Principles of Dynamic Programming

CONTROL AND SYSTEMS THEORY

A Series of Monographs and Textbooks

Volume 1	Discrete Techniques of Parameter Estimation: The Equation Error Formulation, *Jerry M. Mendel*
Volume 2	Mathematical Description of Linear Systems, *Wilson J. Rugh*
Volume 3	The Qualitative Theory of Optimal Processes, *R. Gabasov and F. Kirillova*
Volume 4	Self-Organizing Control of Stochastic Systems, *George N. Saridis*
Volume 5	An Introduction to Linear Control Systems, *Thomas E. Fortmann and Conrad L. Hitz*
Volume 6	Distributed Parameter Systems: Identification, Estimation, and Control, *edited by W. Harmon Ray and Demetrios G. Lainiotis*
Volume 7	Principles of Dynamic Programming. Part I. Basic Analytic and Computational Methods, Part II. Advanced Theory and Applications, *Robert E. Larson and John L. Casti*
Voluem 8	Adaptive Control: The Model Reference Approach, *Yoan D. Landau*
Volume 9	Parameter Estimation: Principles and Problems, *Harold W. Sorenson*
Volume 10	ORACLS: A Design System for Linear Multivariable Control, *Ernest S. Armstrong*
Volume 11	Failure Diagnosis and Performance Monitoring, *L. F. Pau*

Additional Volumes in Preparation

Principles of Dynamic Programming

Part II

Advanced Theory and Applications

by ROBERT E. LARSON
Systems Control, Inc.
Palo Alto, California

JOHN L. CASTI
University of Arizona
Tucson, Arizona

MARCEL DEKKER, INC. New York and Basel

Library of Congress Cataloging in Publication Data (Revised)

Larson, Robert Edward.
Principles of dynamic programming.

(Control and systems theory ; v. 7)
Bibliography: v. 1, p. ; v. 2, p.
CONTENTS: pt. 1. Basic analytic and computational methods.--pt. 2. Advanced theory and applications.
I. Dynamic programming. I. Casti, John, joint author. II. Title.
T57.83.L37 519.7'03 78-15319
ISBN 0-8247-6589-3 (v. 1) AACR2
ISBN 0-8247-6590-7 (v. 2)

MARCEL DEKKER, INC.

270 Madison Avenue, New York, New York 10016

Current printing (last digit):
10 9 8 7 6 5 4 3 2 1

PRINTED IN THE UNITED STATES OF AMERICA

To Richard Bellman

Teacher, Inspiration, and Friend to Us Both

PREFACE

In the first part of our two-part exposition of dynamic programming methodology, the basic theme was determinism and algorithm feasibility, i.e., our attention was centered upon decision processes with no stochastic influences and we developed the basic computational procedures needed for obtaining feedback decision policies.

Taking the material of Part I as our point of departure, in this part we explore many of the extensions and generalizations inherent in the dynamic programming approach to determination of optimal policies. In one direction we make contact with classical variational theory and show how all the standard necessary conditions of the calculus of variations follow in a straightforward way from the basic functional equation of dynamic programming. In a more modern spirit, we also examine a number of questions involving optimal control processes and derive a variety of basic results for determination of optimal control laws using the principle of optimality.

Since uncertainty continually intrudes into most realistic decision processes, a rather lengthy chapter is devoted to a treatment of stochastic and adaptive processes arising in a variety of engineering and management situations. It is at this point in our narrative where the conceptual approach of dynamic programming allows us to treat truly significant applied problems, where the classical variational calculus would be powerless.

As has been well chronicled in the literature, the basic stumbling block standing in the way of a routine use of dynamic programming for most decision processes is the so-called "curse of dimensionality." This computational obstacle is attacked from two standpoints in this part. The first is via the exploitation of problem structure to reduce the original problem to more tractable form. A more or less miscellaneous collection of methods, techniques and subterfuges are presented all of which make use of particular problem structure to cut the standard dynamic programming computational requirements down to a manageable level. The second approach to lifting the "curse" involves a study of new computer architectures, particularly parallelism, and their applicability for dynamic programming calculations. It is seen that recent developments in computer hardware and software have major implications for the feasibility of employing dynamic programming in a wide variety of problems which heretofore had been beyond computational help.

Finally, to demonstrate the many uses of dynamic programming, we present a number of applications in the management and OR areas, together with two extensive case studies involving electric utility operation and planning and nuclear fuel reprocessing plants. These case studies leave no doubt as to the ability of dynamic programming methods to penetrate to the heart of complex decision-making processes involving uncertainty.

It is our hope and expectation that the material provided in Parts I and II of the treatise on dynamic programming will provide the serious system analyst with sufficient tools and confidence to tackle the variety of complex decision-making problems which the modern world continually forces upon us.

ROBERT E. LARSON
JOHN L. CASTI

TABLE OF CONTENTS

CHAPTER 3

Stochastic and Adaptive Systems

CHAPTER 4

Advanced Computational Methods

CHAPTER 5

Dynamic Programming and Operations Research Problems

CHAPTER 6

Two Case Studies

Principles of Dynamic Programming

CHAPTER 1

DYNAMIC PROGRAMMING, OPTIMAL CONTROL THEORY, AND THE CALCULUS OF VARIATIONS

INTRODUCTION

A fundamental problem of modern society, and one to which considerable study has been devoted, is the control or regulation of complex, interdependent systems. These problems occur in all fields of industry and government, and they encompass systems ranging from municipal utility systems to manufacturing plants to large-scale defense systems. Although the specific systems differ greatly in the physical mechanisms of their operation, nevertheless there is a certain commonality in the control problems pertaining to all of them. The purpose of this chapter is to show how dynamic programming can be applied to the analysis of these control problems and to the effective computation of optimal control laws for a variety of systems.

Our concern here will be with what has now become known as feedback control theory, in contrast with previous mathematical analyses of control problems which centered on questions of stability under various conditions. Our analysis will focus upon situations of the type depicted in Figure 1.1. Here we have a system S characterized by its state x. This state is measured by the controlling unit C, which compares the current system performance with a predefined desired mode of operation and generates a controlling input u based upon the current deviation of the system from its desired performance level.

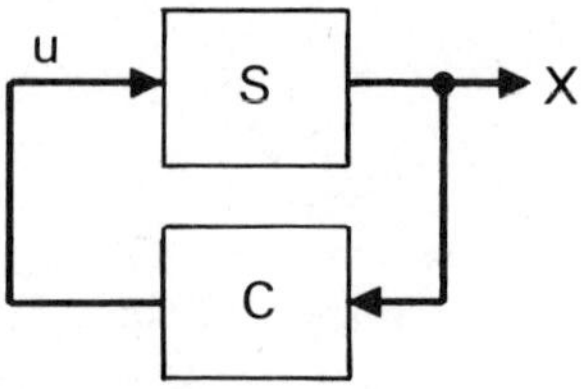

Figure 1.1

An important feature distinguishing the material of this chapter from most of our treatment in Part I, is continuous rather than discrete time. Optimal control problems have typically arisen in mechanical or engineering processes in which the response time of the system to controlling (or perturbing) influences has been quite short as compared with the duration of the process under study. Thus, it has been mathematically convenient, and physically reasonable, to develop the major results in the continuous time framework.

One way to formulate the above situation in mathematical terms is to regard the system S as being governed by the differential equation

$$\frac{dx}{dt} = g(x,u,t), \quad x(0) = c ,$$

where u is to be chosen to minimize some criterion function J which measures the cost of exerting control as well as the cost of deviation of $x(t)$ from its desired level; the form of J is generally assumed to include both an integral term and a final-time term

$$J = \int_0^T L(x,u,t)dt + \varphi[x(T)] .$$

Many of the problems we shall consider in this chapter can be posed as questions within the classical calculus of variations.

The calculus of variations is a rather old and well-studied chapter of mathematics, having its origin in the classical problem of determining the curve of minimum time of descent between two specified points. On the other hand, dynamic programming is a relatively new field of mathematics, whose main development has been stimulated by the recent growth of computational capability and the plethora of important optimization problems which defy analysis by more traditional means. One of the main points we wish to bring out is that dynamic programming provides a very natural framework from which to study both classical and modern variational problems.

DISCRETE CONTROL PROCESSES

To introduce the concepts of this chapter, let us first consider control processes in which decisions are made at a discrete set of times $k = 0,1,2,\ldots,K$. These processes can be treated directly by the methods developed in Part I.

Let us examine the process

$$x(k+1) = g[x(k), u(k)] , \quad x(0) = c , \quad k = 0,1,\ldots,K-1 ,$$

where $x(k)$ is an n-dimensional state vector and $u(k)$ is an n-dimensional control vector. We wish to choose the sequence $\{u(k)\}$ to minimize the criterion functional

$$J = \sum_{k=0}^{K-1} L[x(k), u(k)] + \varphi[x(K)] .$$

Let us denote the minimizing value of J by

$$I(c,a) = \min_{u(a),\ldots,u(k-1)} \sum_{k=0}^{K-1} L[x(k),u(k)] + \varphi[x(K)] ,$$

indicating, as before, that the minimum of J depends only upon c, the initial state, and a, the initial stage of the process.

Employing the principle of optimality and arguing as in Part I, we readily obtain the recurrence formula

$$I(c,a) = \min_u [L(c,u) + I(g(c,u), a+1)] , \quad 0 \le a \le K-1 ,$$

$$I(c,K) = \varphi(c) .$$

EXAMPLE

1.1 Consider the problem of minimizing

$$J[u(0),u(1),\ldots,u(N-1)] = \sum_{k=a}^{N} [x(k) - u(k)] ,$$

over all $u(k)$ subject to the conditions

$$x(k+1) = x(k) + bu(k) , \quad x(a) = c ,$$

$$0 \le u(k) \le x(k) , \quad k = a, a+1, \ldots, N ,$$

where $u(k)$, b and $x(k)$ are scalars. Derive the appropriate recurrence formula for the minimum cost function.

As the result of a decision v in state c, at stage a, the state is transformed from c to $c+bv$, and a cost $c-v$ is incurred. By definition, the optimal continuation from the new state $c+bv$ is $I(c+bv, a+1)$. Thus according to the principle of optimality

$$I(c,a) = \min_{0 \le v \le c} [(c-v) + I(c+bv, a+1)] ,$$

with

$$I(c,N) = \min_{0 \le v \le c} (c-v) = 0 .$$

CONTINUOUS CONTROL PROCESSES

In many problems of practical interest, particularly those associated with physical systems, the problem is given directly in continuous time form. Numerous problems from fields such as process control, mechanical engineering, and aerospace engineering, are specified in this manner. In these cases we wish to minimize

$$\int_a^T L[x(t),u(t),t]\,dt ,$$

over all continuous (say) u, such that

$$\frac{dx}{dt} = g(x,u,t) , \quad x(a) = c .$$

Let us examine the dynamic programming formulation of such a problem, assuming for the moment that x and u are scalar functions.

One way to proceed would be to use finite difference approximations to the integral criterion and differential equation, thereby reducing the problem to the type considered earlier. Under reasonable hypotheses on L and g, this procedure can be justified, and we can assert that the solution to the discrete problem approaches that of the continuous as the discretization step goes to zero. As far as computation of numerical results is concerned, this is the approach which must ultimately be followed, since digital computers are essentially discrete in nature. However, for the moment let us follow another route and stay entirely within the continuous domain in quest of analytical insight into the problem.

Define

$$I(c,a) = \min \int_a^T L(x,u,t)\,dt ,$$

and write

$$\int_a^T L\,dt = \int_a^{a+\Delta} L\,dt + \int_{a+\Delta}^T L\,dt , \tag{1.1}$$

where Δ is an infinitesimal quantity which will be allowed to approach zero. As a result of a decision $u(a)$ at time a, a cost

$$\int_a^{a+\Delta} L[x(t),u(t),t]\,dt ,$$

is incurred. By the Mean-Value theorem for integrals, this may be approximated by

$$\int_a^{a+\Delta} L[x(s),u(s),s]\,ds = L[c,u(a),a]\Delta + 0(\Delta^2) .$$

Also, as a result of the decision $u(a)$, the initial state c is transformed according to the differential equation (5.1). This transformation may be approximated as

$$c \to c + g[x(a),u(a),a]\Delta = c + g[c,u(a),a]\Delta + 0(\Delta^2) .$$

The principle of optimality now asserts that we must proceed from this new state in an optimal manner to have the entire process be optimal. By definition, the cost associated with the optimal continuation is

$$I[c + g(c,u(a),a) , a + \Delta] + 0(\Delta^2) .$$

This corresponds to the cost associated with the second integral in (1.1) along an optimal trajectory. Adding this cost to the initial cost, we have

$$I(c,a) \leq L[c,u(a),a]\Delta + I[c + g(c,u(a),a)\Delta, a + \Delta) + 0(\Delta^2)] , \tag{1.2}$$

for <u>any</u> admissible initial decision $u(a)$, with equality holding for the optimal initial decision. Hence,

$$I(c,a) = \min_{u(a)} \{L[c,u(a),a]\Delta + I[c+g(c,u(a),a)\Delta,a]\} + 0(\Delta^2) .$$

Clearly, when $a = T$,

$$I(c,T) = \varphi(c,T) .$$

To eliminate the term $0(\Delta^2)$, we expand I in a Taylor series about the point (c,a). This gives

$$I[c+g(c,u(a),a)\Delta,\ a+\Delta] = I(c,a) + g[c,u(a),a]\Delta I_c$$
$$+ I_a + 0(\Delta) ,$$

where $I_c = \frac{\partial I}{\partial c}$ and $I_a = \frac{\partial I}{\partial a}$ are partial derivatives of $I(c,a)$ with respect to c and a, respectively. Substituting this result into Eq. (1.2), dividing through by Δ, and letting $\Delta \to 0$, gives the final equation for I as

$$-I_a(c,a) = \min_v [L(c,v,a) + g(c,v,a)I_c(c,a)] , \tag{1.3}$$

where we have let $v = u(a)$ for ease of writing. Equation (1.3) is a nonlinear first order partial differential equation for the optimal value function I.

MULTIDIMENSIONAL PROBLEMS

There is no particular conceptual difficulty in extending the above results when x and u are vector functions. Consider the problem of minimizing

$$\int_a^T L(x,u,t)\, dt ,$$

with

$$\frac{dx}{dt} = g(x,u,t), \quad x(a) = c ,$$

where now x and u are n- and m-dimensional vectors, respectively. Setting

$$I(c,a) = \min \int_a^T L(x,u,t)\, dt ,$$

we readily obtain the partial differential equation

$$-I_a(x,a) = \min_v \,[L(c,v,a) + (g(c,v,a), \text{grad } I)] ,$$

where $(,)$ denotes the usual vector inner product and grad I is the n-dimensional vector whose ith component is $\frac{\partial I}{\partial c_i}$, $i = 1,2,\ldots,n$. The initial condition is again

$$I(c,T) = 0 .$$

GENERALIZED APPROXIMATION IN POLICY SPACE

A common method employed to solve an equation such as (1.3) is the method of successive approximations. After converting the original problem into one of the form

$$I = T(I) ,$$

where T represents a suitable transformation, the sequence of approximants $\{I_n\}$ is generated according to the rule

$$I_{n+1} = T(I_n) ,$$

with I_o chosen in some manner. Under suitable assumptions on T and I_o, it may be established that $I_n \to I$ as $n \to \infty$.

Instead of this approach, let us consider a type of approximation that has no counterpart in classical analysis--approximation in policy space. Let $v^o(c,a)$ be an estimate (guess) of the optimal policy. We compute the associated return function I^o as the solution of the equation

$$-I_a^o(c,a) = L(c,v^o,a) + g^o I_c^o(c,a) . \tag{1.4}$$

(Here we write $g(c,v^o,a) = g^o$.) The initial condition on (1.4) is

$$I^o(c,T) = 0 .$$

Next we determine the new policy $v^1(c,a)$ as the function which minimizes

$$L(c,v,a) + g(c,v,a)\, I_c^o(c,a) .$$

Using this policy, we now compute $I^1(c,a)$ as the solution of the equation

$$-I_a^1(c,a) = L(c,v^1,a) + g^1 I_c^o(c,a) ,$$

$$I^1(c,T) = 0 .$$

In this fashion, we produce a sequence of return functions $\{I^n\}$ and policy functions $\{v^n\}$.

EXAMPLE

1.2 Show that the sequence $\{I^n\}$ is monotonically decreasing in n.

We wish to show that $I^{n+1}(c,a) \leq I^{n}(c,a)$ for all c and a. It will suffice to demonstrate the result for $n=0$, since the mode of proof will make it clear for all $n > 0$.

Observe that the way in which v^1 was obtained implies that

$$-I_a^o(c,a) = L(c,v^o,a) + g^o I_c^o(c,a) \leq L(c,v^1,a) + g^1 I_c^1(c,a) = -I_a^1(c,a) .$$

Hence

$$I_a^o(c,a) \geq I_a^1(c,a) ,$$

which, by the condition $I^o(c,T) = I^1(c,T) = 0$, implies

$$I^o(c,a) \geq I^1(c,a) .$$

It is now clear that this same argument may be employed on each pair of elements I^n, I^{n+1} to show that the sequence $\{I^n\}$ is monotonically decreasing. If we could establish that the sequence $\{I^n\}$ is uniformly bounded from below then, of course, we would have convergence. Unfortunately, this is a somewhat more difficult task and will not be taken up at this time.

The preceding approximation technique suggests that it may be worthwhile to consider an equation of the type

$$u = T(u) ,$$

in the form

$$u = \min_v S(u,v) ,$$

to take advantage of the fact that v may be thought of as a policy. This idea forms the backbone of the theory of quasilinearization. Naturally, the success of this approach hinges upon our ability to find a suitable transformation S.

EXAMPLE

1.3 (a) Consider the scalar Riccati equation $\dot{u} = a(t) + u^2$, $u(0) = 0$. Determine a transformation S such that $\dot{u} = \max_v S(u,v)$.

(b) Obtain a corresponding result for $\dot{u} = g(u)$, $u(0) = c$, where g is uniformly convex in u.

(a) It is easy to verify that

$$u^2 = \max_v [2uv - v^2] .$$

Thus, we can write the equation as

$$\dot{u} = \max_v [2uv - v^2 + a(t)] .$$

Hence, $S(u,v) = 2uv - v^2 + a(t)$.

(b) Since g is uniformly convex in u, g dominates its tangent at any point v (see Figure 1.2). Hence,

$$g(u) \geq g(v) + (u-v)\dot{g}(v) ,$$

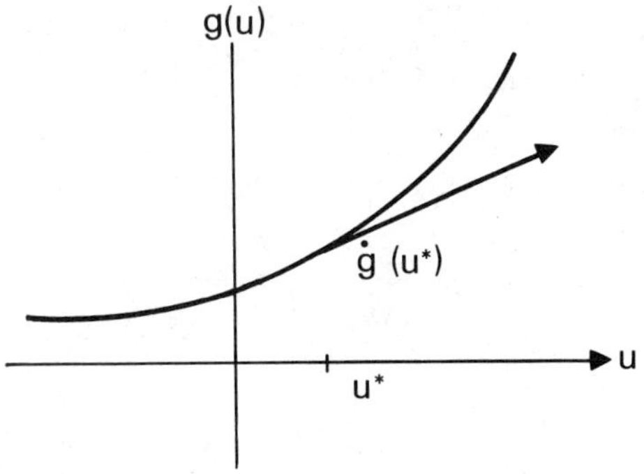

Figure 1.2. A Convex Function g

with equality for $v = u$. Therefore,

$$g(u) = \max_v [g(v) + (u-v)\dot{g}(v)] ,$$

$$= \max_v S(u,v) .$$

CONSTRAINTS

Consider the problem of minimizing

$$J = \int_0^T L(x,u,t)\, dt ,$$

with

$$\frac{dx}{dt} = g(x,u,t) , \quad x(0) = c ,$$

subject now to the additional condition

$$|u(t)| \leq k , \quad 0 \leq t \leq T . \tag{1.5}$$

A constraint of this nature can cause considerable analytic and computational difficulty in the classical treatment of variational problems of this sort. Let us see what it does to the dynamic programming formulation of the problem.

Introduce the function $I(c,a)$ as before. The constraint (1.5) says that our initial decision is not completely free but must be chosen in accordance with the prescribed bound k. After this decision, we must again proceed optimally from the resultant state. Thus, I satisfies the equation

$$-I_a(c,a) = \min_{|v| \leq k} [L(c,v,a) + g(c,v,a)\, I_c(c,a)] , \tag{1.6}$$

$$I(c,T) = 0 .$$

Equation (1.6) is a nonlinear partial differential equation of a rather novel type which is not easy to analyze rigorously in the general case. However, for computational purposes the constraint actually simplifies the solution of the problem since a smaller number of policies need be examined to determine the minimizer. This point will be considered in detail in Chapter 4 on numerical techniques and solutions.

Another type of constraint that often arises is one of a global, rather than local, nature. To illustrate the concept, let us consider an example. In this problem we have the performance criteria

$$\min_u \int_a^T L[x(s),u(s),s]\, ds ,$$

subject to

$$\dot{x}(t) = g[x(t),u(t),t] , \quad x(a) = c , \tag{1.7}$$

and the global constraint

$$\int_a^T u(s)\, ds = M ,$$

where M represents the total decision resource that may be expended on the process. If the functions L and g have certain nonnegativity and monotonicity properties, as is generally the case in practice, equality will hold in this type of side constraint since the more decision resource we have available, the better performance we may achieve.

One way to tackle the foregoing global constraint is to introduce a Lagrange parameter λ. In place of the original problem, we now solve the associated problem

$$\min \int_a^T [L(x(s),u(s),s) + \lambda u(s)]\, ds ,$$

subject to

$$\dot{x}(t) = g[x(t),u(t),t] \ , \qquad x(a) = c \ .$$

For each fixed value of λ, this is a problem of the type considered earlier. The optimal control function u^* will now be a function of both t and λ, i.e., $u^* = u^*(t;\lambda)$. Substituting this function into Eq. (1.7) gives an equation for λ,

$$\int_a^T u^*(s;\lambda) = M \ ,$$

theoretically disposing of the problem. Since we shall have more to say about this approach later, let us now consider an alternate method of treating the constraint.

Let M be considered as an additional state variable. Thus, the minimum value of our criterion now depends upon a, c, and M. Let $I(c,M,a)$ represent this minimum. As a result of the initial decision $u(a)$, c and a transform as before and from Equation (1.7) we see that $M \to M - \Delta u(a)$. Thus, by the principle of optimality I satisfies

$$I(c,M,a) = \min_v \, [L(c,v,a)\Delta + I(c + g(c,v,a),\ a + \Delta,\ M - v\Delta)] \ ,$$

where we again write $v = u(a)$. The initial condition at $a = T$ is

$$I(c,M,T) = \begin{cases} 0 \ , & M = 0 \ , \\ \text{undetermined} \ , & M \neq 0 \ , \end{cases}$$

illustrating a difficulty in use of this procedure; i.e., the precise initial condition may be indeterminate. We shall indicate how to deal with this situation momentarily.

Such a formulation of the global constraint (1.7) has the virtue that it requires no special foresight and/or good fortune in selecting the value of the parameter λ. On the other hand, the computational difficulties of carrying an extra state variable can be considerable, as we have seen in Part I. Which formulation we use will usually depend upon the computing power at our disposal, how much we know about the problem at hand, how accurate our answers need be, and so forth. The important point is to have as many methods as possible at our disposal so that we may remain flexible in our choice of procedure.

EXAMPLE

1.4 (a) Consider now the continuous-time problem of minimizing

$$J = \int_a^T L(u,\dot{u},t)\, dt ,$$

where

$$\int_a^T H(u,\dot{u},t)\, dt = K , \qquad u(a) = c .$$

Let $\min J(u) = I(c,\alpha,a)$. Derive an equation for I.

(b) Consider the same problem using a Lagrange multiplier. Let $\min J(u) = \varphi(c,a;\lambda)$. What is λ in terms of I and what is the physical interpretation of this result?

(a) The initial decision $v = u(a)$ transforms c, a, and α as

$$c \to c + v\Delta, \quad a \to a + \Delta, \quad \alpha \to \alpha - H(c,v,a)\Delta ,$$

to terms $o(\Delta)$. Thus, by the principle of optimality

$$I(c,\alpha,a) = \min_v [L(c,v,a) + I(c + v\Delta, \alpha - H(c,v,a)\Delta, a + \Delta)] + o(\Delta) .$$

Expanding the term in brackets in a Taylor series and letting $\Delta \to 0$ gives

$$-I_a(c,\alpha,a) - \min_v [L(c,v,a) + vI_c - H(c,v,a)I_\alpha] . \quad (1.8)$$

There is some delicacy involved in obtaining an initial condition for I near $a = T$ since, unless $\alpha = 0$, the condition is clearly non-zero. To obtain an approximate condition, let $a = T - \Delta$. Then

$$H(u(T - \Delta), \dot{u}(T - \Delta), T - \Delta) \simeq \alpha/\Delta .$$

Solving for $\dot{u}(T - \Delta)$ gives

$$\dot{u}(T - \Delta) = Z(u(T - \Delta), T - \Delta, \alpha) ,$$

for some appropriate function Z. Substituting this result into the expression for J then yields

$$I(c,T - \Delta,\alpha) \simeq L(c,Z(c,T - \Delta,\alpha), T - \Delta)\Delta .$$

(b) Using a Lagrange multiplier λ, we wish to minimize

$$J = \int_a^T [L(u,\dot{u},t) + H(u,\dot{u},t)] \, dt , \quad u(a) = c .$$

Letting $\varphi(c,a,\lambda) = \min J$, we have

$$\varphi(c,a,\lambda) = \min_{v} [L(c,v,a) + \lambda H(c,v,a)\Delta$$

$$+ \varphi(c + v\Delta, a + \Delta, \lambda)] + o(\Delta) ,$$

or, upon expanding in powers of Δ and letting $\Delta \to 0$,

$$-\varphi_a(c,a,\lambda) - \min_{v} [L(c,v,a) + \lambda H(c,v,a) + v\varphi_c] , \qquad (1.9)$$

$$\varphi(c,T;\lambda) = 0 .$$

Comparing (1.7) and (1.8), we see that the relationship between λ and I is

$$\lambda = -I_\alpha .$$

The physical interpretation of this result is that λ is a price. It trades constraint value for return value, i.e., it represents the change in optimal performance obtained by allowing an infinitesimal change in the value of the constraint.

ASYMPTOTIC CONTROL

In many cases our main concern is not in the initial phases of a process, but rather the steady-state, or "asymptotic" control after the process has been operating for a sufficiently long period of time.

Consider minimizing the functional equation

$$J(x) = \int_0^T [\dot{x}^2(t) + 2g(x)] \, dt ,$$

subject to $x(0) = c$. Suppose that g is convex, ensuring that the minimum of J exists and is furnished by the solution of the Euler equation

$$\ddot{x}(t) - \dot{g}(x) = 0 , \qquad (1.10)$$

$$x(0) = c, \quad \dot{x}(T) = 0 . \qquad (1.11)$$

The minimizing function x exists for all $T > 0$. Hence, we ask about the behavior of J and x as $T \to \infty$.

Let us write

$$I(c,T) = \min_{x} J(x) ,$$

and let $x(t,T)$ denote the solution of Equations (1.10) and (1.11). We shall be interested in the questions:

(a) Does $\lim_{T\to\infty} I(c,T)$ exist and, if so, how may we determine it? Similarly, for $x(t,T)$?

(b) If these limits do exist, is there any connection between the limiting functions and the problem

$$\int_0^\infty [\dot{x}^2(t) + 2g(x)]\, dt , \qquad x(0) = c \ ?$$

(c) Can we use the solution on the infinite interval as the first term in the asymptotic expansion of the solution of the finite interval problem where T is large?

The last question is associated with the goal of using the "steady-state" control associated with the infinite duration problem as an approximation to the optimal control for a finite process of long duration.

The first question regarding the existence of $\lim_{T\to\infty} I(c,T)$ is relatively easy to answer under rather general conditions. Let

(i) $g \geq 0$ and convex,

(ii) $\int_0^\infty g(ce^{-t})\, dt < \infty$, for all c .

First of all, it is clear that (i) implies $I(c,T)$ is monotone increasing in T. Using the trial function $x(t) = ce^{-t}$, we have

$$I(c,T) \leq \int_0^T [c^2 e^{-2t} + 2g(ce^{-t})]\, dt ,$$

$$\leq \int_0^\infty [c^2 e^{-2t} + 2g(ce^{-t})\, dt < \infty ,$$

by (ii). Hence, I is monotone increasing in T and bounded for all c from which it follows that $I(c,T)$ converges as $T \to \infty$. We answer questions (b) and (c) by the following example.

EXAMPLE

1.5 (a) Consider the case where

$$J = \int_0^T [\dot{x}^2(t) + x^2(t)]\, dt , \qquad x(0) = c .$$

Show by direct calculation that $I(c,T)$ and $x(t,T)$ possess limits as $T \to \infty$.

(b) What is the relation of this problem to that of minimizing

$$H = \int_0^\infty [\dot{x}^2(t) + x^2(t)]\, dt , \qquad x(0) = c .$$

(a) It is easily verified for this problem that

$$I(c,T) = c^2 \tanh T ,$$

$$\dot{x}(t,T) = c \frac{\cosh (t-T)}{\cosh T} .$$

Thus, $\lim_{T\to\infty} I(c,T) = c^2$ while $\lim_{T\to\infty} \dot{x}(t,T) = -c$

Writing $x(t,T)$ in the form

$$x(t,T) = \frac{ce^{-t}}{1+e^{-2T}} + \frac{ce^{t-2T}}{1+e^{2T}} ,$$

if $0 \leq t \leq T$, the second term is uniformly bounded by $|c|e^{-T}$. Thus

$$\left| x - \frac{ce^{-t}}{1+e^{-2T}} \right| \leq |c|\, e^{-T} ,$$

which means

$$|\dot{x} - ce^{-t}| \leq 2|c|e^{-T} .$$

From the fact that

$$\dot{x} = c \frac{\cosh (t-T)}{\cosh T} ,$$

in the same way we obtain

$$|\dot{x} + ce^{-t}| \leq 2|c|e^{-T} .$$

Thus, for large T we have the approximate relation

$$\dot{x} \simeq -x , \quad 0 \leq t \leq T .$$

This is an example of the simple type of control law that we hope to obtain from the asymptotic behavior of I and x.

(b) For this problem, the minimum value of H depends only on c. Let

$$\min_{x} H(x) = Z(c) \ .$$

As a result of the initial decision $\dot{x}$, the state c is transformed into $c + \dot{x}\Delta$ and we still face a process of infinite duration from this new state. Applying the principle of optimality gives

$$Z(c) = \min_{\dot{x}} \left[(\dot{x}^2 + c^2)\Delta + Z(c + \dot{x}\Delta) \right] + o(\Delta) \ .$$

Expanding in powers of Δ and letting $\Delta \to 0$ gives the equation for Z,

$$0 = \min_{\dot{x}} \left[\dot{x}^2 + c^2 + \dot{x}\dot{Z}(c) \right] \ .$$

The minimizing $\dot{x}$ is given by

$$\dot{x}_{min} = -1/2 \ \dot{Z}(c) \ . \qquad (1.12)$$

Thus, the resulting ordinary differential equation for Z is

$$\dot{Z}^2(c) = 4c^2 \ ,$$

which has the solution

$$Z(c) = c^2 \ ,$$

in agreement with our earlier result for $I(c,\infty)$. Thus

$$\lim_{T \to \infty} I(c,T) = Z(c) \ .$$

Similarly, from (1.12), we have

$$\lim_{T\to\infty} \dot{x}(t,T) = =1/2\ \dot{Z}(c) = -\ x(t)\ .$$

CALCULUS OF VARIATIONS AND MULTISTAGE PROCESSES

One of the simplest problems that arises in the calculus of variations is the determination of a function $u(t)$ which minimizes the integral

$$\int_0^T \ell[u(s),\ \dot{u}(s)]\ ds\ ,$$

where ℓ is a prescribed function. In the classical theory one regards u as an element in a space of candidate functions, and we then employ various conditions to be discussed below to isolate the optimizing u. Form this viewpoint, the basic function is u and it is chosen over the interval $0 \leq t \leq T$ at one stroke.

In place of determining u as a function of t, an alternate approach would be to prescribe $\dot{u}$ as a function of u and t. That is, we can regard a choice of $\dot{u}$ as a choice of policy, where the decision to be made depends upon the state of the system at the time the decision is made. In this way, we can regard the type of variational process described above as a multistage design process.

In geometric terms, the classical approach to the calculus of variations regards a curve as a locus of points while the dynamic programming approach regards a curve as an envelope of tangents. The fundamental point-line duality of euclidean geometry ensures the equivalence of these two approaches. As a result, for deterministic processes we have a choice as to which approach to use. However, in the case of a stochastic process no choice exists; we must use the dynamic programming approach.

SOME CLASSICAL RESULTS

In this section, we present a brief discussion of the fundamental variational approach. We shall derive the Euler equation as well as a few of the more celebrated necessary conditions of Weierstrass, Legendre, and others. In addition, we shall indicate some of the computational difficulties inherent in the classical approach. For readers interested in more detailed discussion of these matters, we refer to the excellent texts cited in the references.

We now illustrate the basic idea of the classical approach by formally deriving the Euler equation associated with the problem of minimizing

$$J(u,v) = \int_0^T F[u(t), v(t)]\, dt ,$$

subject to the conditions that

$$\frac{du}{dt} = G[u(t), v(t)] , \quad u(0) = c . \tag{1.13}$$

Let u^*, v^* be a pair of functions which yield the minimum (assuming such a minimum exists). Write

$$u(t) = u^*(t) + \varepsilon w(t) , \tag{1.14}$$

$$v(t) = v^*(t) + \varepsilon z(t) , \tag{1.15}$$

where ε is a small perturbation parameter and w and z are arbitrary functions of t, $0 \leq t \leq T$.

Substituting (1.14) and (1.15) into J gives

$$J(u^* + \varepsilon w, v^* + \varepsilon z) = \int_0^T F(u^* + \varepsilon w, v^* + \varepsilon z)dt ,$$

or, upon expanding the integral in a Taylor series up to terms of order ε^2,

$$J(u^* + \varepsilon w,\ v^* + \varepsilon w) = J(u^*, v^*) + \int_0^T \left[w \frac{\partial F}{\partial u^*} + z \frac{\partial F}{\partial v^*} \right] dt + 0(\varepsilon^2) \ . \tag{1.16}$$

A similar substitution and expansion in (1.13) yields

$$G(u^* + \varepsilon w,\ v^* + \varepsilon z) = G(u^*, v^*) + \varepsilon \left[\frac{\partial G}{\partial u^*} w + \frac{\partial G}{\partial v^*} \right] z \ + 0(\varepsilon^2) \ . \tag{1.17}$$

In (1.16) and (1.17), the partial derivatives of F and G that occur are evaluated along the curves u^* and v^* and, therefore, are particular functions of t, once F, G, u^*, and v^* are specified.

Setting the coefficient of ε equal to zero in (1.16) (justification for this is the fact that if it were not zero, we could choose ε of opposite sign, thereby reducing the value of $J(u^* + \varepsilon w,\ v^* + \varepsilon w)$ below that of $J(u^*, v^*)$, contradicting the minimality of the pair (u^*, v^*)), and equating like powers of ε in (1.17), we obtain

$$\int_0^T \left[w \frac{\partial F}{\partial u^*} + z \frac{\partial F}{\partial v^*} \right] dt = 0 \ , \tag{1.18}$$

$$\frac{dw}{dt} = w \frac{\partial G}{\partial u^*} + z \frac{\partial G}{\partial v^*} \ , \qquad w(0) = 0 \ , \tag{1.19}$$

which hold for all admissible functions w and z.

Solving for z in (1.19) and substituting into (1.18) gives

$$\int_0^T \left[w \frac{\partial F}{\partial u^*} + \frac{\partial F}{\partial v^*} \frac{dw}{dt} - w \frac{\partial G}{\partial u^*} \Big/ \frac{\partial G}{\partial v^*} \right] dt = 0 \ , \tag{1.20}$$

for all w.

Eliminating the term involving $\frac{dw}{dt}$ through integration by parts, (1.20) becomes

$$w \left. \frac{\frac{\partial F}{\partial v^*}}{\frac{\partial G}{\partial v^*}} \right|_{t=0}^{t=T} - \int_0^T \left[\frac{\partial F}{\partial u^*} - \frac{\frac{\partial F}{\partial v^*}\frac{\partial G}{\partial u^*}}{\frac{\partial G}{\partial v^*}} - \frac{d}{dt}\left(\frac{\frac{\partial F}{\partial v^*}}{\frac{\partial G}{\partial v^*}} \right) \right] dt = 0 \ . \tag{1.21}$$

In the integrated term of (1.21), the part at $t = 0$ vanishes because $w(0) = 0$, but the term at $t = T$ gives the additional condition

$$\left. \frac{\frac{\partial F}{\partial v^*}}{\frac{\partial G}{\partial v^*}} \right|_{t=T} = 0 \ . \tag{1.22}$$

The Fundamental Lemma of the calculus of variations now ensures that the bracketed term in the integral of (1.21) vanishes, leading to the Euler equation

$$\frac{\frac{\partial F}{\partial u^*}\frac{\partial G}{\partial v^*} - \frac{\partial F}{\partial v^*}\frac{\partial G}{\partial u^*}}{\frac{\partial G}{\partial v^*}} - \frac{d}{dt}\left(\frac{\frac{\partial F}{\partial v^*}}{\frac{\partial G}{\partial v^*}} \right) = 0 \ , \tag{1.23}$$

with associated boundary condition (1.22). Equations (1.13), (1.22), and (1.23) comprise the two-point boundary value problem whose solution theoretically disposes of the original problem.

We say theoretically since the fact that conditions are specified at the points $t = 0$ <u>and</u> $t = T$ introduces considerable computational complexity into the foregoing theoretical considerations. We shall return to this point below.

EXAMPLE

1.6 In the case $G(u,v) = v$, the Euler equation simplifies to $\frac{d}{dt}\left(\frac{\partial F}{\partial \dot{u}}\right) = \frac{\partial F}{\partial u}$ with the boundary conditions $u(0) = c$, $\left.\frac{\partial F}{\partial \dot{u}}\right|_{t=T} \quad 0.$

In what follows, we shall assume that $G(u,v) = v$ unless otherwise indicated. The foregoing discussion of the Euler equation provides us with necessary conditions which a candidate extremal curve must satisfy.

Several additional necessary conditions have been discovered which enable us to eliminate as candidate extremals many curves which satisfy the Euler equations. Since a detailed discussion of the origin and restrictions surrounding these tests would take us too far afield, we content ourselves with giving only statements of the tests, leaving the interested reader to consult the appropriate references for full details.

The three most celebrated additional necessary conditions are:

(1) Legendre Condition:

$$\frac{\partial^2 F}{\partial \dot{u}^2} \geq 0 \ ;$$

(2) Jacobi Condition:

For a problem with a fixed left endpoint and free right endpoint, the solution $p(t)$ of the linear second order differential equation

$$\left(\frac{\partial^2 F}{\partial \dot{u}^2}\right) \frac{d^2 p(t)}{dt^2} + \frac{d}{dt}\left(\frac{\partial^2 F}{\partial \dot{u}^2}\right) \frac{dp(t)}{dt}$$

$$+ \frac{d}{dt}\left(\frac{\partial^2 F}{\partial u \partial \dot{u}} - \frac{\partial^2 F}{\partial u^2}\right) p(t) = 0 \ ,$$

must have no zeros in the interval $0 \leq t \leq T$ except at the point $t = 0$;

(3) Weierstrass Condition:

$$F(u,\dot{U}) - F(u,\dot{u}) - (\dot{U} - \dot{u}) \frac{\partial F(u,u)}{\partial \dot{u}} \geq 0 ,$$

for all functions $U \neq u$ if u is to be the minimizing curve.

In some cases, there may be freedom in the choice of endpoints as, for example, when we seek the shortest path from a point to a specified curve where the exact point on the curve is not given and is to be selected optimally. In these situations we can determine necessary conditions which must be satisfied if the candidate curve is to be better than neighboring curves with neighboring endpoints. These are called transversality conditions.

The general transversality condition states that if the terminal condition for a problem is the intersection of the solution curve $u(t)$ with a given terminal curve $h(t)$ which is continuously differentiable in $0 < t < T$, then the minimizing curve must satisfy

$$F(u,\dot{u}) + (\dot{h} - \dot{u}) \left. \frac{\partial F}{\partial \dot{u}} \right|_{t=T} = 0 .$$

Implicit in our derivation of the Euler equation was that it holds only in a region where the solution function has a continuous derivative. The complete solution to a given problem may consist of segments, each satisfying the Euler equation between corners, where we define a corner as a point at which the solution curve has a discontinuous derivative. A set of conditions that must be satisfied at a corner if the trial function is to be unimprovable within a neighborhood are the so-called Weierstrass-Erdmann corner conditions, which are given by

(a) $\left.\frac{\partial F}{\partial \dot{u}}\right|_{t_o^-} = \left.\frac{\partial F}{\partial \dot{u}}\right|_{t_o^+}$,

(b) $F - \dot{u}\left.\frac{\partial F}{\partial \dot{u}}\right|_{t_o^-} = F - \dot{u}\left.\frac{\partial F}{\partial \dot{u}}\right|_{t_o^+}$,

where $|_{t_o^-}$ means evaluated in terms of the left-hand limit of $\dot{u}$ at the corner point $(t_o, u(t_o))$, and analogously for $|_{t_o^+}$ from the right side.

DYNAMIC PROGRAMMING

Our brief excursion into the realm of classical variational theory completed, let us turn to our primary objective of detailing the use of dynamic programming for the solution of variational problems.

Consider the problem of minimizing the functional

$$J(u) = \int_a^T F(t,u,\dot{u})\, dt \ , \tag{1.24}$$

with $u(a) = c$. Note that we now use a to measure the time interval and regard T as being fixed. Define the function

$$I(c,a) = \min_u J(u) \ .$$

To determine an equation for I, write the integral (1.24) as

$$\int_a^T = \int_a^{a+\Delta} + \int_{a+\Delta}^T \ ,$$

where Δ is a small positive quantity. The principle of optimality now yields the equation

$$I(c,a) = \min_{u} \left[\int_{a}^{a+\Delta} F(t,u,\dot{u})\,dt + I(u(a+\Delta),\ a+\Delta) \right] + o(\Delta) .$$

where the minimum is over all functions u defined on the interval $a \le t \le a+\Delta$, $u(a) = c$.

Under the assumptions that the minimizing function u (if it exists) possesses at least two continuous derivatives, we observe that a choice of u over $[a, a+\Delta]$ is equivalent to a choice of $\dot{u}(a)$. Let us denote $\dot{u}(a)$ by $v(a)$. Then (1.25) becomes

$$I(c,a) = \min_{v(a)} \left[\int_{a}^{a+\Delta} F(t,u,\dot{u})\,dt + I(u(a+\Delta),\ a+\Delta) \right] + o(\Delta) .$$

Making the further approximation

$$\int_{a}^{a+\Delta} F(t,u,\dot{u})\,dt = F[a,u(a),\dot{u}(a)]\Delta = F(a,c,v)\Delta ,$$

we have

$$I(c,a) = \min_{v(a)} [F(a,c,v)\Delta + I(c+v(a)\Delta,\ a+\Delta)] + o(\Delta) ,$$

or, upon expanding in powers of Δ and letting $\Delta \to 0$,

$$-\frac{\partial I}{\partial a} = \min_{v} \left[F(a,c,v) + v\frac{\partial I}{\partial c} \right] , \tag{1.26}$$

$$I(c,T) = 0 .$$

Equation (1.26), which is basic for relating dynamic programming to classical variational theory, is just a special case of the multi-dimensional control problem treated earlier.

To illustrate the use of dynamic programming in problems from the variational calculus, we consider a few foundational examples.

EXAMPLE

1.7 Obtain the: (a) Euler equation, (b) Legendre Condition, and (c) Weierstrass Condition from Eq. (1.26).

(a) Equation (1.26) is equivalent to the two equations

$$\frac{\partial F}{\partial v} + \frac{\partial I}{\partial c} = 0 \ , \tag{1.27}$$

obtained by taking the partial derivative with respect to v in (1.26) and

$$F(a,c,v) + \frac{\partial I}{\partial a} + v \frac{\partial I}{\partial c} = 0 \ , \tag{1.28}$$

valid for a, c, v related by (1.27). Differentiating (1.27) with respect to a gives

$$\frac{d}{da}\frac{\partial I}{\partial v} + \frac{\partial^2 I}{\partial c^2} v + \frac{\partial^2 I}{\partial a \partial c} = 0 \ , \tag{1.29}$$

while partial differentiation of (1.28) with respect to c gives

$$\frac{\partial F}{\partial c} + \frac{\partial F}{\partial v}\frac{\partial v}{\partial c} + \frac{\partial^2 I}{\partial a \partial c} + \frac{\partial^2 I}{\partial c^2} v + \frac{\partial I}{\partial c}\frac{\partial v}{\partial c} = 0 \ . \tag{1.30}$$

Combining Eqs. (1.29) and (1.30), together with relation (1.27), yields

$$\frac{d}{da}\left(\frac{\partial F}{\partial v}\right) = \frac{\partial F}{\partial c} \ ,$$

which is the classical Euler equation after making the identifications $a \to t$, $c \to u$, $v \to \dot{u}$.

(b) Equation (1.27) is the first derivative condition for the stationary points of the right-hand side of (1.26). For a minimum, the second derivative of the right side of (1.26) must be nonnegative, which leads to the inequality

$$\frac{\partial^2 F}{\partial v^2} \geq 0 ,$$

which is precisely the Legendre Condition.

(c) For an absolute minimum, we must have

$$f(a,c,v) + \frac{\partial I}{\partial a} + v \frac{\partial I}{\partial c} \leq F(a,c,V) + \frac{\partial I}{\partial a} + V \frac{\partial I}{\partial c} ,$$

for all functions $V(c,a)$. This is equivalent to

$$F(a,c,V) - F(a,c,v) + (V-v) \frac{\partial I}{\partial c} \geq 0 ,$$

which, using Eq. (1.27), yields

$$F(a,c,V) - F(a,c,v) - (V-v) \frac{\partial F}{\partial v} \geq 0 ,$$

the Weierstrass Condition.

This example shows that the rather non-intuitive necessary conditions of the classical calculus of variations appear in a very natural form when obtained from the basic dynamic programming equation (1.26). Let us now examine how the transversality and corner conditions follow in an equally painless fashion.

EXAMPLE

1.8 Derive the: (a) Transversality, and (b) Weierstrass-Erdmann corner conditions from Eqs. (1.27) and (1.28) of the previous example.

(a) Suppose the optimal curve must begin somewhere on a given curve $h(t)$. For the <u>optimal curve</u>, the change in the optimal value function I as the initial point varies along h must be zero, i.e., at the initial point

$$\frac{\partial I}{\partial a} + \dot{h}(a) \frac{\partial I}{\partial c} = 0 .$$

From Eq. (1.28) of the last example,

$$F(a,c,v) + v \frac{\partial I}{\partial c} - \dot{h}(a) \frac{\partial I}{\partial c} = 0 ,$$

which, with Eq. (1.27), yields

$$F(a,c,v) + [\dot{h}(a) - v] \frac{\partial F}{\partial v} = 0 ,$$

the classical transversality condition relating the initial slope v to the initial point (a,c) and the slope $\dot{h}(a)$.

(b) Inspection of (1.27), together with the continuity of $\frac{\partial I}{\partial c}$, shows that $\frac{\partial F}{\partial v}$ is a continuous function across any discontinuity in the function v. Equation (1.28) now shows that

$$F(a,c,v) - v \frac{\partial F}{\partial v} ,$$

is a continuous function across any discontinuity in v. These are the corner conditions given earlier.

SOLVED PROBLEMS

1.1 Introduce the "marginal return" function $I_c(c,a)$. Show that we can determine the optimal policy in terms of the sequence of marginal return functions $I_c(c,a)$, without requiring knowledge of the optimal return function $I(c,a)$.

From the discrete-time version of (1.3), we assume that g and L are continuously differentiable with respect to u, the minimizing u is determined as the solution of the equation

$$L_u + g_u I_c(g(c,u),\ a+1) = 0 \ . \tag{1.31}$$

Thus, (1.3) gives

$$I(c,a) = L(c,u) + I(g(c,u),\ a+1) \ , \tag{1.32}$$

with u determined by (1.31). Differentiating both sides of (1.32) with respect to c gives

$$\begin{aligned} I_c(c,a) &= L_c + L_u u_c + I_c(g(c,u),\ a+1)f_c \\ &\qquad + I_c(g(c,u),\ a+1)g_u u_c \ , \\ &= L_c + I_c(g(c,u),\ a+1)g_c \ , \end{aligned} \tag{1.33}$$

by virtue of (1.31).

Hence, using (1.31) and (1.33), we may determine the optimal policy u without making use of the optimal value function $I(c,a)$.

1.2 To apply discrete dynamic programming to a continuous time system, it is necessary to use discrete approximations to the state differential equations and performance criteria.

Determine discrete approximations to use for the system

$$\dot{x}_1(t) = \dot{x}_2(t) ,$$

$$\dot{x}_2(t) = -x_1(t) + [1 - x_1^2(t)]\, x_2(t) + u(t) ,$$

which is to be controlled so as to minimize the criteria

$$J = [x_1(T) - 5]^2 + \int_b^T \{x_2^2(t) + 20[x_1(t) - 5]^2 + u^2(t)\} dt .$$

Let the final time $T = 10$, and use an interval of $\Delta t = 0.1$.

By Taylor's theorem

$$x_1(t + \Delta t) \simeq x_1(t) + \Delta t\, x_2(t) ,$$

$$x_2(t + \Delta t) \simeq x_2(t) + \Delta t[-x_1(t) + (1 - x_1^2(t))x_2(t)] + \Delta t\, u(t) .$$

Collecting terms and letting $x_1(t) = x_1(k)$, $x_2(t) = x_2(k)$, $x_1(t + \Delta t) = x_1(k+1)$, etc., we have

$$x_1(k+1) = x_1(k) + 0.1\, x_2(k) ,$$

$$x_2(k+1) = 0.1\, x_1(k) + [1 + 0.1(1 - x_1^2(k))x_2(k)] + 0.1\, u(k) ,$$

$$J = (x_1(T) - 5)^2 + \int_{a\Delta t}^{N\Delta t} \{x_2^2(t) + 20[x(t) - 5]^2 + u^2(t)\}\, dt ,$$

$$J = (x_1(T)-5)^2 + \int_{a\Delta t}^{(a+1)\Delta t} [\ldots]dt + \int_{(a+1)\Delta t}^{N\Delta t} [\ldots]dt ,$$

$$= (x_1(N)-5)^2 + 0.1 \sum_{k=a}^{N-1} \{x_2^2(k) + 20[x_1(k)-5]^2 + u^2(k)\} ,$$

where $N = 10/0.1 = 100$, $a\Delta t = b$.

1.3 Assuming L is continuously differentiable in v in Eq. (1.3), obtain a first order partial differential equation for the minimizing control v when $g(c,v,a) = v$.

Equation (1.3) is equivalent to the two equations

$$L_v + I_c = 0 , \tag{1.34}$$

expressing the fact that v minimizes, and

$$-I_a = L(c,v,a) + v\, I_c , \tag{1.35}$$

for the value of v determined by (1.34). We can use (1.34) to rewrite (1.35) as

$$-I_a = L(c,v,a) = v\, L_v . \tag{1.35'}$$

Differentiate with respect to a. This gives

$$L_{vv} v_a + I_{ca} = 0 . \tag{1.36}$$

Similarly, differentiation of (1.35') with respect to c gives

$$-I_a = L_c + L_v v_c - v_c L_v - vL_{vv} v_c - vLv_c . \tag{1.37}$$

Equating the terms I_{ac} and I_{ca} in (136) and (137), we obtain the final result

$$L_{vv}v_a = L_c - vv_cL_{vv} - vL_{vc} \tag{1.38}$$

Note that whereas (1.3) is a nonlinear partial differential equation for I, Equation (1.38) is only a quasilinear equation for v, i.e., it is linear in v_a and v_c.

1.4 (a) Write the equations for I and v when $L(c,v,a) = c^2+v^2$, $g(c,v,a) = v$.

(b) Use the fact that, in this case, I has the form

$$I(c,a) = c^2R(a)$$

to find closed form solutions for I and v.

(a) Substituting into (1.3), we have

$$-I_a = \min_v [c^2 + v^2 + vI_c] \tag{1.39}$$

while from problem (1.3) we see that the minimizing v is given by

$$v_{min} = -1/2\ I_c\ ,$$

Using this result in (1.39) gives

$$-I_a = c^2 - 1/4\ I_c^2 \tag{1.40}$$

To determine the relevant equation for v, form the partial derivatives

$$L_{vv} = 2 ,$$

$$L_{vc} = 0 ,$$

$$L_c = 2c .$$

Substituting these results into Equation (1.38) of Problem (1.3) gives

$$v_a = c - vv_c . \tag{1.41}$$

(b) Since $I = c^2R(a)$, we have

$$I_a = c^2\dot{R}(a) ,$$

$$I_c = 2cR(a) ,$$

which, upon substituting into (1.40), yields

$$-c^2\dot{R}(a) = c^2 - c^2R^2(a) .$$

Cancellation of c^2 gives the Riccati equation for R

$$\dot{R}(a) = R^2(a) - 1 .$$

From the definition of $I(c,a)$, we have the condition

$$R(T) = 0 .$$

An elementary quadrature yields

$$R(a) = \tanh (T-a) .$$

Thus,

$$I(c,a) = c^2 \tanh (T-a) .$$

The minimizing v is now

$$v_{min} = -c \tanh (T-a) .$$

1.5 Consider the problem of minimizing

$$\int_0^T (x-u)\, dt ,$$

subject to

$$\frac{dx}{dt} = g(u),\ x(0) = c,\quad 0 \leq u \leq x .$$

Assume g satisfies the conditions:

(a) $g(0) = 0$,

(b) $g'(u) > 0$,

(c) $g''(u) < 0$,

(d) $g'(0) = \infty$,

(e) $g'(u) \to 0$ as $u \to \infty$.

(A simple function with these properties is $\sqrt{u}$.)

Show that there exists a time t_1 such that the optimal control policy has the form

$$u(x) = \begin{cases} x , & 0 \leq t \leq t_1 , \\ 0 < u < x , & t_1 < t < T , \end{cases}$$

and that the first part of the curve, where $u = x$, will appear only if T is sufficiently large. If T is small, the solution will consist only of the second part where $0 < u < x$.

The basic equation for this problem is

$$I_T(c,T) = \max_{0 \le v \le c} [c - v + g(v) I_c(c,T)] ,$$

$$I(c,0) = 0 .$$

Assume $I_c > 0$. (The nature of the constraint and the type of criterion function involved guarantee this assumption holds.) Let $K(v) = c - v + g(v) I_c$. To maximize $K(v)$, we see that

$$K'(v) = -1 + g'(v) I_c(c,T) ,$$

is positive for small v, negative for large v, and zero for only one value of v. Let us further assume that $I_c = 0$ at $T = 0$ and monotone increasing thereafter as T increases.

If we allow v to run through the interval $0 \le v < \infty$, there will always be a solution of $K'(v) = 0$. However, if v is constrained by $0 \le v \le c$, then if I_c is large (i.e., if T is sufficiently large), $K'(v) > 0$ for $0 \le v \le c$. This means the maximum will be at $v = c$ (or $u = x$) for T large compared to c.

The point of this example is to show that when appropriate structure is present the qualitative nature of optimal controls can often be determined even for constrained problems.

1.6 Consider the continuous control problem of minimizing

$$\max_{0 \le t \le T} |x(t) - p| ,$$

over all functions $u(t)$ satisfying $-1 \le u(t) \le 1$, and where

$$\frac{dx}{dt} = g(x,u) , \qquad x(0) = c .$$

(a) Define a corresponding discrete problem.

(b) Obtain recurrence relations for the solution of the discrete problem.

(a) A corresponding discrete time problem is

$$\min_{\{u(k)\}} \max_{0 \leq k \leq N} |x(k) - p| ,$$

where

$$x(k+1) = x(k) + f[x(k), u(k)]\Delta ,$$

$$x(0) = c ,$$

with

$$u(k) \triangleq u(k\Delta) , \quad \Delta = T/N , \quad x(k) \triangleq x(k\Delta) .$$

(b) Define

$$I(c,N) = \min_{\{u(k)\}} \max_{0 \leq k \leq N} |x(k) - p| .$$

Clearly, the maximum deviation either comes from the initial state c, or it occurs at some future time after the initial control has been applied. Taking these two possibilities into account and applying the principle of optimality gives

$$I(c,N) = \max\{|c-p| , \min_{|v| \leq 1} [I(c+g(c,v)\Delta, N-1]\} ,$$

$$N \geq 1 .$$

When $N = 0$, we have

$$I(c,0) = |c-p| .$$

1.7 Using Hamilton's principle that a particle moves so as to minimize the Lagrangian $\int L(t,q,\dot{q})dt$, show that the classical Hamilton-Jacobi equation may be obtained using the principle of optimality.

Let the vector q describe the state of the mechanical system and let $\dot{q}$ be the decision variable. We wish to transfer the system to the state $\bar{q}$ at time t so that the functional $\int L(t_1,q,\dot{q})dt_1$ is minimized.

Define

$$I(q,t;\bar{q},t_o) = \text{minimum value of } \int_{t_o}^{t} L(t_1,q,\dot{q})dt_1, \quad \text{subject}$$

to the above boundary conditions.

Regarding both ends of the interval $[t_o,t]$ as variable, application of the principle of optimality at each end yields the relations

$$I(q,t;\ \bar{q},t_o) = \min_{\dot{\bar{q}}(t_o)} [L(t_o,\bar{q},\dot{\bar{q}})\Delta + I(q,t;\ \dot{\bar{q}}\Delta,\ t_o+\Delta)]\ ,$$

$$I(q,t;\ \bar{q},t_o) = \min_{\dot{q}(t)} [L(t,q,\dot{q})\Delta + I(q-\dot{q}\Delta,\ t-\Delta;\ \bar{q},t_o)]\ . \tag{1.42}$$

At time t_o, we have

$$0 = \min_{\dot{\bar{q}}} \quad L + \dot{\bar{q}}\frac{\partial I}{\partial \bar{q}} + \frac{\partial I}{\partial t_o}\ ,$$

which, upon defining $\frac{\partial L}{\partial \dot{\bar{q}}} = p$, gives

$$p + \frac{\partial I}{\partial \bar{q}} = 0\ .$$

At a general time t, expansion of (1.42) gives two equations

$$p = L - \dot{q}\frac{\partial I}{\partial q} - \frac{\partial I}{\partial t}, \qquad (1.43)$$

$$\frac{\partial L}{\partial \dot{q}} = -\frac{\partial I}{\partial q} = p .$$

Define the Hamiltonian function $H(q,p,t) = pq - L$; Eq. (1.43) becomes

$$-\frac{\partial I}{\partial t} = H\left(q, \frac{\partial I}{\partial q}, t\right),$$

the Hamilton-Jacobi partial differential equation.

1.8 Consider the discrete variational problem

$$\min J(\{u_k\}) = \sum_{k=0}^{N} \ell(u_k, x_k)\Delta ,$$

$$x_{k+1} = x_k + u_k\Delta , \qquad u_o = c ,$$

where

$$|u_k| \leq m , \qquad k = 0,1,\ldots,N .$$

Show that under suitable hypotheses on ℓ, the solution of this discrete problem converges to the solution of the continuous problem

$$\min \int_0^T \ell(u,v)\, dt ,$$

$$\frac{dx}{dt} = u , \qquad x(0) = c ,$$

$$|u| \leq m ,$$

as $\Delta \to 0$.

Since the details are rather onerous, we sketch the necessary ingredients for proof, leaving the interested reader the option of consulting the literature for a complete account.

Using by now familiar arguments, the relevant functional equation for this problem is

$$I(c,N) = \min_{|u_N| \le m} [\ell(c,u_N)\Delta + I(c+u_N\Delta, N-1)] , \quad N \ge 1 ,$$

$$I(c,0) = 0 ,$$

where $I(c,N) = \min J$.

The first step entails showing that the sequence $\{I(c,N)\}$ satisfies a Lipschitz condition in a suitable c interval, i.e.,

$$|I(c,N) - I(c_1,N)| \le T\, m_2\, |c - c_1|^a ,$$

for suitable constants m_2 and a determined by imposing a similar condition on ℓ:

$$|\ell(c,u) - \ell(c_1,u)| \le m_2\, |c - c_1|^a ,$$

for $a > 0$, $|c|$, $|c_1| \le m_1$, $|u| \le m$.

Next, replace Δ by 2Δ. This yields a sequence of functions $\{P(c,N)\}$ determined by the relation

$$P(c,N+1) = \min_{|u| \le m} [\ell(c,u)2\Delta + P(c+2u\Delta,N)] , \quad N \ge 1 ,$$

$$P(c,0) = 0 .$$

We wish to compare $P(c,N)$ with $I(c,2N)$. To simplify this comparison, introduce the intermediate process with step Δ by modifying the original process by allowing only suboptimal policies in which the same choice of u is made two stages at a time. This process gives rise to the sequence of functions $h(c,N)$ defined by

$$h(c,2N) = \min_{|u| \leq m} [\ell(c,u)\Delta + \ell(c+u\Delta,u)\Delta + h(c+2\Delta, 2N-2)] , \quad N \geq 1 ,$$

$$h(c,0) = \min_{|u| \leq m} [\ell(c,u)\Delta + \ell(c+u\Delta,u)\Delta] .$$

Clearly, $I(c,2N) \leq h(c,2N)$. We now compare $P(c,N)$ and $h(c,2N)$ and obtain

$$|P(c,N) - h(c,2N)| \leq T\, m_4\, \Delta^a ,$$

where m_4 is a constant determined by the allowable c-region.

Now define the sequence of return functions

$$I(c,N,\Delta/2^r) ,$$

where $I(c,N,\Delta)$ denotes the function obtained using the time step Δ. Using the relation

$$I(c,2N) \leq h(c,2N) \leq P(c,N) + T\, m_4\, \Delta^a ,$$

we show that

$$u_{r+1}(c) \leq u_r(c) + T\, m_4 \Delta^a\, 2^{-ra} .$$

Finally, we conclude convergence of the sequence $\{u_r(c)\}$ by use of the lemma:

Lemma: If $\{a_n\}$ is a bounded sequence such that

$$a_{n+1} \leq a_n + b_n , \quad n = 0,1,\ldots ,$$

where $\Sigma b_n < \infty$ then a_n approaches a limit as $n \to \infty$.

1.9 Determine an appropriate functional equation for the problem of maximizing

$$J = \int_0^T \min\,[x(t), y(t)]\,dt$$

where

$$\frac{dx}{dt} = ax + h \ , \qquad x(0) = c_1 \ ,$$

$$\frac{dy}{dt} = by + g \ , \qquad y(0) = c_2 \ ,$$

and the functions h and g satisfy $h(t) + g(t) = 1$, $h, g \geq 0$, $0 \leq t \leq T$.

Let $\max J = I(c_1, c_2, T)$ and let α, β denote the initial choice of h and g, respectively. Note that the constraints on h and g imply $0 \leq \alpha$, β, and $\alpha = 1 - \beta$. As a result of a choice of α, the initial conditions c_1 and c_2 are transformed as $c_1 \to c_1 + (ac_1 + \alpha)\Delta$, $c_2 \to c_2 + [bc_2 + (1 - \alpha)]\Delta$ and a return $[\min\,(c_1, c_2)]\Delta$ is achieved, where $\Delta \ll 1$.

By the principle of optimality, we have

$$I(c_1, c_2, T) = \max_{0 \leq \alpha \leq 1} [\min(c_1, c_2)\Delta + I(c_1 + (ac_1 + \alpha)\Delta \ ,$$

$$c_2 + (bc_2 + (1 - \alpha))\Delta, T - \Delta] \ , \ T > 0 \ ,$$

$$I(c_1, c_2, 0) = 0 \ .$$

In the limit as $\Delta \to 0$, this becomes the nonlinear partial differential equation

$$I_T = \max_{0 \le \alpha \le 1} [\min(c_1, c_2) + (ac_1 + \alpha) I_{c_1} + (bc_2 + 1 - \alpha) I_{c_2}] .$$

1.10 Consider the differential equation

$$\ddot{x} = g(x, \dot{x}, u) , \quad x(0) = c_1 , \quad \dot{x}(0) = c_2 ,$$

where $u(t)$ is a function to be determined such that $|u| \le K$, for all t and u forces the initial state (c_1, c_2) to $(0,0)$ in minimum time. Derive a functional equation for determining u.

Let the function $I(c_1, c_2, N)$ be defined as the square of the minimum distance from the origin after N decisions have been made. Then $I(c_1, c_2, 0) = c_1^2 + c_2^2$ and

$$I(c_1, c_2, N) = \min_{|u| \le K} I(c_1 + c_2\Delta, \ c_2 + g(c_1, c_2, u)\Delta, N) , \qquad N \ge 1 .$$

To determine the minimum time needed to reach the equilibrium position, we compute the sequence $\{I(c_1, c_2, N)\}$ until we find the first value of N for which $I(c_1, c_2, N) = 0$. The time $N\Delta$ will clearly be the minimum time.

1.11 Consider the problem of finding the extreme values of the quadratic form

$$Q(x_1,x_2,\ldots,x_N) = \sum_{i=1}^{N} \sum_{j=1}^{N} a_{ij}\, x_i\, x_j ,$$

subject to the constraint $\sum_{i=1}^{N} x_i^2 = 1$, where $a_{ij} = a_{ji}$. Show that any stationary point is a characteristic vector of the matrix $A = [a_{ij}]$.

The Lagrange function is

$$F(x_1,x_2,\ldots,x_N) = Q(x_1,x_2,\ldots,x_N) + \lambda \left[\sum_{i=1}^{N} x_i^2 - 1 \right] .$$

Then

$$\frac{\partial F}{\partial x_i} = 2 \sum_{j=1}^{N} a_{ij}\, x_j + 2\, \lambda x_i , \qquad i=1,2,\ldots,N ,$$

or

$$\sum_{j=1}^{N} a_{ij}\, x_j = - \lambda x_i .$$

Thus, the stationary points are characteristic vectors and the Lagrange multiplier (or more properly its negative) is a characteristic value for the symmetric matrix A.

1.12 Let $u(t)$ be a function yielding the absolute minimum of the functional

$$J = \int_0^T G(u,\dot{u})\, dt + \lambda \int_0^T H(u,\dot{u})\, dt ,$$

and let

$$\int_0^T H(u,\dot{u})\, dt = k ,$$

for this function u. Prove that u yields the minimum value of the functional

$$J = \int_0^T G(u,\dot{u})\, dt ,$$

subject to the constraint

$$\int_0^T H(u,\dot{u})\, dt = k .$$

Assume $v(t)$ is a function such that

$$\int_0^T G(v,\dot{v})\, dt < \int_0^T G(u,\dot{u})\, dt ,$$

while satisfying

$$\int_0^T H(v,\dot{v})\, dt = k .$$

Then

$$\int_0^T G(v,\dot{v})\,dt + \lambda \int_0^T H(v,\dot{v})\,dt = \int_0^T G(v,\dot{v})\,dt + \lambda k$$

$$< \int_0^T G(u,\dot{u})\,dt + \lambda k ,$$

$$= \int_0^T G(u,\dot{u})\,dt + \int_0^T H(u,\dot{u})\,dt ,$$

which contradicts the absolutely minimizing properties of u. Thus, no such function v exists, thereby establishing the desired result.

1.13 Consider the problem of minimizing

$$J = \int_0^T F(u,v)\,dt ,$$

subject to

$$\frac{du}{dt} = H(u,v) , \quad u(0) = c ,$$

$$\int_0^T v^2 dt = K .$$

Introduce a Lagrange multiplier and then derive a functional equation for $I(c,T)$, the minimizing value of a modified functional $\tilde{J}$. Obtain a relationship between the Lagrange multiplier and the function $\frac{\partial I}{\partial K}$.

We consider minimizing the functional

$$\tilde{J} = \int_0^T F(u,v)\, dt - \lambda \int_0^T v^2\, dt ,$$

subject to

$$\frac{du}{dt} = H(u,v) , \qquad u(0) = c .$$

Letting $I(c,T)$ denote the minimum of J, we have in the usual manner

$$I(c,T) = \min_v \, [F(c,v)\Delta - \lambda v^2 \Delta + I(c + H(c,v)\Delta, T - \Delta)] .$$

Letting $\Delta \to 0$, gives the partial differential equation

$$\frac{\partial I}{\partial T} = \min_v \left[F(c,v) - \lambda v^2 + H(c,v) \frac{\partial I}{\partial c} \right] . \tag{1.44}$$

Comparing (1.44) with the corresponding equation in the text, we see that $\lambda = \frac{\partial I}{\partial K}$. This makes the physical interpretation of the multiplier clear. As noted earlier, it represents a "price" in the usual sense that it trades constraint value for return value.

SUPPLEMENTARY PROBLEMS

1.1 Obtain a functional equation for the problem

$$\min_u \int_0^T [(x - c)^2 + ku^2]\, dt ,$$

$$\frac{dx}{dt} = -ax + u , \quad x(0) = c .$$

1.2 Determine the minimum of

$$\int_0^T (1 - x)^2 ,$$

$$\frac{dx}{dt} = -x + u , \quad x(0) = 1 , \quad 0 \le u \le M , \quad M > 1 ,$$

$$\int_0^T u \, dt \le a < T .$$

1.3 Consider the problem of minimizing

$$\max_{0 \le t \le T} |x(t) - a(t)| ,$$

where

$$\frac{dx}{dt} = f(x,u) , \qquad x(0) = c , \qquad \int_0^T h(x,u)dt = K ,$$

and $a(t)$ is a known function.

1.4 Consider the equation

$$\frac{d^2x}{dt^2} + a^2x = u(t) ,$$

$$u(0) = c_1 , \quad \dot{u}(0) = c_2 , \quad a^2 < 1 .$$

We wish to choose $u(t)$, subject to $|u(t)| \le 1$, so as to reduce x to zero in minimum time. What is the appropriate functional equation?

1.5 Show that the Legendre condition is a corollary of the Weierstrass condition.

1.6 Consider the minimum distance problem

$$\min \int_0^1 \sqrt{1 + \dot{u}^2}\, dt\ , \qquad u(0) = 0\ , \qquad u(1) = 1\ .$$

(a) Derive the Euler equation for the minimizing arc;

(b) Apply the Legendre, Jacobi, and Weierstrass conditions to test the Euler equation solution as a possible minimizer.

1.7 (a) Show that the condition (1.22) is a special case of the transversality condition.

(b) What does the above transversality condition imply about the intersection of the minimizing curve with a given terminal curve for the minimum distance problem of Supplementary Problem 1.6?

1.8 Verify that for Supplementary Problem 1.6, the corner conditions (a) and (b) imply that an optimal path can have no corners.

1.9 Find a curve $u(t)$ connecting a point P with a given curve C such that the mean value of its slope is minimized. Formulate as a dynamic programming problem.

REFERENCES

Control Processes: [A-21], [B-22], [B-63], [B-65], [B-87], [B-108], [H-16], [P-11].

Calculus of Variations: [B-101], [B-107], [C-3], [H-15], [H-17], [M-26], [Y-4].

Dynamic Programming and Calculus of Variations: [B-13], [B-15], [B-20], [D-12], [L-12], [M-11], [M-32], [R-22].

CHAPTER 2

QUADRATIC CRITERIA — LINEAR EQUATIONS

INTRODUCTION

Many of the worked examples of the preceding volume have had the general structure of minimizing a quadratic form in the state and decision variables subject to a linear system difference equation. The current chapter is devoted to a systematic examination of this extremely important class of optimization problems.

In general, control problems do not possess closed form solutions, and recourse must be taken to numerical methods for their solution in most cases of practical interest. However, it is fortunate for both the mathematician and the engineer that there exists a class of problems for which explicit solutions may often be obtained. This class is important to the engineer, since it corresponds to the way control problems sometimes arise in practice; and it is of interest to the mathematician, since the explicit solutions provide a firm jumping-off point for detailed analysis of the properties to look for in the solution of more complex problems. In addition, this special class of problems often forms the basis for various numerical successive approximations schemes and perturbation procedures.

The class of problems we have in mind are those control processes in which the criterion function is quadratic and the governing system dynamics are linear. The reason for the favored position of this type of problem in the mathematical scheme of things is that the associated Euler equation is linear, thereby allowing us to employ the vast body of results available in the theory of linear differential equations. As we shall attempt to point out during the course of this chapter, the linearity of the Euler equation may be profitably employed in our dynamic programming formulation of control problems to obtain both significant reductions in dimensionality and, especially in the case of problems with constant coefficients, interesting analytic results.

DISCRETE PROBLEMS

To fix our ideas without worrying about the delicate subtleties of the variational calculus, let us first investigate the control of a system whose state is described at a discrete set of times $0,1,2,\ldots$. Let $x(k)$ denote the state at time k. Assume the dynamics of the system are

$$x(k+1) = fx(k) + gu(k), \qquad k = 0,1,\ldots,N-1 .$$

Suppose we wish to minimize the quadratic form

$$J = \sum_{k=0}^{N-1} [q(k)x^2(k) + r(k)u^2(k)] + q(N)x^2(N) ,$$

$$q(k) \geq 0,\ k = 0,1,\ldots,N-1 , \qquad q(N) > 0 ,$$

$$r(k) > 0,\ k = 0,1,\ldots,N-1 .$$

As before, we observe that the minimum value of J depends upon c, the initial state, and on a, the initial stage. Let

$$I(c,a) = \min_{u(a),u(a+1),\ldots,u(N)} \left\{ \sum_{k=a}^{N-1} [q(k)x^2(k) + r(k)u^2(k)] + q(N)x^2(N) \right\} ,$$

where $x(a) = c$.

An easy application of the principle of optimality yields

$$I(c,a) = \min_{u(a)} \left\{ [q(a)c^2 + r(a)u^2(a)] + I[f(a)c + g(a)u(a), a+1] \right\},$$

$$a < n, \tag{2.1}$$

$$I(c,N) = q(N)c^2. \tag{2.2}$$

EXAMPLE

2.1 Assume that $q(k) = 1 = r(k)$, $k = 0,1,\ldots,N$. Let us show that

(a) an actual minimum value of J exists for all c

(b) the function $I(c,a)$ possesses the form $I(c,a) = c^2 s(a)$, where $s(a)$ is a positive quantity independent of c.

(a) Let us take $u(k) = 0$ for all $k = 0,1,\ldots,N$. The corresponding value of J, call it J^0, is

$$J^0 = c^2 \sum_{k=0}^{N} (f)^{2k}.$$

Thus,

$$\min J \leq J^0 = c^2 \sum_{k=0}^{N} (f)^{2k}.$$

Since

$$\sum_{k=0}^{N} u^2(k) < \sum_{k=0}^{N} [x^2(k) + u^2(k)],$$

it suffices to minimize J over the finite region

$$\left\{u(k) \ : \ \sum_{k=0}^{N} u^2(k) \le c^2 \sum_{k=0}^{N} f^{2k}\right\}.$$

The minimum of J is assumed over this closed, bounded region since J is a continuous function of its arguments.

(b) This contention is most easily proved by mathematical induction. For $k = N$, we see from Eq. (2) that

$$I(c,N) = \operatorname*{Min}_{u(N)} \{c^2(N) + u^2(N)\} .$$

Clearly, the minimizing value is $\hat{u}(c,N) = 0$, so that

$$I(c,N) = c^2 ,$$

which has the desired form. Now, assume that $I(c,a+1)$ has the form

$$I(c,a+1) = c^2 s(a+1) ,$$

where $s(a+1)$ is positive. Then, applying the principle of optimality, we have

$$I(c,a) = \operatorname*{Min}_{u(a)} \{c^2 + u^2(a) + [f(a)c + g(a)u(a)]^2 s(a+1)\} .$$

A necessary condition for $u(a)$ to minimize the quantity in the brackets can be obtained by differentiating with respect to $u(a)$ and setting the result equal to zero. The result is

$$u(c,a) = -\frac{f(a)g(a)s(a+1)}{1 + g^2(a)s(a+1)} c ,$$

where the denominator is guaranteed to be non-zero because $s(a+1)$ is positive, by hypothesis. This condition is also sufficient, and the second derivative of the expression in brackets is

$$2 + 2g^2(a)\ s(a+1) ,$$

which is again guaranteed to be positive. If this result is substituted back into the equation for $I(c,a)$, we obtain

$$\begin{aligned} I(c,a) &= c^2 + \frac{f^2(a)g^2(a)s^2(a+1)}{[1 + g^2(a)s(a+1)]^2}\, c^2 \\ &\quad + \left[f(a) - g(a)\ \frac{f(a)s(a)s(a+1)}{1 + g^2(a)s(a+1)}\right]^2 s(a+1)c^2 \\ &= c^2 + \frac{f^2(a)g^2(a)s^2(a+1)}{[1+g^2(a)s(a+1)]^2}\, c^2 + \frac{f^2(a)s(a+1)}{[1+g^2(a)s(a+1)]^2}\, c^2 \\ &= c^2 + \frac{f^2(a)s(a+1)}{1+s^2(a)s(a+1)}\, c^2 \\ &= c^2 \left[\frac{1 + [f^2(a) + g^2(a)]s(a+1)}{1+g^2(a)s(a+1)}\right] . \end{aligned}$$

Thus, if $I(c,a+1)$ has the form $c^2 s(a+1)$, $s(a+1)$ positive, then $I(c,a)$ has the form $c^2 s(a)$, where

$$s(a) = \frac{1 + [p^2(a) + s^2(a)]s(a+1)}{1 + g^2(a)s(a+1)} ,$$

which is also clearly positive. This completes the proof by induction.

MULTIDIMENSIONAL VERSION

Since most control problems of interest have several state and decision variables, let us employ a small amount of vector-matrix theory to handle multidimensional versions of the type of control processes discussed in the previous section. Let $x(k)$ and $u(k)$ be $n \times 1$, $m \times 1$ vectors, respectively, with F and G, $n \times n$, $n \times m$ constant real matrices, respectively. Assume the system dynamics are linear

$$x(k+1) = Fx(k) + Gu(k) \ , \qquad x(0) = c \ .$$

We want to choose the $u(k)$ to minimize the criterion function

$$J = \sum_{k=0}^{N-1} [(x(k),\ Qx(k)) + (u(k),\ Ru(k))] + (x(N),\ Mx(N)) \ ,$$

where $M \geq 0$, $Q \geq 0$, $R > 0$ are matrices of sizes $n \times n$, $n \times n$, $m \times m$, respectively.

Setting

$$I(c,a) = \min J \ ,$$

we obtain the recurrence relation

$$I(c,a) = \min_u [(c,Qc) + (u,Ru) + I(Fc + Gu,\ a+1)] \ ,$$

$$a = N-1, N-2, \ldots \ , \tag{2.3}$$

$$I(c,N) = [c,Mc) \ . \tag{2.4}$$

We shall later prove that $I(c,a)$ is a pure quadratic form in c, i.e.,

$$I(c,a) = (c,P(a)c) \ ,$$

where $P(a)$ is an $n \times n$ matrix function, independent of c. Thus, we may write

$$(c,P(a)c) = \min_v \left[(c,Qc) + (v,Rv) + (Fc+Gv,P(a+1)(Fc+Gv))\right] ,$$

with

$$P(N) = M .$$

Using calculus, we obtain the minimizing v as

$$v_{min} = -(R + G'P(a+1)G)^{-1}G'P(a+1)Fc , \tag{2.5}$$

where we have used the fact that each $P(a)$ is symmetric and positive definite by the assumptions on Q and R. Substituting this expression into (2.3), doing some algebra, and equating coefficients of like powers of c, we obtain the minimizing value of J by calculating the sequence $\{P(i)\}$ according to the recurrence formula

$$P(i) = A + F'P(i+1)F - F'P(i+1)G(G'P(i+1)G + R^{-1}G'P(i+1)F ,$$

$$P(N) = M , \qquad i = N-1, N-2, \ldots, a .$$

The minimizing value of J is given by

$$J_{min} = I(c,a) = (c,P(a)c) .$$

The optimal control law (2.5) is seen to be a linear function of the state, i.e.,

$$v_{min}(i,c) = -K_i c , \qquad i = N-1, N-2, \ldots, a ,$$

where

$$K(i) = [R + G'P(i+1)G]^{-1}G'P(i+1)F .$$

This quantity is called the gain function. Thus we see that the minimizing control is a linear feedback law.

The following alternate derivation of the above results uses a completion-of-squares argument.

EXAMPLE

2.1 Letting

$$I(x,k) = \min J ,$$

we have

$$I(x,k) = \min_u [x^TQx + u^TRu + I(Fx+Gu, k+1)] ,$$

$$I(x,N) = x^TMx .^*$$

Since $I(x,k)$ has the structure

$$I(x,k) = x^TP(k)x , \quad P(k) \geq 0 ,$$

we see that

$$I(x,i) = \min_u [x^TQx + u^TRu + (Fx+Gu)^TP(k+1)(Fx+Gu)]$$

$$= \min_u [u^T(R+G^TP(k+1)G)u + 2u^TG^TP(k+1)Fx + x^T(Q+G^TP(k+1)F)x .$$

We can now write

$$I(x,k) = \min_u [(u+Sx)^TH(u+Sx) + x^T(Q+F^TP(k+1)F - S^THS)x] ,$$

where

$$S = [R+G^TP(k+1)G]^{-1} G^TP(k+1)F$$

$$H = R + G^TP(k+1)G .$$

---

*Here we adopt the alternate notation $x^TAx = (x,Ax)$, where A^T denotes the transpose of A.

Clearly, to minimize we choose

$$u(k) = -Sx(k)$$

$$= -[R(k) + G^T(k)P(k+1)G(k)]^{-1} G^T(k)P(k+1)F(k)x$$

(assuming, as always, that $R(k) > 0$ for all k).

We now have

$$I(x,k) = x^T[Q + F^T P(k+1)F - S^T HS]x$$

$$= x^T[Q + F^T P(k+1)F - F^T P(k+1)G(R + G^T P(k+1)G)^{-1} P(k+1)F]x .$$

But since $I(x,k) = x^T P(k)x$, we see that

$$P(k) = Q(k) + F^T(k)P(k+1)F(k) - F^T(k)P(k+1)G(k)[R(k) + G^T(k)P(k+1)G(k)]^{-1} G(k)P(k+1)F(k)$$

$$= Q(k) + F^T P(k+1)[F(k) - G(k)S(k)] .$$

$$P(N) = M .$$

We note that if the matrices F, G, H, Q, R are functions of i, i.e., they are time-varying, the foregoing equations carry over mutatis mutandis simply by attaching the subscript i to each quantity.

CONTINUOUS CONTROL PROCESSES

Many significant optimization problems, particularly those arising in the calculus of variations, naturally occur in a continuous, rather than discrete, time setting. While it is conceptually simpler to consider the discrete version, many interesting analytical properties do not make themselves apparent until we pass to the limit and consider the continuous problem. Also, certain simplifications of the relevant control equations occur as a result of letting time be continuous.

Let us begin by investigating the scalar problem of minimizing

$$J(x) = \int_a^T [\dot{x}^2(t) + x^2(t)]\, dt ,$$

over all $x(t)$ such that $\dot{x}(t)$ is square-integrable on $[a,T]$ with $x(a) = c$. Dynamic programming may be used to deal with such problems by writing the integral as

$$\int_a^T = \int_a^{a+\Delta} + \int_{a+\Delta}^T ,$$

where Δ is an infinitesimal. Let

$$I(c,a) = \min J .$$

Then application of the mean-value theorem for integrals gives

$$I(c,a) = \min_v \, [(c^2 + v^2)\Delta + I(c + v\Delta,\ a + \Delta)] + o(\Delta) , \qquad (2.6)$$

where $\lim_{\Delta \to 0} \frac{0(\Delta)}{\Delta} = 0$. The initial condition at $a = T$ is

$$I(c,T) = 0 .$$

Assuming requisite smoothness of I, the limiting form of (2.6) as $\Delta \to 0$ is

$$-\frac{\partial I}{\partial a}(c,T) = \min_v \left[c^2 + v^2 + v \frac{\partial I}{\partial c}(c,a) \right] ,$$

a nonlinear first order partial differential equation of rather novel form. Carrying out the minimization with respect to v yields

$$v_{min} = -1/2\left(\frac{\partial I}{\partial c}\right).$$

Hence,

$$-\frac{\partial I}{\partial a} = c^2 - 1/4\left(\frac{\partial I}{\partial c}\right)^2, \tag{2.7}$$

$$I(c,T) = 0 . \tag{2.8}$$

By the substitution $x(t) = cw(t)$, or otherwise, we see that the structure of I is quadratic in c. i.e.,

$$I(c,a) = c^2 r(a) ,$$

where $r(a)$ does not depend upon c. Substituting into (2.7) and equating like powers of c, we obtain the Riccati equation for r as

$$-\dot{r}(a) = 1 - r^2(a) , \qquad a < T ,$$

$$r(T) = 0 .$$

The minimal value of J and the optimal control are given by

$$I(c,a) = c^2 \tanh(T-a) ,$$

$$v(c,a) = -c \tanh(T-a) .$$

Now consider the more general multidimensional problem of minimizing over u

$$J = \int_a^T [(x,Qx) + (u,Ru)]dt + (x(T), Mx(T)) ,$$

where x and u are connected by the differential equation

$$\frac{dx}{dt} = Fx + Gu , \qquad x(a) = c .$$

The vectors x, u and the matrices F, G, M, Q, R are assumed to be of the same dimensions and satisfy the same restrictions as for the discrete time case.

A chain of reasoning completely analogous to that employed in the discrete time case gives rise to the result that

$$J_{min}(a,c) = (c,P(a)c) , \qquad a < T ,$$

$$u_{min}(a,c) = -R^{-1}G'P(a)c , \qquad a \leq t \leq T ,$$

where $P(a)$ is a solution of the matrix Riccati equation

$$\dot{P}(a) = - Q - F'P - PF + PGR^{-1}G'P , \tag{2.9}$$

$$P(T) = M . \tag{2.10}$$

EXAMPLE

We derive the above results using a variant of the completion-of-squares argument employed previously.

We first demonstrate an important preliminary result. For any differentiable trajectory $x(t)$, we have

$$\int_a^T [(x,\dot{P}(t)x) + (\dot{x},P(t)x) + (x,P(t)\dot{x})]dt - (x,P(t)x) \Big|_a^T = 0 ,$$

where F, G, P are given matrix functions with $P(t)$ being differentiable on $a \leq t \leq T$. Substitution of $F(t)x + G(t)u$ for $\dot{x}$ in the above expression gives the important identity

$$J = \int_a^T [(x,PGu) + (x,GPu) + (\dot{P} + F'(t)P + PF(t))x]dt$$

$$- (x,Px) \Big|_a^T + (x(T), Mx(T)) .$$

Collecting terms in the above integral and using the Riccati equation (2.9) satisfied by $P(t)$, we have

$$J = \int_a^T ||R^{\frac{1}{2}}u + R^{-\frac{1}{2}}G'Px||^2 dt + (c,P(a)c) + (x(T),Mx(T)) .$$

This last expression clearly indicates the form of the optimal control and the optimal cost.

Formally, we can also derive the continuous-time result by a passage to the limit in the discrete-time case.

EXAMPLE

Let us use the discretization

$$x(a + k\Delta t) \triangleq x(k) ,$$

$$u(a + k\Delta t) \triangleq u(k) .$$

Then the continuous criterion J is associated with the discrete criterion

$$J = \sum_{k=0}^{N-1} (x(k), Q(k)x(k))\Delta t + (u(k), R(k)u(k))\Delta t ,$$

with the dynamics

$$x(k+1) = [I + F(k)\Delta t]x(k) + G(k)\Delta t u(k) .$$

Substituting into the earlier discrete-time results, we obtain

$$K(t) = [R\Delta t + G'P(k+1)G(\Delta t)^2]^{-1} G'\Delta t P(k+1)[I + F\Delta t]$$

$$= (R\Delta t)^{-1}[G'P(k+1)\Delta t + O((\Delta t)^2)]$$

$$= R^{-1}G'P(k+1) + O(\Delta t) .$$

Thus, as $\Delta t \to 0$, we have

$$K(t) = R^{-1}G'P(t) .$$

Similarly,

$$P(k) = Q\Delta t + (I + F\Delta t)'P(k+1)[I + F\Delta t - GR^{-1}G'P(k+1)\Delta t]$$

$$= P(k+1) + Q\Delta t + F'P(k+1)\Delta t + P(k+1)F\Delta t$$

$$- P(k+1)GR^{-1}G'P(k+1)\Delta t ,$$

which, as $\Delta t \to 0$, gives Eq. (2.9).

Again, all relevant matrices may be replaced by their time-dependent counterparts and all conclusions and equations remain unchanged.

CONTROLLABILITY, OBSERVABILITY, AND STABILITY

Physically, the foregoing quadratic criteria control processes may be interpreted in the following sense: we have a system governed by the linear dynamical equation $\dot{x} = Fx$, whose initial condition has been perturbed from the origin. Consequently, we wish to exert a controlling influence on the system in order to force the state x back to the origin. On the one hand, there is a cost $(u(t), Ru(t))$ associated with applying the restoring influence $u(t)$ at time t. On the other hand, there is a cost $(x(t), Qx(t))$ of deviation of the state from the origin at time t. Thus, our previous problem (with $M=0$) represents an attempt to optimally balance these two different costs and, since we want to return the system to the origin, we are concerned with a regulator system.

The considerations of the previous section give us the best possible control over a finite interval of time $[a,T]$. However, it is both theoretically important and practically interesting to know whether or not it is possible to drive the system to the origin, given an arbitrary amount of time to exert control, i.e., letting $T \to \infty$ (or $a \to -\infty$). It is by no means obvious that the optimal control laws of the preceding sections possess this property, i.e., loosely speaking, that they are asymptotically stable laws. In fact, as we shall see, it is possible to obtain optimal control laws which are not asymptotically stable laws. These considerations lead immediately to the concept of a completely controllable and completely observable linear system. What we will be interested in is that the system to be regulated is completely controllable and that the associated optimal control law is completely observable.

The first condition, that of complete controllability, means roughly that all state variables can be affected by some suitable choice of the control function $u(t)$. The second condition, that of complete observability, means that the control function with constant gain K

$$u = -Kx$$

(conditions for a constant law to be optimal are given later) can be identically zero only if the state x is identically zero, i.e., the law K is chosen in such a way that any change from zero in the state variables is counteracted by some control action.

Mathematically speaking, the first of these conditions is expressed as rank $[G, FG, \ldots, F^{n-1}G] = n$, where the quantity inside the brackets is regarded as an nxnm matrix whose columns are those of $G, FG, \ldots, F^{n-1}G$, respectively. We shall defer the mathematical statement of the second condition until later.

We are concerned with the problem of determining what conditions must be imposed on a regulator problem in order to insure that the optimal control law K is constant, i.e., not explicitly dependent on t (if the interval length is finite, we have a non-stationary situation and the optimal control will not be constant). Thus, we consider $T = \infty$ and assume that all matrices that appear are constant, i.e., we are looking at the "steady-state" control law.

Under our positive definiteness assumptions on Q and R, we may write

$$Q = Z'Z \ ,$$

$$R = S'S \ ,$$

where Z is a $p \times n$ matrix (p = rank Q), and S is an $n \times m$ matrix. The cost function may then be written in the form

$$\int_0^\infty [(Zx,Zx) + (Su,Su)]\, dt \ . \qquad (2.11)$$

Under the complete controllability assumption, it can be established that the Riccati matrix equation (2.9) possesses a limiting solution

$$P = \lim_{t\to\infty} P(t) \ ,$$

which is unique. The corresponding optimal control law is given as

$$u(t) = -R^{-1}G'P\,x(t) \ .$$

To see that optimality does not imply stability, let $\chi(x,t)$ be any control law. Define the integral of the cost function to be zero when $u = -\chi(x,t)$ and positive for all other values of u.

(Here we depart from a purely quadratic integral and consider the general situation.) Thus, the loss is identically zero if, and only if, the law χ is used. Since the integral is non-negative, the optimal cost is also non-negative for any initial state c. It follows that the law χ, which is entirely arbitrary, is optimal and even unique.

To understand this situation more carefully, let us consider the physical significance of the integral being identically zero. (We now return to the quadratic integral situation of (2.11).) Evidently, this can happen only when $u = 0$.

Now let X_1 be the subspace consisting of those elements mapped to zero by the transformation

$$\theta = \begin{bmatrix} Z \\ \hline ZF \\ \hline \vdots \\ \hline ZF^{n-1} \end{bmatrix},$$

i.e., $X_1 = \ker \theta$. In this case, if $x \varepsilon X_1$, then $x(t) = e^{Ft}x_0$ is also contained in X_1 for all t. This follows from the identity $e^{F(t+\tau)} = e^{Ft} e^{F\tau}$, valid for all F, t, τ or from the Cayley-Hamilton theorem. Thus, X_1 is invariant under the control law $u = 0$; in fact, X_1 is the largest subspace of the state space such that the cost is identically zero when $x_0 \varepsilon X_1$. Hence, $u = 0$ is the optimal control for all states in X_1.

It is now reasonably clear that there is no loss of generality in demanding that

$$\text{dimension } X_1 = 0 .$$

This requirement is equivalent to the control-theoretic statement that the pair (F,Z) is completely observable. Mathematically, this is identical to the condition that

$$\text{rank } \Theta = n .$$

In more prosaic language, the above condition insures the non-degeneracy of the statement of the optimal control problems.

Finally, with regard to stability we may state that if the pair (F,G) is completely controllable and the pair (F,Z) is completely observable, then P_∞ is positive definite and the optimal control law K is asymptotically stable (i.e., all the characteristic roots of F-GK have negative real parts).

SOLVED PROBLEMS

2.1 Consider the problem defined by

$$J = \int_0^T (x^2 + u^2)dt ,$$

$$\frac{dx}{dt} = u .$$

(a) Solve for the minimum cost function and optimum decision policy as continuous functions of T.

(b) Compute these quantities for $T = \infty$.

(c) Let $T = 1$. Compute and sketch the optimum trajectory in state space and the optimum decision function corresponding to $x(0) = 1$ as continuous functions of t for $0 \leq t \leq 1$.

(d) Compute and sketch the trajectory in state space and the decision function corresponding to $x(0) = 1$ if the decision policy corresponding to $T = \infty$ is implemented over the interval $0 \leq t \leq 1$.

(a) We use the continuous time Riccati equation from the text. We can directly identify $F = 0$, $Q = 1$, $R = 1$, $G = 1$, then,

$$A = P \ ,$$

$$P = -(1 - P^2) \ , \qquad P(T) = 0 \ ,$$

$$\frac{dP}{1-P^2} = -dt \ ,$$

or

$$\tanh^{-1}P = c_o - t \ .$$

Thus,

$$P(t) = \tanh(T-t) \ ,$$

$$I(x,t) = P(t)x^2 = x^2\tanh(T-t) \ ,$$

$$\hat{u}(x,t) = -x \tanh(T-t) \ .$$

(b) For $T = \infty$, $\tanh(T-t) = 1$. Thus, $P = 1$ and

$$P = 1, \quad A = 1 \ ,$$

$$I(x,t) = x^2 \ ,$$

$$\hat{u}(x,t) = -x \ .$$

(c) and (d) In order to specify $\hat{u}(x,t)$ in either case, it is necessary to compute $x(t)$.

(i) For part (c), let $\sigma = T - t$. Then

$$\frac{dx(\sigma)}{dt} = \frac{dx(\sigma)}{d\sigma} = -x(\sigma) \tanh(\sigma) \ ,$$

$$\frac{dx}{x} = \tanh\sigma \, d\sigma \ ,$$

$$\log x = \log(\cosh\sigma) + c_1 \ ,$$

$$x(\sigma) = c_2 \cosh\sigma \ ,$$

when

$$t = 0 \ , \quad \sigma = T \ .$$

Hence,

$$x(0) = x_2 \cosh T \ ,$$

$$c_2 = \frac{x(0)}{\cosh T} \ .$$

For $x(0) = 1, T = 1$

$$x(1-t) = 0.648 \cosh(1-t) \ .$$

(ii) For part (d), where $T = \infty$, $\hat{u}(x,t) = -x$

$$\dot{x}(t) = -x(t) \ ,$$

$$x(t) = x(0)e^{-t} \ .$$

Thus, for $x(0) = 1$, $x(t) = e^{-t}$.

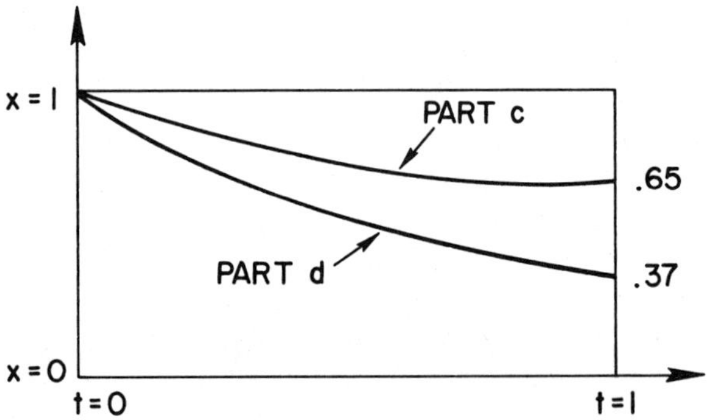

Now it is possible to obtain $u(t)$ in both cases.

(i) For part (c)

$$u(\sigma) = -x(\sigma)\ \tanh(\sigma) = -0.648\ \cosh\sigma\ \tanh\sigma$$

$$= -0.648\ \sinh\sigma\ ,$$

$$u(1-t) = -0.648\ \sinh(1-t)\ .$$

(ii) For part (d)

$$u(t) = -x(t) = -e^{-t} = u(t)\ .$$

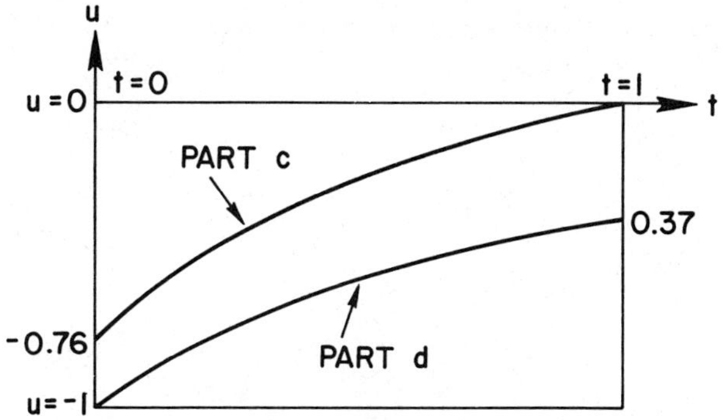

2.2 Consider the problem defined by

$$J = \int_0^T (x^2 + u^2)dt\ ,$$

$$\frac{dx}{dt} = u\ , \qquad 0 \le x \le 4\ , \quad -4 \le u \le 0\ .$$

(a) Using the solution to the previous problem, find the exact closed-form solution.

(b) Convert the problem to the discrete-time case using an increment size $\Delta t = 1$. Use the performance criterion

$$J = \sum_{k=0}^{5} [x^2(k) + u^2(k)] ,$$

and the system equation $x(k+1) = x(k) + u(k)$. Compute the minimum cost and optimal control as <u>closed-form</u> functions of x and of k, $k = 0,1,2,\ldots,5$.

(c) Solve the discrete-time problem in part (b) by the conventional dynamic programming procedure with $X = \{0,1,2,3,4\}$ and $U = \{-4,-3,-2,-1,0\}$.

(d) For parts (a)-(c), plot the optimal trajectory from $x(0) = 4$.

6.3 (a) Using the results of Problem 2.2,

$$\hat{u}(x,t) = -x \tanh(6-t) , \quad 0 \leq x \leq 4 ,$$

$$x(t) = \frac{x(0)\cosh(6-t)}{\cosh(6)} ,$$

$$I(x,t) = x^2 \tanh(6-t) .$$

Since $0 \leq \tanh(6-t) \leq 1$, the constraint $-4 \leq u \leq 0$ is always satisfied whenever $0 \leq x \leq 4$. Thus, this solution is exact.

(b) The discrete-time equations are

$$x(k+1) = x(k) + u(k) ,$$

$$J = \sum_{k=0}^{5} [x^2(k) + u^2(k)] .$$

The optimal gain matrix for this problem is

$$K(k) = \frac{P(k+1)}{1 + P(k+1)} ,$$

while the Riccati equation becomes

$$P(k) = \frac{1 + 2P(k+1)}{1 + P(k+1)} .$$

Solution of these equations yields

$$I(x,6) = 0 ,$$

$$I(x,5) = x^2 , \qquad u(x,5) = 0 ,$$

$$I(x,4) = \frac{3}{2} x^2 , \qquad u(x,4) = -\frac{1}{2} x ,$$

$$I(x,3) = \frac{8}{5} x^2 , \qquad u(x,3) = -\frac{3}{5} x ,$$

$$I(x,2) = \frac{21}{13} x^2 , \qquad u(x,2) = -\frac{8}{13} x ,$$

$$I(x,1) = \frac{55}{34} x^2 , \qquad u(x,1) = -\frac{21}{34} x ,$$

$$I(x,0) = \frac{144}{89} x^2 , \qquad u(x,0) = -\frac{55}{89} x , \quad 0 \le x \le 4 .$$

A check shows that $0 \le x \le 4$ implies that $-4 \le \hat{u}(x,k) \le 0$.

(c) Solution by conventional method.

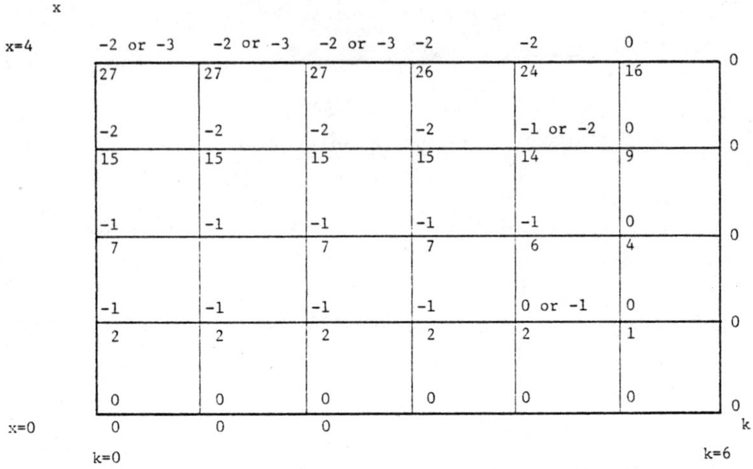

(d) The solutions for $x(0) = 4$ for parts (a), (b) and (c) are shown below.

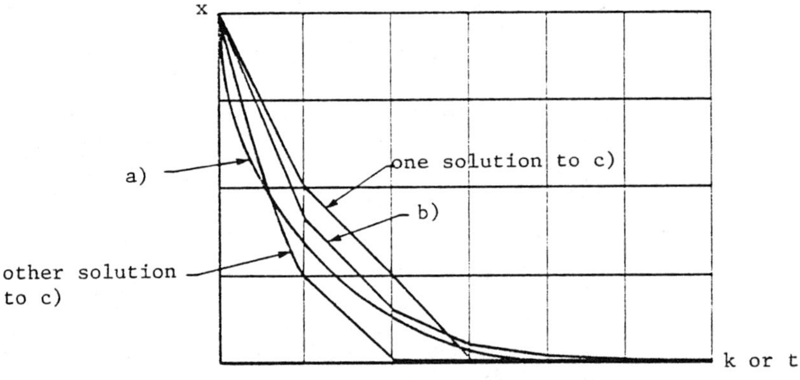

2.4 The Fibonacci sequence is a well-known series of numbers that turns up in numerous diverse areas, ranging from function minimization to Greek sculpture. The sequence is defined by each term being the sum of the two previous terms; the sequence is thus {1, 1, 2, 3, 5, 8, 13, 21, 34, 55, 89, ...}. Recall that for the problem with performance criterion

$$J = \sum_{k=0}^{N-1} [x^2(k) + u^2(k)] + x^2(N) ,$$

system equation

$$x(k+1) = x(k) + u(k) ,$$

and no constraints, the solution can be written in closed-form as

$$I(x,N) = \frac{1}{1} x^2 ,$$

$$I(x,N-1) = \frac{3}{2} x^2 , \qquad \hat{u}(x,N-1) = -\frac{1}{2} x ,$$

$$I(x,N-2) = \frac{8}{5} x^2 , \qquad \hat{u}(x,N-2) = -\frac{3}{5} x ,$$

$$I(x,N-3) = \frac{21}{13} x^2 , \qquad \hat{u}(x,N-3) = -\frac{8}{13} x ,$$

$$I(x,N-4) = \frac{55}{34} x^2 , \qquad \hat{u}(x,N-4) = -\frac{21}{34} x .$$

Note that the coefficients multiplying $-x$ in $u(x,k)$ are fractions of the form A/B, where A and B are integers. If the numerator and denominator of the fractions are written consecutively, the Fibonacci sequence is obtained. Similarly, the coefficients of x^2 in $I(x,k)$ are fractions that form the Fibonacci sequence, except that the denominator is taken <u>before</u> the numerator each time. The problem is to explain both of these phenomena quantitatively. (Hint: the Riccati equation for this problem may be helpful.)

As noted in the previous problem, the gain matrix is given by

$$K(k) = \frac{P(k+1)}{1 + P(k+1)} ,$$

while the Riccati equation is

$$P(k) = \frac{1 + 2P(k+1)}{1 + P(k+1)} .$$

We first write $P(k)$, the coefficient of $x^2 = I(x,k)$, as a fraction of $A(k)/B(k)$. We then show that

$$\frac{A(k)}{B(k)} = \frac{1 + 2\,\frac{A(k+1)}{B(k+1)}}{1 + \frac{A(k+1)}{B(k+1)}}$$

$$= \frac{B(k+1) + 2A(k+1)}{B(k+1) + A(k+1)} .$$

We see that

$$B(k) = B(k+1) + A(k+1) ,$$

$$A(k) = B(k+1) + 2A(k+1)$$

$$= B(k) + A(k+1) .$$

As noted, the Fibonacci sequence is defined by

$$F(k+2) = F(k+1) + F(k) .$$

Thus, we see that the consecutive numbers $B(k)$, $A(k)$ in reverse order of k obey this equation. Since $A(N) = B(N) = 1$ and $F(1) = F(0) = 1$, the initial conditions also correspond, and the result is obtained.

A similar analysis can be performed for $K(k)$, the coefficient of $-x$ in $\hat{u}(x,k)$.

2.5 Consider the problem with performance criterion

$$J = \sum_{k=0}^{4} [x_1^2(k) + x_2^2(k) + u^2(k)] + x_1^2(5) + x_2^2(5) ,$$

system equations

$$x_1(k+1) = x_1(k) + 2x_2(k) ,$$

$$x_2(k+1) = 2x_2(k) + u(k) ,$$

and no constraints.

(a) Calculate the minimum cost function and optimum decision policy as an explicit function of x and k.

(b) Compute the optimum trajectory in state space and the optimum decision sequence from the initial state $x_1(0) = 0$, $x_2(0) = 1$. Calculate the corresponding cost.

(c) Compute the trajectory in state space from the initial state $x_1(0) = 0$, $x_2(0) = 1$ corresponding to the decision sequence $u(0) = u(1) = u(2) = u(3) = u(4) = 0$. Calculate the corresponding cost. How does it compare to the cost in part (b)?

(d) Calculate the <u>steady-state</u> minimum cost function and optimum decision policy.

(e) Compute the trajectory in state space that results from using the <u>steady-state</u> optimum decision policy at each stage from the initial state $x_1(0) = 0$, $x_2(1) = 1$. Calculate the corresponding cost. How does it compare to the costs in parts (b) and (d)?

(a) We recall that

$$\hat{u}(x,k) = K(k)x ,$$

$$I(x,k) = x^T P(k)x ,$$

where $K(k)$ and $P(k)$ satisfy the basic recurrence equations

$$K(k) = [R + G^T P(k+1)G]^{-1} G^T P(k+1)F ,$$

$$P(k) = Q + F^T P(k+1)[F-GK(k)] .$$

We identify

$$Q = \begin{bmatrix} 1 & 0 \\ 0 & 1 \end{bmatrix}, \quad R = [1] , \quad F = \begin{bmatrix} 1 & 2 \\ 0 & 2 \end{bmatrix}, \quad G = \begin{bmatrix} 0 \\ 1 \end{bmatrix} .$$

We can then solve for the components of $K(k)$ and $P(k)$. We see that $P(k)$ is a symmetric 2x2 matrix

$$P = \begin{bmatrix} p_{11} & p_{12} \\ p_{12} & p_{22} \end{bmatrix},$$

while $K(k)$ is a 1x2 matrix

$$K = [k_1 \quad k_2] .$$

We note that

$$[R + G^T P(k+1)G]^{-1} = [1 + p_{22}(k+1)]^{-1} ,$$

and then obtain

$$k_1(k) = \frac{p_{12}(k+1)}{1 + p_{22}(k+1)} ,$$

$$k_2(k) = \frac{2p_{12}(k+1) + 2p_{22}(k+1)}{1 + p_{22}(k+1)} .$$

Further algebraic manipulation results in

$$p_{11}(k) = \frac{1}{[1+p_{22}(k+1)]}\,[1+p_{11}(k+1)+p_{22}(k+1)$$
$$+\,p_{11}(k+1)p_{22}(k+1)-p_{12}^2(k+1)]\ ,$$

$$p_{12}(k) = \frac{1}{[1+p_{22}(k+1)]}\,[2p_{11}(k+1)+2p_{12}(k+1)$$
$$+\,2p_{11}(k+1)p_{22}(k+1)-2p_{12}^2(k+1)]\ ,$$

$$p_{22}(k) = \frac{1}{[1+p_{22}(k+1)]}\,[1+4p_{11}(k+1)+8p_{12}(k+1)$$
$$+\,5p_{22}(k+1)+4p_{11}(k+1)p_{22}(k+1)$$
$$-\,4p_{12}^2(k+1)]\ .$$

We begin with

$$P(5) = \begin{bmatrix} 1 & 0 \\ 0 & 1 \end{bmatrix}\ .$$

We then obtain

$$K(4) = [0 \quad 1]\ , \qquad P(4) = \begin{bmatrix} 2 & 2 \\ 2 & 7 \end{bmatrix}\ ,$$

$$K(3) = [0.25 \quad 2.25]\ , \qquad P(3) = \begin{bmatrix} 2.5 & 3.5 \\ 3.5 & 12.5 \end{bmatrix}\ ,$$

$$K(2) = [0.262 \quad 2.382]\ , \qquad P(2) = \begin{bmatrix} 2.593 & 3.704 \\ 3.704 & 13.148 \end{bmatrix}\ ,$$

$$K(1) = [0.264 \quad 2.387]\ , \qquad P(1) = \begin{bmatrix} 2.623 & 3.770 \\ 3.770 & 13.305 \end{bmatrix}\ ,$$

$$K(0) = [0.264 \quad 2.387]\ , \qquad P(0) = \begin{bmatrix} 2.629 & 3.786 \\ 3.786 & 13.346 \end{bmatrix}\ .$$

(b) We note that

$$u(k) = -K(k)x(k) = -k_1(k)x_1(k) - k_2(k)x_2(k) .$$

Using the expressions derived in part (a), we obtain the optimum solution as shown in Table 2.1. The corresponding cost can be calculated directly. Alternatively, we observe that

$$I(x,k) = x^T P(k) x ,$$

so that

$$I(0,1,0) = p_{22}(0) .$$

In either case

$$I(0,1,0) = 13.346 .$$

(c) The results of applying the decision sequence $u(0) = u(1) = u(2) = u(3) = u(4) = 0$ are shown in Table 2.2.

Table 2.1. Optimum Solution

k	$\hat{x}_1(k)$	$\hat{x}_2(k)$	$\hat{u}(k-1)$
0	0	1	
1	2	-0.387	-2.387
2	1.226	-0.376	0.398
3	0.474	-0.178	0.574
4	0.118	-0.074	0.282
5	-0.030	-0.074	0.074

Table 2.2. Solution Using Zero Control

k	x_1	x_2	u
0	0	1	0
1	2	2	0
2	6	4	0
3	14	8	0
4	30	16	0
5	62	32	

The corresponding cost can be calculated as 6,345, a great increase over the 13.346 corresponding to the optimal solution. The benefits of applying a non-zero control to this system, which is inherently unstable, are thus seen to be very great.

(d) Because the gain matrix K(0) and the cost matrix P(0) have elements that are close to the corresponding elements of K(1) and P(1), we suspect that they are very close to the steady-state values. We can thus try to solve the steady-state Riccati equation

$$P = Q + F^T PF - F^T PG[R + G^T PG]^{-1} G^T PF$$

using these values as initial guesses and then use approximation in function space. The resulting steady-state values are

$$P = \begin{bmatrix} 2.629 & 3.789 \\ 3.789 & 13.365 \end{bmatrix} .$$

The steady-state minimum cost is computed as

$$I(x_1, x_2) = 2.629 x_1^2 + 7.578\ x_1 x_2 + 13.356\ x_2^2 .$$

The corresponding steady-state gain matrix, K, is calculated directly from P as

$$K = [0.264 \quad 2.388] .$$

The steady-state optimum decision policy is

$$u(x_1, x_2) = -0.264x_1 - 2.388x_2 .$$

(e) The solution using the optimum steady-state decision policy is shown in Table 2.3. Note that this solution is very close to the true optimum and that the cost is 13.354, compared to the true optimum cost of 13.346. Thus, the degradation in system performance due to using the steady-state solution rather than the true optimum solution is negligible. On the other hand, the constant gains of the optimum steady-state solution are much easier to implement than the time-varying gains of the true optimum solution. The steady-state solution is thus very attractive from an implementation standpoint.

Table 2.3. Solution Using Steady-State Decision Policy

k	x_1	x_2	u
0	0	1	2.388
1	2	-0.388	0.399
2	1.224	-0.377	0.577
3	0.470	-0.177	0.299
4	0.116	-0.055	0.101
5	0.006	-0.009	

2.6 Consider the problem with performance criterion

$$J = \sum_{k=0}^{4} [x_1^2(k) + x_2^2(k) + x_3^2(k) + u^2(k)] + 4x_1^2(5) + 2x_2^2(5) + x_3^2(5) ,$$

system equations

$$x_1(k+1) = x_1(k) + 0.5x_2(k)$$

$$x_2(k+1) = x_2(k) + 0.5x_3(k)$$

$$x_3(k+1) = 0.5x_1(k) - 0.5x_2(k) + 1.5x_3(k) + u(k)$$

and no constraints.

(a) Calculate the minimum cost function and optimum decision policy as an explicit function of x and k.

(b) Compute the optimum trajectory in state space and the optimum decision sequence from the initial state $x_1(0) = 0$, $x_2(0) = 0$, $x_3(0) = 1$. Calculate the corresponding cost.

(c) Compute the trajectory in state space corresponding to the decision sequence $u(0) = u(1) = u(2) = u(3) = u(4) = 0$. Calculate the corresponding cost. How does it compare to the cost in part (b)?

(a) The minimum cost function is given as $I(x,k) = x^T P(k)x$, while the optimum decision policy is $\hat{u}(x,k) = -G(k)x$.

$P(k)$ is a 3x3 matrix, and $G(k)$ is a 1x3 vector. They obey the recursive equations

$$K(k) = [R + G^T P(k+1) G]^{-1} G^T P(k+1) F \ ,$$

$$P(k) = Q + F^T P(k+1)[F - GK(k)] \ ,$$

where

$$Q = \begin{bmatrix} 100 \\ 010 \\ 001 \end{bmatrix}, \quad R = 1 \ , \quad F = \begin{bmatrix} 1 & 0.5 & 0 \\ 0 & 1 & 0.5 \\ 0.5 & -0.5 & 1.5 \end{bmatrix}, \quad G = \begin{bmatrix} 0 \\ 0 \\ 1 \end{bmatrix}$$

Numerical values are given below.

$$P(5) = \begin{bmatrix} 4 & 0 & 0 \\ 0 & 2 & 0 \\ 0 & 0 & 1 \end{bmatrix}, \qquad K^T(5) = \begin{bmatrix} 0 \\ 0 \\ 0 \end{bmatrix},$$

$$P(4) = \begin{bmatrix} 5.125 & 1.875 & 0.375 \\ 1.875 & 4.125 & 0.625 \\ 0.375 & 0.625 & 2.625 \end{bmatrix}, \qquad K^T(4) = \begin{bmatrix} 0.25 \\ -0.25 \\ 0.75 \end{bmatrix},$$

$$P(3) = \begin{bmatrix} 6.371 & 4.233 & 1.647 \\ 4.233 & 8.056 & 2.211 \\ 1.647 & 2.211 & 3.892 \end{bmatrix}, \qquad K^T(3) = \begin{bmatrix} 0.466 \\ -0.138 \\ 1.172 \end{bmatrix},$$

$$P(2) = \begin{bmatrix} 7.352 & 6.340 & 2.959 \\ 6.340 & 12.582 & 4.621 \\ 2.959 & 4.621 & 5.232 \end{bmatrix}, \qquad K^T(2) = \begin{bmatrix} 0.734 \\ 0.223 \\ 1.419 \end{bmatrix},$$

$$P(1) = \begin{bmatrix} 7.632 & 7.162 & 3.600 \\ 7.162 & 15.023 & 6.265 \\ 3.600 & 6.265 & 6.289 \end{bmatrix}, \qquad K^T(1) = \begin{bmatrix} 0.895 \\ 0.559 \\ 1.630 \end{bmatrix},$$

$$P(0) = \begin{bmatrix} 7.563 & 7.085 & 3.636 \\ 7.085 & 15.269 & 6.629 \\ 3.636 & 6.629 & 6.639 \end{bmatrix}, \qquad K^T(0) = \begin{bmatrix} 0.925 \\ 0.675 \\ 1.724 \end{bmatrix}.$$

(b) The optimal decision sequence and optimal trajectory in state space from $x^T(0) = [0\ 0\ 1]$ are found by solving the equations

$$\hat{u}[\hat{x}(k),k] = -K(k)\hat{x}(k) ,$$

$$\hat{x}(k+1) = F\hat{x}(k) - GK\hat{x}(k) ,$$

where $\hat{x}^T(0) = [0\ 0\ 1]$. The solution is shown in Table 2.4.

The minimum cost can be calculated directly as

$$J[x(0),0] = x^T(0)P(0)x(0)$$

$$= [0\ 0\ 1] \begin{bmatrix} 7.563 & 7.085 & 6.636 \\ 7.085 & 15.269 & 6.629 \\ 3.636 & 6.629 & 6.638 \end{bmatrix} \begin{bmatrix} 0 \\ 0 \\ 1 \end{bmatrix}$$

$$= 6.638 .$$

The same result can be found by directly calculating

$$\sum_{k=0}^{4} [\hat{x}_1^2(k) + \hat{x}_2^2(k) + \hat{x}_3^2(k) + \hat{u}^2(k)] + 4\hat{x}_1^2(5)$$

$$+ 2\hat{x}_2^2(5) + \hat{x}_3^2(5) ,$$

Table 2.4. Optimal Trajectory

k	$x_1(k)$	$x_2(k)$	$x_3(k)$	u(k)
0	0.000	0.000	1.000	-1.724
1	0.000	0.500	-0.224	0.086
2	0.250	0.388	-0.501	0.440
3	0.444	0.138	-0.379	0.256
4	0.513	-0.051	-0.160	-0.210
5	0.488	-0.131	0.210	0.000

using the above values. The same value, 6.638, is obtained.

(c) If the decision is set to zero at each stage, the system is essentially operating without any restraining force. Since the system is unstable, one might expect the state to grow rapidly under this policy. The results are shown in Table 2.5.

The cost must be evaluated directly at

$$J = \sum_{k=0}^{4} [x_1^2(k) + x_2^2(k) + x_3^2(k) + u^2(k)] + 4x_1^2(5)$$

$$+ 2x_2^2(5) + x_3^2(5) .$$

This value is 168.49, as compared to 6.638 for the minimum cost. The benefit of the optimum feedback decision policy over zero decisions is thus seen to be staggering for this unstable system.

Table 2.5. Uncontrolled Trajectory

k	$x_1(k)$	$x_2(k)$	$x_3(k)$	$u(k)$
0	0.000	0.000	1.000	0.000
1	0.000	0.500	1.500	0.000
2	0.250	1.250	2.000	0.000
3	0.875	2.250	2.500	0.000
4	2.000	3.500	3.063	0.000
5	3.750	5.031	3.844	0.000

2.7 For the functional

$$J = \sum_{k=a}^{N-1} [(x(k), Qx(k)) + (u(k), Ru(k))] + (x(N), Hx(N)) ,$$

and the differential equation

$$x(k+1) = Fx(k) + Gu(k), \quad x(a) = c ,$$

if $I(c,a) = \min J$, show that $I(c,a)$ has the structure

$$I(c,a) = (c, Z(a)c) ,$$

where $Z(a)$ is a matrix independent of c.

From the text, I satisfies the equation

$$I(c,a) = \min_{v} [(c,Qc) + (v,Rv) + I(Fc + Gv, a+1)] , \quad a < N ,$$

$$I(c,N) = (c,Hc) .$$

Clearly, the stated result holds if $a = N$. Assume that

$$I(Fc + Gv, a+1) = [Kc, Z(a+1)Kc] ,$$

for an appropriate K (note that K, in general, depends on a but not c). Thus,

$$I(c,a) = (c,Qc) + (Bc, RBc) + [Kc, Z(a+1)Kc] ,$$

where $v_{min} = Bc$. This is clearly a pure quadratic form in c

$$I(c,a) = (c, Z(a)c) ,$$

where $Z(a)$ is determined by Q, R, $B(a)$, $Z(a+1)$. Hence, the result is established for all a by induction.

2.8 Consider the system of linear algebraic equations

$$Ax = y$$

where A is symmetric, real, and positive definite. Using the fact that the solution x minimizes the quadratic form

$$Q(x) = (x,Ax) - 2(x,y) ,$$

derive a recurrence formula for determining the minimum value of Q and the solution vector x.

Introduce the function of n variables

$$I_n(y_1,y_2,\ldots,y_n) = \min_{\{x_i\}} \sum_{i=1}^{n} \sum_{j=1}^{n} a_{ij}x_i x_j - 2 \sum_{i=1}^{n} x_i y_i$$

where x_i, y_i are the ith components of x and y, respectively. Write

$$Q(x) = a_{nn}x_n^2 - 2x_n y_n + \sum_{i=1}^{n-1} \sum_{j=1}^{n-1} a_{ij}x_i x_j - 2 \sum_{i=1}^{n-1} x_i(y_i - a_{in}x_n).$$

From this expression, we readily obtain

$$I_n(y_1,y_2,\ldots,y_n) = \min_{x_n} \begin{cases} a_{nn}x_n^2 - 2x_ny_n \\ + I_{n-1}(y_1-a_{1n}x_n, y_2-a_{2n}x_n, \ldots \\ \ldots, y_{n-1}-a_{n-1,n}x_n) \\ n \geq 2 . \end{cases} \quad (1)$$

Clearly,

$$I_1(y_1) = \frac{-y_1^2}{a_{11}} .$$

Notice that at each stage, the number of state variables is increased by one to account for the fact that we are increasing the "size" of the system by one more variable.

2.9 Using the results of Problem 2.8 and the fact that $I_n(y_1,y_2,\ldots,y_n)$ has the form

$$I_n(y_1,y_2,\ldots,y_n) = \sum_{i=1}^{n} \sum_{j=1}^{n} c_{ij}(n)y_iy_j ,$$

show how to derive relations linking

$$\{c_{ij}(n)\} \text{ with } \{c_{ij}(n-1)\} .$$

From Eq. (1) of Problem 2.8,

$$\sum_{i=1}^{n}\sum_{j=1}^{n} c_{ij}(n)y_i y_j = \min_{x_n} \left\{ \begin{aligned} & a_{nn}x_n^2 - 2x_n y_n \\ & + \sum_{i=1}^{n-1}\sum_{j=1}^{n-1} c_{ij}(n-1) \\ & \cdot (y_i - a_{in}x_n)(y_j - a_{jn}x_n) \end{aligned} \right\}$$

$$= \min_{x_n} \left\{ \begin{aligned} & x_n^2 \quad a_{nn} + \sum_{i=1}^{n-1}\sum_{j=1}^{n-1} a_{in}a_{jn}c_{ij}(n-1) \\ & -2x_n \quad y_n + \sum_{i=1}^{n-1}\sum_{j=1}^{n-1} a_{in}y_j c_{ij}(n-1) \\ & + \sum_{i=1}^{n-1}\sum_{j=1}^{n-1} y_i y_j c_{ij}(n-1) \end{aligned} \right\} . \tag{1}$$

Carrying out the minimization, we have the minimizing x_n satisfying

$$x_n = \frac{y_n + \sum_{i=1}^{n-1}\sum_{j=1}^{n-1} a_{in}y_j c_{ij}(n-1)}{a_{nn} + \sum_{i=1}^{n-1}\sum_{j=1}^{n-1} a_{in}y_{jn}c_{ij}(n-1)} .$$

The minimum value of I_n is given by substituting this result back into (1). This yields

$$\sum_{i=1}^{n}\sum_{j=1}^{n} c_{ij}(n)y_i y_j = \frac{-\left[y_n + \sum_{i=1}^{n-1}\sum_{j=1}^{n-1} a_{in}y_j c_{ij}(n-1)\right]^2}{a_{nn} + \sum_{i=1}^{n-1}\sum_{j=1}^{n-1} a_{in}a_{jn}c_{ij}(n-1)}$$

$$+ \sum_{i=1}^{n-1}\sum_{j=1}^{n-1} c_{ij}(n-1)y_i y_j .$$

Equating coefficients of $y_i y_j$ in (2) gives appropriate recurrence relations linking the sequence

$$\{c_{ij}(n)\} \text{ with } \{c_{ij}(n-1)\} .$$

2.10 Consider the case when A is a positive definite Jacobi matrix, i.e.,

$$a_{ij} = 0 , \qquad |i-j| \geq 2 .$$

Show that a recurrence formula involving only functions of one variable may be obtained for the solution of

$$Ax = y .$$

We wish to minimize

$$Q(x) = (x,Ax) - 2(x,y) .$$

Writing this expression out, we have

$$Q(x) = a_{11}x_1^2 + 2a_{12}x_1x_2 + a_{22}x_2^2 + 2a_{23}x_2x_3 + \dots +$$

$$2a_{n-1n}x_{n-1}x_n + a_{nn}x_n^2 - 2y_1x_1 - 2y_2x_2 - \dots - 2y_nx_n .$$

We consider the problem of minimizing

$$Q_k(x,z) = a_{11}x_1^2+2a_{12}x_1x_2+a_{22}x_2^2+2a_{23}x_2x_3 + \dots$$

$$+ 2a_{k-1,k}x_{k-1}x_k+a_{kk}x_k^2-2y_1x_1-2y_2x_2- \dots -2zx_k ,$$

$$k = 1,2,\dots,n .$$

Clearly,

$$Q_1(x,z) = a_{11}x_1^2 - 2zx_1 .$$

Let

$$I_k(z) = \min_x Q_k(x,z) .$$

The usual arguments now give

$$I_k(z) = \min_{x_k} [a_{kk}x_k^2 - 2zx_k + I_{k-1}(y_{k-1} - a_{k-1,k}x_k)] ,$$

$$k = 2,3,\ldots,n ,$$

$$I_1(z) = \frac{z^2}{a_{11}} .$$

It is easily established by induction, or otherwise, that each $I_k(z)$ is quadratic in z, i.e.,

$$I_k(z) = u_k z^2 + w_k z + v_k .$$

Substituting this expression into (1) and performing the required minimization yields recurrence relations for the sequences $\{u_k\}$, $\{w_k\}$, and $\{v_k\}$.

2.11 Consider the minimization of

$$Q_{m,n}(x) = \sum_{k=0}^{n} \left(b_k - \sum_{i=0}^{m} x_i a_{k-i} \right)^2 , \quad n > m \geq 0 ,$$

over the x_i, where $\{a_i\}$ and $\{b_i\}$ are given. Derive a recurrence relation for obtaining the minimum of $Q_{m,n}$.

Introduce the quadratic form

$$I_{n,m}(b_0,b_1,\ldots,b_n) = \min_{x_0,x_1,x_2,\ldots,x_m} \left\{ \begin{array}{l} (b_0-x_0a_0)^2 \\ + (b_1-x_0a_1-x_1a_0)^2+\ldots \\ \ldots+(b_n-x_0a_n-x_1a_{n-1}-\ldots \\ \ldots-x_ma_{n-m})^2 \end{array} \right\} .$$

Some elementary algebra shows that

$$\sum_{i=0}^{n} a_i^2 \; I_{n,0}(b_0,b_1,\ldots,b_n) = \sum_{i=0}^{n} a_i^2 \sum_{i=0}^{n} b_i^2 - \left(\sum_{i=0}^{n} a_ib_i\right)^2 .$$

Applying our previous arguments shows that

$$I_{n,m}(b_0,b_1,\ldots,b_n) = \min_{x_0} \; (b_0-x_0a_0)^2$$

$$+I_{n-1,m-1}(b_1-x_0a_1,b_2-x_0a_2,\ldots,b_n-x_0a_n) .$$

2.12 Consider the problem

$$\max_{\{u(k),v(k)\}} \; [b_1x(n) + b_2y(n)] ,$$

where

$$x(k+1) = a_{11}u(k) + a_{12}v(k),\; x(0) = c_1 ,$$

$$y(k+1) = a_{21}u(k) + a_{22}v(k),\; y(0) = c_2 ,$$

with the constraints

$$0 \leq u(k) \leq x(k) \ , \quad 0 \leq v(k) \leq y(k) \ , \quad k = 0, ., \ldots, n-1 \ .$$

Let $I_n(c_1,c_2)$ denote the maximizing value.

(a) Derive a recurrence relation for the sequence $\{I_n(c_1,c_2)\}$.

(b) Show that the sequence of functions of two variables $\{I_n(c_1,c_2)\}$ may be calculated by introducing two functions of one variable.

(a) By definition

$$I_n(c_1,c_2) = \max \ [b_1 x(n) + b_2 y(n)] \ .$$

As a result of any initial decision $[u(0), v(0)]$, the states c_1 and c_2 are transformed as

$$c_1 \to a_{11}u(0) + a_{12}v(0) \ ,$$

$$c_2 \to a_{21}u(0) + a_{22}v(0) \ .$$

For ease of writing, let $u = u(0)$, $v = v(0)$, be the optimal initial decisions for an n-stage process. Let R be the region in u-v space given by $0 \leq u \leq c_1$, $0 \leq v \leq c_2$. An application of the principle of optimality now gives

$$I_n(c_1,c_2) = \max_R \ [I_{n-1}(a_{11}u + a_{12}v, \ a_{21}u + a_{22}v)] \ , \qquad n \geq 1 \ ,$$

$$I_0(c_1,c_2) = b_1 c_1 + b_2 c_2 \ .$$

Let us now establish the result that each element of the sequence $I_n(c_1,c_2)$ is a homogeneous function of the first degree in both c_1 and c_2. Our proof is by induction on n. First of all, the result is clearly true for $n=0$. Assume $I_{n-1}(c_1,c_2)$ possesses the homogeneity property, i.e.,

$$I_{n-1}(\alpha c_1,\alpha c_2) = \alpha c_1 I_{n-1}(1,c_2) = \alpha c_2 I_{n-1}(c_1,1) \quad ,$$

α a scalar.

Define the new variables $u^1 = \frac{u}{c_1}$, $v^1 = \frac{v}{c_2}$. Then

$$I_n(c_1,c_2) = \max_{0\le u^1,v^1\le 1} [I_{n-1}(a_{11}c_1u^1 + a_{12}c_c v^1,$$
$$a_{21}c_1u^1 + a_{22}c_2v^1)] \; .$$

Applying the induction hypothesis and the homogeneity property, this becomes

$$I_n(c_1,c_2) = \max_{u^1,v^1} \left[\left(c_1 a_{11}u^1 + a_{12}\frac{c_2}{c_1}v^1 \right) \cdot I_{n-1}\left(1, \frac{a_{21}u^1 + a_{22}\frac{c_2}{c_1}v^1}{a_{11}u^1 + a_{12}\frac{c_2}{c_1}v^1} \right) \right] \tag{1}$$

$$= c_1 I_n\left(1, \frac{c_2}{c_1}\right) , \tag{2}$$

which completes the proof.

(b) Define the new functions

$$g_n(t) = I_n(1,t) \ ,$$

$$h_n(t) = I_n(t,1) \ , \qquad 0 \le t \le 1 \ .$$

Turning to (1), we have

$$g_n(t) = \max_{\substack{0 \le u^1 \le 1 \\ 0 \le v^1 \le 1}} \left[(a_{11}u^1 + a_{12}v^1 t) g_{n-1}\left(\frac{a_{21}u^1 + a_{22}v^1 t}{a_{11}u^1 + a_{12}v^1 t} \right) \right] ,$$

$$0 \le t \le 1 \ , \ n \ge 1 \ ,$$

with a similar expression for the sequence $\{h_n(t)\}$. Thus, instead of dealing with a single sequence of functions of two variables which involves the expanding grid phenomena previously illustrated, we may calculate the two sequences $\{g_n(t)\}$ and $\{h_n(t)\}$ over the fixed interval, $0 \le t \le 1$, and then use (2) to obtain $I_n(c_1,c_2)$, for any $c_1,c_2 \ge 0$. As pointed out in an Part I, it is generally computationally desirable to deal with many functions of one variable rather than one function of many variables, so the above technique represents a significant dimensionality reduction tool.

2.13 Consider the problem of minimizing

$$J = \int_0^T (\dot{x}^2 + x^2) dt \ , \qquad x(0) = c \ .$$

We have seen that J has the structure

$$J_{min} = c^2 R(T) .$$

Show that $R(T) = 1 + 0(e^{-2T})$ as $T \to \infty$, where $0(x)$ is a function such that

$$\lim_{x \to \infty} \frac{0(x)}{x} = \text{constant} .$$

Write $R(T) = 1 + s(T)$. Since $R(T)$ satisfies the equation

$$\dot{R} = 1 + R^2(T), \quad R(0) = 0 ,$$

$$\dot{s}(T) = -2s(T) - s^2(T) , \quad s(0) = -1 .$$

Solving for s, we obtain

$$s(T) = -e^{-2T} - \int_0^T e^{-2(T-y)} s^2(y)dy .$$

Since $0 \leq R(T) \leq 1$, we have $|s(T)| \leq 1$. Thus,

$$|s(T)| \leq e^{-2T} + \int_0^T e^{-2(T-y)}dy$$

$$= 1/2e^{-2T} [1 + e^{2T}]$$

$$= 1/2[e^{-2T} + 1] = 0(e^{-2T}) .$$

2.14 Consider the problem of minimizing

$$J(x) = \int_0^T [\dot{x}^4(t) + x^4(t)]dt \quad ,$$

where $x(0) = c$. Let $I(c,T) = \min J$.

(a) Show that $I(c,T) = c^4 R(T)$.

(b) Show that $\dot{R}(T) = 1 - 3R^{4/3}(T)$, $R(0) = 0$.

(c) Show that $v(c,T) = g(T)c$ and determine the function g.

(a) Let $x(t) = cw(t)$, where $w(0) = 1$ and $\dot{w} \varepsilon L^4[0,T]$. Then

$$J(x) = J(cw) = \int_0^T c^4[\dot{w}^4 + w^4]dt$$

$$= c^4 R(T) \quad . \tag{1}$$

(b) The usual arguments give the relation for I:

$$\frac{\partial I}{\partial T}(c,T) = \min_v \left[c^4 + v^4 + v\frac{\partial I}{\partial c}\right] . \tag{2}$$

Hence,

$$v_{min} = \left[-1/4 \frac{\partial I}{\partial c}\right]^{1/3} . \tag{3}$$

Substituting (1) into (2), we see that

$$c^4\dot{R}(T) = c^4 + c^4R^{4/3} - 4c^4R^{4/3} \quad ,$$

or

$$\dot{R}(T) = 1 - 3R^{4/3} , \qquad R(0) = 0 .$$

(c) From (1) and (3) we see

$$v_{min} = \left(-1/4 \frac{\partial I}{\partial c}\right)^{1/3}$$

$$= -R^{1/3}(T)c .$$

Thus,

$$g(T) = -R^{1/3}(T) .$$

2.15 Let

$$J(x) = \int_0^T [\dot{x}^2(t) + x^2(t)]dt .$$

Write $I(c,T) = \min_{x(0)=c} J(x)$ and let $\tilde{x}(t,c,T)$ denote the minimizing function.

(a) Show that

$$I(c,T) \leq \int_0^T [\dot{\tilde{x}}(t,c,T_1)^2 + \tilde{x}(t,c,T_1)^2]dt ,$$

for all $T_1 \geq T$;

(b) Show that

$$I(c,T_1) \leq I(c,T) + (T_1 - T)\tilde{x}^2(T,c,T) .$$

This problem shows that the functional I is always Lipschitz continuous in T, provided the minimizing function stays bounded at the right endpoint of the interval of consideration.

(a) By definition,

$$I(c,T) = \min_{x(0)=c} \int_0^T [\dot{x}(t)^2 + x(t)^2]dt .$$

Let $\tilde{x}(t,c,T_1)$ be the minimizing function when the interval length is $T_1 \geq T$. Clearly, the restriction of $\tilde{x}(t,c,T_1)$ to the interval $0 \leq t \leq T$ is an admissible function. Hence,

$$I(c,T) \leq \int_0^T [\dot{\tilde{x}}(t,c,T_1)^2 + \tilde{x}(t,c,T_1)^2]dt .$$

(b)

$$I(c,T_1) = \min \int_0^{T_1} [\dot{x}^2(t) + x^2(t)]dt$$

$$= \min_{x(0)=c} \left\{ \int_0^T [\dot{x}^2(t) + x^2(t)]dt + \int_T^{T_1} [\dot{x}^2(t) + x^2(t)]dt \right\} .$$

Since $\tilde{x}(t,c,T)$ is the minimizing function on $[0,T]$,

$$I(c,T_1) \leq \min_{x(0)=c} \left\{ \int_0^T [\dot{\tilde{x}}(t,c,T_1)^2 + \tilde{x}(t,c,T_1)^2]dt + \int_T^{T_1} [\dot{x}^2(t) + x^2(t)]dt \right\} .$$

Since the minimization over x does not involve the first integral, we may write

$$I(c,T_1) \leq \int_0^T [\dot{\tilde{x}}(t,c,T_1)^2 + \tilde{x}(t,c,T_1)^2]dt + \min_{x(0)=c} \int_T^{T_1} [\dot{x}^2(t) + x^2(t)]dt$$

$$= I(c,T) + \min_{x(0)=c} \int_T^{T_1} [\dot{x}^2(t) + x^2(t)]dt .$$

Choosing the admissible function

$$x(t) = \tilde{x}(T,c,T)$$

for the integral term, and applying the mean-value theorem for integrals, we see that

$$I(c,T_1) \leq I(c,T) + (T_1-T)\tilde{x}^2(T,c,T) .$$

2.16 Consider the problem of minimizing (over u)

$$J = \int_0^T [(x,Qx) + (u,Ru)] ,$$

where $\dot{x}(t) = Fx + Gu, \quad x(0) = c.$

Show that this problem may be brought into the form

$$\min_{\bar{u}} \int_0^T [(\bar{x},\tilde{A}\bar{x}) + (\bar{u},\bar{u})]dt ,$$

$$\dot{\bar{x}} = \tilde{C}\bar{x} + \bar{u} , \quad \bar{x}(0) = \bar{c} ,$$

by a change of variable in both x and u, provided G is non-singular and R positive definite. (Note that the non-singularity of G is a very severe restriction.)

Define the new variables

$$\bar{x} = R^{1/2}G^{-1}x ,$$

$$\bar{u} = R^{1/2}u .$$

Thus,

$$(x,Qx) = (\bar{x}, R^{1/2}G^TQGR^{-1/2}\bar{x})$$

$$(u,Ru) = (R^{-1/2}\bar{u}, R^{1/2}\bar{u})$$

$$= (\bar{u},\bar{u}) , \text{ since } R > 0 .$$

Also,

$$\dot{x} = GR^{-1/2}\dot{\bar{x}} = FGR^{-1/2}\bar{x} + GR^{-1/2}\bar{u} ,$$

or

$$\dot{\bar{x}} = R^{1/2} G^{-1} FGR^{-1/2} \bar{x} + \bar{u} ,$$

$$\bar{x}(0) = R^{1/2} G^{-1} c .$$

Consequently, the original problem has been transformed into that of minimizing

$$J = \int_0^T [(\bar{x},\bar{Q}\bar{x}) + (\bar{u},\bar{u}) ,$$

$$\dot{\bar{x}}(t) = \overline{F\bar{x}} + \bar{u} , \quad \bar{x}(0) = \bar{c} ,$$

with

$$\bar{Q} = R^{-1/2} G^T QGR^{-1/2} ,$$

$$\bar{F} = R^{1/2} G^{-1} FGR^{-1/2} ,$$

$$\bar{c} = R^{1/2} G^{-1} c .$$

2.17 Let L be a bounded linear functional defined on the space $L^2[0,T]$, the square-integrable functions on the interval $[0,T]$.

Using the fact that the quadratic function

$$Q(f) = \int_0^T f \, dt - 2L(f) ,$$

possesses a unique minimizing function g, establish the Riesz Representation Theorem (weak form)

$$L(h) = \int_0^T gh \, dt \quad \text{for \underline{all}} \quad h \in L^2[0,T] .$$

Consider the function $u(\varepsilon) = Q(g+\varepsilon h)$ as a function of ε. Since g is the unique minimizing function, u has an absolute minimum at $\varepsilon = 0$. But

$$u(\varepsilon) = Q(g+\varepsilon h) = \int_0^T (g+\varepsilon h)^2 dt - 2L(g+\varepsilon h)$$

$$= \int_0^T g^2 dt + 2\varepsilon \int_0^T gh\, dt + \varepsilon^2 \int_0^T h^2\, dt$$

$$- 2L(g) - 2\varepsilon L(h)\ .$$

Differentiating with respect to ε, we have

$$\dot{u}(\varepsilon) = 2\left[\int_0^T gh\, dt + \varepsilon \int_0^T h^2\, dt - L(h)\right] .$$

Thus, the minimizing ε is

$$\varepsilon_{min} = \frac{L(h) - \int_0^T gh\, dt}{\int_0^T h^2\, dt} .$$

But $\varepsilon_{min} = 0$ by the quadratic nature of Q, which concludes our proof.

2.18 Consider the problem of minimizing the quadratic functional

$$J = \int_0^T [(\dot{x},\dot{x}) + (x,Ax)]dt ,$$

over all x such that

$$\int_0^T (x,f)dt = c .$$

Here x and f are n-dimensional vectors and A is a positive definite matrix. Show that the minimizing function x and the minimum value of J can be determined by solving the equation

$$\ddot{y} - Ay = -f , \quad \dot{y}(0) = \dot{y}(T) = 0 .$$

Define the new inner product of two vectors as

$$\langle x,y \rangle = \int_0^T [(\dot{x},\dot{y}) + (x,Ay)]dt .$$

Since J is a positive definite quadratic form, we have

$$J(\alpha x + \beta y) = \alpha^2 \langle x,x \rangle + 2\alpha\beta \langle x,y \rangle + \beta^2 \langle y,y \rangle \geq 0 ,$$

for any two real scalars α, β. Thus, the Cauchy Schwarz inequality gives

$$\langle x,x\rangle \langle y,y\rangle \geq 2 \langle x,y\rangle ,$$

with equality if, and only if, $x = \lambda y$, for some scalar λ. Note that

$$\langle x,\dot{y}\rangle = (x,\dot{y}) \Big|_0^T + \int_0^T (x,\ Ay - \ddot{y})dt .$$

Thus, if we choose y to be a solution of

$$Ay - \ddot{y} = f,\ \dot{y}(0) = \dot{y}(T) = 0 ,$$

we obtain the minimum value of J as

$$\min_x \langle x,x\rangle = \min_x J = \frac{2c}{(y,y)} .$$

The minimizing function is given by

$$x = \lambda y ,$$

where λ is determined from

$$\int_0^T (\lambda y,f)dt = c ,$$

or

$$\lambda = \frac{c}{\int_0^T (y,f)dt} .$$

2.19 Consider the problem of minimizing

$$\int_0^\infty [(x,Qx) + (u,Ru)]dt ,$$

with

$$\frac{dx}{dt} = Fx + Gu \quad , \quad x(0) = c \ .$$

Let $Q = I_{4x4}$ (the 4x4 identity matrix), $R = I_{2x2}$. Assume F and G are given by

$$F = \begin{bmatrix} -3 & -3 & 0 & 1 \\ 26 & 36 & -3 & -25 \\ 30 & 49 & -2 & -27 \\ 30 & 43 & -3 & -32 \end{bmatrix} , \quad G = \begin{bmatrix} 3 & 3 \\ -2 & -1 \\ 0 & 0 \\ 0 & 1 \end{bmatrix} .$$

Is the steady-state control law

$$K = -R^{-1}G'P_\infty$$

$$= -G'P_\infty \ ,$$

asymptotically stable?

For K to be asymptotically stable, it suffices to have the pair (F,G) completely controllable since $Q > 0$. Thus, we must check whether the condition

$$\text{rank } (G, \ FG, \ F^2G, \ F^3G) = 4$$

is satisfied.

With the above F and G this matrix is

$$\begin{bmatrix} 3 & 3 & -3 & -3 & 3 & 3 & -3 & -3 \\ -2 & -1 & 6 & 8 & 2 & 6 & 14 & 22 \\ 0 & 3 & 12 & 18 & 12 & 24 & 36 & 60 \\ 0 & 1 & 4 & 6 & 4 & 8 & 12 & 20 \end{bmatrix} .$$

The rank of this matrix is 2, implying that (F,G) is not completely controllable. Thus, K is optimal, but may not be stable and it may not be possible to drive this system to the origin for all initial disturbances.

2.20 Consider a system governed by the linear dynamics

$$\dot{x} = Fx + gu ,$$

where x and g are n-dimensional vectors, F is an nxn matrix, and F and g are constant. Assume that the pair (F,g) is completely controllable. Show that given n arbitrary complex numbers $\{\lambda_i\}$, there exists a constant 1xn matrix k' such that the characteristic roots of the matrix $F - gk'$ are the set $\{\lambda_i\}$.

(This is the scalar "pole-shifting theorem." Since we may interpret k as a linear feedback law through the control $u = -k'x$, this result shows that there is no need to design stability into a system since we may always induce whatever stability behavior we desire by a suitable feedback control law.)

Writing the characteristic polynomial of F as

$$\chi_F(z) = z^n + \alpha_1 z^{n-1} + \dots + \alpha_n ,$$

introduce the new vectors

$$e_i = F^{n-1}g + \alpha_1 F^{n-2}g + \dots + \alpha_{n-1}g ,$$
$$\vdots$$
$$e_{n-1} = Fg + \alpha_1 g ,$$
$$e_n = g .$$

Since (F,g) is completely controllable, the set $e_1,\dots,e_n$ forms a basis for R^n. In this basis F and G have the matrix representations

$$F = \begin{bmatrix} 0 & 1 & 0 & \cdots & 0 \\ 0 & 0 & 1 & \cdots & 0 \\ \vdots & \vdots & \vdots & & \vdots \\ 0 & 0 & 0 & & 1 \\ -\alpha_n & -\alpha_{n-1} & -\alpha_{n-2} & & -\alpha_1 \end{bmatrix}, \quad g = \begin{bmatrix} 0 \\ \vdots \\ 0 \\ 1 \end{bmatrix}.$$

Form the desired characteristic polynomial

$$\chi_{F-gk'} = (z-\lambda_1)(z-\lambda_2)\ \dots\ (z-\lambda_n)$$
$$= z^n + \beta_1 z^{n-1} + \dots + \beta_n .$$

We see that the control gain matrix

$$k' = [\alpha_n - \beta_n, \alpha_{n-1} - \beta_{n-1}, \dots, \alpha_1 - \beta_1] ,$$

is the desired feedback law since we may use the special basis $\{e_i\}_{i=1}^n$ and the fact that the characteristic polynomial of a matrix is invariant under a change in basis.

SUPPLEMENTARY PROBLEMS

2.21 Consider the problem of minimizing

$$J = \sum_{n=0}^{N} [(x_n, x_n) + (u_n, u_n)] ,$$

where

$$x_{n+1} = Fx_n + u_n , \quad x_o = c .$$

Let $I(c,N) = \min J$. It has been shown that I has the structure

$$I(c,N) = (c, R(N)c) ,$$

where the matrices $\{R(N)\}$ are independent of c. Show that

$$I \leq R(N-1) < R(N) \leq (I + FF') ,$$

(in the ordering $\leq$ induced on the set of non-negative definite matrices by $A \leq B$ if and only if $B - A \geq 0$) and thus that as $N \to \infty$, $R(N) \to R_\infty$, where R_∞ satisfies

$$R_\infty = (I + FF') - F'(I + R_\infty)^{-1}F .$$

2.22 Consider the problem of minimizing

$$J(x,\lambda) = \int_0^T [(\dot{x},\dot{x}) + (x,Bx)]dt + \lambda(x(T)-d, x(T)-d) ,$$

$\lambda > 0$, where $x(0) = c$. Discuss the limiting behavior of $x = x(t,\lambda)$ as $\lambda \to \infty$.

2.23 Consider the problem of minimizing the quadratic function

$$Q = \int_0^T \left[f(t) - \sum_{k=1}^{N} \alpha_k \, g_k(t) \right]^2 dt \; .$$

Let $I(f,N) = \min Q$. Show that

$$I(f,1) = \min_{\alpha_1} \int_0^T [f(t) - \alpha_1 \, g_1(t)]^2 dt \; ,$$

$$I(f,N) = \min_{\alpha_n} [I(f - \alpha_N g_N, N-1)] \; , \qquad N \geq 2 \; .$$

Also, show that

$$I(f,N) = \int_0^T f^2 dt - \int_0^T \int_0^T k_N(s,t) f(s) f(t) ds \, dt \; ,$$

and use the above relations to obtain a recurrence relation linking the functions k_N and k_{N-1}.

2.24 Consider the problem of minimizing

$$J = \int_0^T (\dot{x}^2 + x^2) dt \; ,$$

where $x(0) = c$. Without finding the explicit solutions, show that $|x| \leq |c|g(T)$ for an appropriate function g.

2.25 Consider the constant linear system

$$x = Fx + gu \ ,$$

where

$$F = \begin{bmatrix} 2 & & & & 0 \\ 1 & 2 & & & \\ & 0 & 1 & & \\ & & 1 & 1 & \\ 0 & & & 0 & 4 \end{bmatrix}, \quad g = \begin{bmatrix} a \\ 0 \\ b \\ 1 \\ c \end{bmatrix} .$$

Determine necessary and sufficient conditions on a, b, c such that (F,g) is completely controllable.

2.26 With reference to the linear system of the previous problem, show that a necessary and sufficient condition for (F,g) to be completely controllable is that no characteristic vector of F is orthogonal to g. (Actually, this result holds in the general case.)

2.27 The functional equation

$$I(c) = \min_{v} [c^2 + v^2 + I(c + bv)] \ ,$$

possesses a solution of the form $I(c) = kc^2$. Does it have any other strictly convex solution?

2.28 Obtain a functional equation for the quantity

$$\min_u \int_0^T [(x-c)^2 + ku^2]\, dt ,$$

$$\frac{dx}{dt} = -fx + u , \qquad x(0) = c .$$

2.29 Determine the minimum of

$$J = \int_0^T (1 - x)^2 ,$$

when

$$\frac{dx}{dt} = -x + u , \qquad x(0) = 1 ,$$

$$0 \le u \le M , \qquad M > 1 ,$$

$$\int_0^T u\, dt \le a < T .$$

REFERENCES

Discrete Problems: [A-14], [B-65], [C-18], [F-21], [K-22], [T-4].

Multidimensional Problems: [B-41], [K-27], [L-8], [S-18], [W-22].

Continuous Problems: [B-22], [B-87], [J-1], [L-4].

Controllability, Observability, and Stability: [K-3], [K-5], [K-7], [L-22], [Z-4].

CHAPTER 3

STOCHASTIC AND ADAPTIVE SYSTEMS

INTRODUCTION

In Part I, our investigations have taken us along paths that have, by and large, stayed within the confines of deterministic variational theory. Although we occasionally strayed from the classical problems to illustrate certain attractive features of the dynamic programming theory, many of the results presented could have been obtained using more standard arguments from the variational calculus. Thus, the utility of dynamic programming would be greatly diminished were its applicability confined only to the problem areas covered in the first volume.

A key feature distinguishing the treatment in Volume I was determinism. We made the convenient assumption (often not justified in practice), that there is a definite cause and effect relation governing all the systems dealt with and that this relationship between the control we exert and the resulting state of the system is known to as great a degree of accuracy as we desire. When put in such a bold fashion, it is clear just how serious (and optimistic) this deterministic assumption is. In fact, the only thing that saves the day is that many physical systems of interest exhibit a high degree of stability with respect to unknown influences which enables us to get away with using the mathematically desirable, but generally unrealistic, deterministic assumption.

Unfortunately, when faced with many significant practical problems we can no longer afford the luxury of neglecting unknown effects if we wish to have the system operate in an acceptable fashion. Thus, we must explicitly take into account such features as unknown dynamics, noisy observations, uncertain criteria, and so forth. That is, we must account for stochastic influences which affect the performance of the system under investigation. If we are fortunate, the probability distributions for these random influences will be known in which case we are in the realm of stochastic control theory. Often we do not even know these distributions a priori and must try to estimate them on the basis of information obtained about the system as the process unfolds. This leads us into the study of adaptive control processes.

Fortunately, the theory of dynamic programming extends in a natural fashion to these more realistic types of processes. In fact, it is the only truly general theory available for studying such problems. In contrast to the deterministic situation, where the euclidean point-line duality extends to show the duality between dynamic programming and the calculus of variations, there appears to be no such duality present in the stochastic case. The two approaches give entirely different results as we shall indicate in the following pages, and it is only the dynamic programming concept which seems to extend in a natural fashion to obtain optimal policies in the presence of random effects.

RANDOM VARIABLES, DISTRIBUTION FUNCTIONS, AND INDEPENDENCE

Before studying the application of dynamic programming to stochastic systems, let us briefly review some of the elementary aspects of probability theory and stochastic processes that we shall require. The reader desiring more details in these areas is referred to one of several excellent books cited in the bibliography. We shall confine our attentions to the bare essentials of the theory and then illustrate with examples to clarify the notions that we will use in the sequel.

The theory of probability begins with a triple $(\Omega, \mathscr{A}, \mathscr{P})$ where Ω is a set, $\mathscr{A}$ is a certain collection of subsets of Ω, and $\mathscr{P}$ is a function defined on $\mathscr{A}$ which has real values in the range 0 to 1. The elements $w \varepsilon \Omega$ are called elementary events. The set $\mathscr{A}$ is assumed to contain Ω and be closed under a countable number of the set operations $\cap$ (set intersection), $\cup$ (set union), and c (set complement), i.e., $\mathscr{A}$ is a σ-field. Since the elements of $\mathscr{A}$ are themselves events, this assumption assures us that events may be combined to obtain new events by using the logical set operations of intersection, union, and complement.

The following axioms are assumed concerning the function $\mathscr{P}$:

(a) $\mathscr{P}(\Omega) = 1$,

(b) $\mathscr{P}\left(\bigcup_{i=1}^{\infty} \mathscr{A}_i\right) = \sum_{i=1}^{\infty} \mathscr{P}(\mathscr{A}_i) \qquad \text{if } \mathscr{A}_i \cap \mathscr{A}_j = \{\phi\} \text{ , } i \neq j$.

To introduce the important idea of a random variable, consider the following game. A coin is tossed twice; if heads come up on a flip, you receive a dollar, and if tails appear you must pay fifty cents. Define the elementary events:

$$w_1 = \text{H T} ,$$
$$w_2 = \text{H H} ,$$
$$w_3 = \text{T T} ,$$
$$w_4 = \text{T H} , \qquad \text{H = heads , T = tails} .$$

Then $\Omega = \{w_1, w_2, w_3, w_4\}$. Choose $\mathscr{A}$ to be the 16-element set of all subsets of Ω, i.e.,

$$\mathscr{A} = \left\{\begin{array}{l} w_1,\ w_2,\ w_3,\ w_4,\ \{w_1,\ w_2\},\ \{w_1,\ w_3\},\ \{w_1,\ w_4\}, \\ \{w_2,\ w_3\},\ \{w_2,\ w_4\},\ \{w_3,\ w_4\},\ \{w_1,\ w_2,\ w_3\}, \\ \{w_1,\ w_3,\ w_4\},\ \{w_2,\ w_3,\ w_4\},\ \{w_1,\ w_2,\ w_4\}, \\ \{w_1,\ w_2,\ w_3,\ w_4\},\ \phi \end{array}\right\} .$$

may be specified by assigning four nonnegative numbers, p_1, p_2, p_3, p_4, satisfying $\sum_{i=1}^{4} p_i = 1$ and $\mathscr{P}(A) = \sum_{w_i \varepsilon A}$, $A \varepsilon \mathscr{A}$. A quantity of interest would be the gain G received; it is a <u>function</u> with domain Ω and range the real numbers given by $G(w_1) = \$0.50$, $G(w_2) = \$2$, $G(w_3) = -\$1$, $G(w_4) = \$0.50$. This function G is an example of <u>random variable</u>. Thus, we see that a random variable is just a real-valued function defined on the space of elementary events Ω.

Assume for a moment that Ω is a finite set. We define the <u>expected value</u> of the random variable X(w) to be

$$E(X) = \sum_{w \varepsilon \Omega} X(w) \mathscr{P}(w) ,$$

i.e., E(X) is just the weighted average of the values of X, weighted according to the probability of occurrence of the event w. The expected value operation E has the important mathematical property of linearity. Thus, if X and Y are two random variables and α and β are real scalars, then

$$E(\alpha X + \beta Y) = \alpha E(X) + \beta E(Y) .$$

EXAMPLE

3.1 Calculate the expected value of the random variable G defined in the coin-tossing game above.

The elements of Ω are the events w_1, w_2, w_3, and w_4. Assume that the coin is fair (prob. heads = prob. tails = 1/2). Then $\mathscr{P}(w_i) = 1/4$, $i = 1,2,3,4$.

Thus,

$$E(G) = \sum_{i=1}^{4} G(w_i)\,\mathscr{P}(w_i) ,$$

$$= [1/2 + 2 - 1 + 1/2](1/4) ,$$

$$= 1/2 .$$

Thus, we may expect to gain, on the average, \$0.50 each time we play the game.

Of interest are the so-called "central moments" of the random variable X. These are defined by the relation

$$E(X^n) = \sum_{w \varepsilon \Omega} X^n(w)\,\mathscr{P}(w) , \quad n = 1,2,\ldots .$$

Of particular importance is the case $n = 2$, the second moment. It is used to compute the <u>variance</u> of X, which measures the deviation of X from its expected value; i.e., the variance gives a quantitative estimate of whether most of the values of X are close to the expected value or whether the values of X regularly depart significantly from the expected value. The definition is

$$\mathrm{Var}(X) = E\{[X - E(X)]^2\} ,$$

$$= E(X^2) - [E(X)]^2 ,$$

where use has been made of the linearity of E in obtaining the last relation.

EXAMPLE

3.2 Let us calculate the variance of the random variable G in the coin-tossing game.

From the previous example, we know $E(G) = 1/2$. Also,

$$E(G^2) = \sum_{i=1}^{4} G^2(w_i)\,\mathscr{P}(w_i) ,$$

$$= [1/4 + 4 - 1 + 1/4]\ 1/4 ,$$

$$= 7/8 .$$

Thus, $\text{Var}(G) = 7/8 - 1/4 = 5/8$. Hence, most payoffs will be in the range $E(G) \pm \sqrt{\text{Var}G} = 0.50 \pm 0.79$.

In the case where Ω is an uncountable set, say the real numbers, then we must suitably extend the foregoing definitions by proper interpretation of the limiting process. Thus, we define the function

$$F_X(z) = \mathscr{P}[X(w) \leq z] ,$$

where $w \varepsilon \Omega$, $-\infty \leq z \leq \infty$. F_X is called the cumulative distribution function of the random variable X. In terms of F_X we may define the moments of X as

$$E(X^n) = \int_{-\infty}^{\infty} X^n \, dF_X(z) ,$$

where dF_X represents the Stieltjes measure associated with $\mathscr{P}$. For our purposes, it may be assumed that

$$dF_X(z) = f(z)\, dz ,$$

for an appropriate continuous function f, which is often called the probability density function, although it will be well worth the reader's effort to spend some time learning the Riemann-Stieltjes notation since it allows us to consider both discrete and continuous problems with a single notation. Since our objectives are primarily numerical, all problems must ultimately be approximated by discrete distributions anyway, so we will have no occasion to make anything other than formal use of continuous distributions.

A concept inherent in probability theory which distinguishes it from an exercise in measure theory is the idea of independence. Intuitively, we feel that two events are independent if the occurrence or non-occurrence of one of the events has no influence on the occurrence or non-occurrence of the other. In terms of random variables, this basic idea may be stated by the formula

$$E(X)\,E(Y) = E(XY)\ ,$$

for any two random variables X and Y which are independent. Alternate definitions are possible, but this seems to be the one most useful in practice so we will employ it as the situation dictates.

DISCRETE STOCHASTIC PROCESSES

One of the most attractive features of dynamic programming is that it allows us to analyze, in a systematic fashion, problems containing stochastic variables. Because of the importance of these problems in many areas of application, we introduce here the notion of a stochastic multistage process. This concept will enable us to introduce the dynamic programming concepts for stochastic problems in a fashion similar to that of Part I for deterministic processes.

Until now, we have assumed that the state of our system at time $k+1$ is completely determined by the state at time k and the transformations T_k. Now let us introduce an element of uncertainty into the discussion by relaxing our assumptions to the point where we know only that the state $x(k+1)$ belongs to a set of new states, the particular state being determined by a random parameter r having the probability distribution function $dG_k(r)$. For simplicity, let us first assume that the sequence $[r(k), r(k+1), \ldots]$ consists of independent random variables with <u>distribution</u> functions $dG_k[r(k)]$. What we are saying is that now the transformation T_k is a function not only of time, but also of the random parameter $r(k)$; formally,

$$x(k+1) = T_k[x(k),\ r(k)], \quad k = a, a+1, \ldots, N \ .$$

Thus, the sequence of elements $[x(a), x(a+1), \ldots, x(N)]$ forms a discrete stochastic multistage process.

The associated criterion functions are now stochastic variables themselves. For mathematical simplicity and practical utility, let us assume that we will only be interested in their expected values. Suppose that we wish to determine the expected value of the performance function

$$L_a[x(a)] + L_{a+1}[x(a+1)] + \ldots + L_N[x(N)] \ .$$

We observe that the expected value depends upon the initial state $x(a)$ and initial stage a. Thus, we write

$$I[x(a),a] = \underset{r(a),\ldots,r(N)}{E} \{L_a[x(a)] + L_{a+1}[x(a+1)] + \ldots$$

$$\ldots + L_N[x(N)]\} \ ,$$

$$= L_a[x(a)] + \underset{r(a),\ldots,r(N)}{E} \{L_{a+1}[x(a+1)] + \ldots$$

$$\ldots + L_N(x(N)]\} \ ,$$

where $\underset{r(a),\ldots,r(N)}{E}$ denotes the expectation taken with respect to the random variables $r(a)$, $r(a+1)$, ..., $r(N)$. The assumed independence of the $[r(k)]$ allows us to write

$$\begin{aligned} I[x(a),a] &= L_a[x(a)] + \underset{r(a)}{E}\ \{L_{a+1}[T_a(x(a),r(a))] \\ &\quad + \underset{r(a+1),\ldots,r(N)}{E}\ [L_{a+2}(x(a+2)) + \ldots \\ &\quad \ldots + L_N(x(N))]\} , \\ &= L_a[x(a)] + \underset{r(a)}{E}\ \{I[T_a(x(a),r(a),a+1)]\} , \end{aligned}$$

with

$$I[x(N),N] = L_N[x(N)] .$$

EXAMPLE

3.3 Show that if the distribution function for $r(k)$ depends on $r(k-1)$ (i.e., the random variables $r(k)$, $r(k-1)$ are correlated), an appropriate recurrence relation for the associated function above may still be obtained.

Let the joint distribution for $r(k)$ and $r(k-1)$ be given by $dG_k[r(k),r(k-1)]$. Let us attempt to establish a recurrence relation for the quantity $I[x(a),r(a-1),a]$, where $x(a)$ is the present state and $r(a-1)$ is the previously observed random variable. Then we may define

$$\begin{aligned} I[x(a),r(a-1),a] = \underset{r(a),\ldots,r(N)}{E}\ &\{L_a[x(a)] + L_N[x(a+1)] + \ldots \\ &\ldots + L_N[x(N)]\} , \end{aligned}$$

where the distribution of $r(a)$ is uniquely determined for a given $r(a-1)$ from $dG_a[r(a),r(a-1)]$. Proceeding as before, we see that

$$\begin{aligned} I[x(a),r(a-1),a] &= L_a[x(a)] + \{L_{a+1}[T_a(x(a)]\} \\ &\quad + \underset{r(a+1),\ldots,r(N)}{E} \{L_{a+2}[x(a+2)] + \ldots \\ &\quad \ldots + L_N[x(N)]\} \, dG_a[r(a),r(a-1)] , \\ &= L_a[x(a)] + I\{T_a[x(a),r(a)],r(a),a+1\} \\ &\quad dG_a[r(a),r(a-1)] . \end{aligned}$$

Thus, we see that the effect of the correlation is to make the function I at each stage a depend on two variables rather than one.

In order to start the recursion, we observe that

$$I[x(N),r(N-1),N] = L_N[x(N)] .$$

EXAMPLE

3.4 Consider the situation in which there are noisy observations on the state of the system. Thus, when the actual state of the system is $x(k)$, we observe $z(k) = x(k) + r(k)$ where $r(k)$ is a random variable with known distribution. Show that this situation may be handled by considering the sequence $\{z(k)\}$ to be a process of the type considered in Example 3.3 (assume $z(a) = x(a)$).

Consider the two processes $[x(a),x(a+1),\ldots]$ and $[z(a), z(a+1),\ldots]$, where the x process represents the actual state of the system and the z process the information we actually possess. Then the two processes unfold in the following fashion:

$$\begin{array}{ll} x(a) & z(a) \\ x(a+1) = T[x(a)] & z(a+1) = x(a+1) + r(a+1) \\ x(a+2) = T[x(a+1)] & z(a+2) = x(a+2) + r(a+2) \\ \vdots & \vdots \end{array}$$

Thus, we see that

$$z(a+1) = T[x(a)] + r(a+1) = T[z(a)] + r(a+1) ,$$

$$z(a+2) = T[x(a+1)] + r(a+2) = T[z(a+1) - r(a+1)] + r(a+2) ,$$

$$z(a+3) = T[x(a+2)] + r(a+3) = T[z(a+2) - r(a+2)] + r(a+3) ,$$

$$\vdots$$

and so on. This demonstrates that the elements of the sequence $\{z(k)\}$ depend upon the random variables $r(k)$ and $r(k-1)$, which shows that the sequence $\{z(k)\}$ is of the same sort as the correlated sequence considered in Example 3.3.

STOCHASTIC MULTISTAGE DECISION PROCESSES

Since we now admit that the outcome of any particular decision does not result in a uniquely determined change of state, we are faced with the problem of choosing a sequence of decisions which is in some way "best" or "optimal." As we no longer have complete control over the process, it is impossible to obtain a maximum or minimum return from the process and we must therefore agree to accept a statistical measure of the performance of our system. To keep the functional equations which describe our stochastic processes at the simplest possible level, we will agree to use the expected value in the discussion and examples of this chapter and to denote the operation of taking the expected value of a particular random variable X by $E(X)$.

Consider the following stochastic version of a one-dimensional allocation process. We have an amount of resource x which is to be allocated over K activities. Assume that the result of dividing x into the two quantities v and $x-v$ is a return of $g(v)$ with probability p and a return $h(v)$ with probability $q=1-p$. Further, assume that in the first case the amount of resource remaining is $a(x-v)$, $a > 0$, and in the second event the amount remaining is $b(x-v)$, $0<a<b\leq 1$. If we assume that $g(v)>h(v)$, the question is how to allocate x over the K activities to maximize the expected return.

To attack this problem, we write

$I(x,k)$ = expected return from an allocation to $K-k+1$ activities when the total resource available is x and an optimal allocation policy is used, $k=1,\ldots,K$.

On the basis of the principle of optimality, we may argue that

$$I(x,k) = \max_{0\leq v\leq x} \left\{ \begin{array}{l} p[g(v) + I(a(x-v),k+1)] \\ \quad + q[h(v) + I(b(x-v),k+1)] \end{array} \right\} .$$

For one activity we clearly have

$$I(x,K) = \max_{0\leq v\leq x} [pg(v) + qh(v)] .$$

These equations have the same analytic structure as those we obtained in the deterministic case, the only difference being the minor modification necessary to include the stochastic state transition. The extension of the above treatment to include the case when there may be more than two possible outcomes from a given decision is straightforward and is left to the reader. It is also evident that the above formulation allows us to deal with a time-dependent stochastic process by allowing the probabilities p and q to be functions of k, i.e., $p = p(k)$, $q = q(k)$.

Now let us examine another type of stochastic decision process which can be treated by the preceding techniques. Assume that we have a system which may be in one of two possible states: acceptable, denoted by A, or unacceptable, denoted by U. To begin with, we do not know which state the system is in but we may perform two types of operations on the system to determine its state. The first type of operation R reduces the probábility that the system is in state U by a factor $0 < \alpha < 1$, while the second type of operation E consists of an examination of the system to determine its true state. If each operation takes one unit of time, our desire is to determine a sequence of R, E operations which will transform the system to the state A with complete certainty in a minimum amount of time.

To deal with the above problem, let

x = probability the system is in state U,

$I(x)$ = the expected time required to transform the system state A with complete certainty, give the current probability of state U is x and an optimal policy is followed.

As a result of a decision R, the state x is transformed as $x \rightarrow \alpha x$, while a decision E transforms $x \rightarrow 1$ if the system is in state U and $x \rightarrow 0$ if it is in state A. The principle of optimality then gives

$$I(x) = \min \begin{cases} R : 1 + I(\alpha x) \\ R : 1 + xI(1) \end{cases}, \tag{3.1}$$

$$I(0) = 0 \tag{3.2}$$

To solve Eq. (3.1) we note that there exists a value x^*, such that we choose E if $x < x^*$ and R if $x > x^*$. Also, for $x = x^*$ we have

$$1 + x^* I(1) = 1 + I(\alpha x^*) .$$

Since $\alpha x^* < x^*$, we must have

$$I(\alpha x^*) = 1 + \alpha x^* I(1) ,$$

which means

$$2 + \alpha x^* I(1) = 1 + x^* I(1) .$$

Consequently, for the critical value x^* we obtain

$$x^* = \frac{1}{(1-\alpha)I(1)} .$$

We must now calculate $I(1)$ in terms of the parameter α. This is accomplished by noting that there exists a positive integer M such that

$$\begin{aligned} I(1) &= 1 + I(\alpha) , \\ I(\alpha) &= 1 + I(\alpha^2) , \\ &\vdots \\ I(\alpha^{M-1}) &= 1 + I(\alpha^M) , \\ I(\alpha^M) &= 1 + \alpha^M I(1) . \end{aligned}$$

Then

$$I(1) = M + 1 + \alpha^M I(1) ,$$

or,

$$I(1) = \frac{M+1}{1-\alpha^M} . \tag{3.3}$$

The unknown value of M must be that positive integer which minimizes the right side of (3.3). Thus,

$$I(1) = \min_{M=1,2,\ldots} \frac{M+1}{1-\alpha^M} .$$

In this way, the threshold x^* is determined and the optimal decision rule is found.

EXAMPLE

3.5 Assume the parameter $\alpha = 0.9$ and calculate the optimal x^*.

Calculation of the minimizing M in (3.3) shows that $M_{min} = 5$, giving the value $I(1) \cong 14.6$. Hence, employing the representation for x^*, we obtain

$$x^* = \frac{1}{(1-0.9)(14.6)} \cong 0.685 .$$

Thus, as soon as the probability of the system being in the acceptable state U drops below 0.685, the optimal policy is to then directly examine the system.

A type of problem which exhibits many of the features inherent in other stochastic decision processes is that of communicating information through a noisy communication channel. Although this problem has several aspects including origination of the data, encoding, decoding and so forth, to illustrate our methods we shall confine our attention to a process in which the only noise present is in the communication channel itself.

Consider a gambler (investor) who receives advance information regarding the outcome of certain independent sporting events over a noisy communication channel. Assume the outcome of each event is a result between two evenly matched teams, and that p is the probability of correct transmission of the result.

Assuming that the gambler begins with x units of capital and bets to maximize his expected value at the end of K states of play, what is his optimal betting policy? To answer this question, let

$I(x,k)$ = the expected value of the gambler's capital at the end of k stages of play when he uses an optimal betting policy and he begins with x units of capital.

Then,

$$I(x,k) = \max_{0 \le y \le x} [pI(x+y,k+1) + I(x-y,k+1)] , \quad k<K , \tag{3.4}$$

$$I(x,K) = \max_{0 \le y \le x} [py - qy] , \quad (q = 1-p) . \tag{3.5}$$

It is not difficult to show by induction that Eqs. (3.4)-(3.5) yield the solution that, to maximize his expected gain, the gambler should bet his entire fortune x when $p > 1/2$, and bet nothing when $p \le 1/2$. This is a very risky policy since there is an appreciable chance of ruin before the process is completed.

Thus, it may be better to adopt a criterion which enables us to escape total ruin over a sequence of wagers. For example, let us assume the gambler plays so as to maximize the expected value of the logarithm of his total amount of capital at the end of his wagering. To see how this modification affects his betting policy, let

$I(x,K)$ = the expected value of the logarithm of the gambler's final capital after a K-stage wagering process, starting at stage k with the amount x and using an optimal policy.

As before, we obtain the recurrence relation

$$I(x,k) = \max_{0 \le y \le x} [pI(x+y,k+1) + qI(x-y,k+1)] , \quad k<K , \tag{3.6}$$

with

$$I(x,K) = M + \log x , \tag{3.7}$$

where M is the expected value of the logarithm of the amount of money the gambler wins after one wager. To determine this quantity, we see that for an initial wager y, the expected return is

$$E(y) = p \log (x+y) + q \log (x-y) .$$

The maximum is attained at $y = (2p-1)x$, if $p > 1/2$, and $y = 0$ for $p \leq 1/2$. Let us assume $p > 1/2$. Then

$$E_{max} = \log x + \log 2 + p \log p + q \log q \,.$$

Thus

$$I(x,K) = \begin{cases} \log x + \log 2 + p \log p + q \log q \,, & p < 1/2 \\ \log x \,, & p \geq 1/2 \end{cases} \,.$$

EXAMPLE

3.6 (a) Show that the functions $I(x,k)$ possess the analytic structure

$$I(x,k) = \log x + (K - k+1)\, M \,.$$

(b) What does this structure tell us about the structure of the optimal wagering policy?

The postulated result is obviously true for $k = K$. Assume it holds for $k+1$. Then

$$I(x,k) = \max_{0 \leq y \leq x} \{p[\log(x+y)+M(K-k)]+q[\log(x-y)+M(K-k)]\} \,,$$

$$= \max_{0 \leq y \leq x} \{p[\log(x+y)] + q[\log(x-y)]\} \,,$$

$$= M(K-k) + \log x + M \,,$$

$$= \log x + (K - k+1)\, M \,.$$

From the second equation, we see that the optimal policy at <u>any</u> stage is

$$y = \begin{cases} (2p-1)x \,, & p > 1/2 \\ 0 \,, & p \leq 1/2 \end{cases} \,.$$

Thus, the structure of the optimal betting policy is <u>independent</u> of k and x and is determined only by the properties of the communication channel. The bet is a fraction (2p-1) of the available capital if $p > 1/2$, and nothing otherwise.

The foregoing problem admits generalization and extension in several directions. In the first place, we may assume the probability of correct transmission is time-dependent. Other extensions are to assume the successive signals are not independent or that the channel may transmit any of several different signals, a limiting case being where a continuum of signals is possible. Another type of generalization is also possible if we assume that p itself is a random variable subject to a known, or perhaps unknown, probability distribution. We shall examine some of these extensions in the Solved Problems.

STOCHASTIC CONTROL PROCESSES

Let us now indicate the rather surprising fact that dynamic programming allows us to treat many stochastic control processes of discrete type with little more effort than is required to treat the deterministic version of the same problem. We say "surprising fact" since, as already mentioned, classical variational techniques generally don't extend in any fashion to stochastic problems.

Consider the process defined by

$$x(k+1) - f[x(k),\ u(k),\ r(k)]\ , \qquad x(0) = c\ ,$$

where we wish to minimize the criterion

$$J = \underset{\{r(0),\ldots,r(N-1)\}}{E} \sum_{k=0}^{N-1} L[x(k),u(k),r(k)] + \varphi[x(N)]\ ,$$

where E denotes the expected value taken over the random variables $\{r(k)\}$. We assume for simplicity that the random variables $\{r(k)\}$ are independent and identically distributed with common distribution function $dG(r)$.

We note the crucial point that this control process is not completely specified by the defining equations and criterion function. We must, in addition, specify the information that is available to the decision maker at each stage where a control variable is to be chosen. The two extremes of information are:

(1) The decision maker can examine the actual state of the system at each point where a decision is required.

(2) The initial state, c, is known, but the system is not observable at any subsequent time.

In the first case, a policy function $u(c,k)$ represents the solution. If we write

$$I(c,a) = \underset{\{u(k)\}}{\text{Min}} \; J(\{u(k)\}) ,$$

there is no difficulty in obtaining the functional equation

$$\begin{aligned} I(c,a) &= \min_{v} \left\{ \underset{\{r(k)\}}{E} \sum_{k=a}^{N-1} L[x(k),u(k),r(k)] \right\} , \\ &= \min_{v} \left\{ \underset{\{r(a)\}}{E} [L(x(a),u(a),r(a))] \right. \\ &\qquad \left. + \underset{r(k)}{E} \sum_{k=a+1}^{N-1} L[x(k),u(k),r(k)] \right\} , \\ &= \min_{v} \left\{ \int [L(c,v,r) + I(f(c,v,r),a+1)] \, dG(r) \right\} , \end{aligned} \tag{3.8}$$

where the notation $u(a) = v$, $r(a) = r$ has been adopted. Clearly,

$$I(c,N) = \varphi(c) .$$

In the second case, a set of control values $u(a), u(a+1), \ldots, u(N)$ represents the solution, and no corresponding functional equation exists. In the deterministic case, these two different conceptual approaches, event orientation versus time orientation, turned out to be equivalent. In the stochastic case they are different, both conceptually and analytically; the difference lies in the information available to the decision maker at each stage. Since we can have all possibilities of partial information at each stage, we see that the theory of stochastic control processes is infinitely richer than the deterministic theory.

It is important to note that Eq. (3.8) for the stochastic problem has precisely the same form as that given earlier for the deterministic case. Thus, the functional equation approach of dynamic programming serves to unify the treatment of both the stochastic and deterministic processes, once we agree to suitably interpret what we mean by the state of the system. In one case, the state is a definite quantity, while in the other the state is a stochastic quantity determined by the amount of information we have at our disposal concerning the behavior of the system.

In a numerical calculation, the distribution function $dG(r)$ will be taken to be discrete, so that the relation (3.8) will become

$$I(c,a) = \min_{v} \sum_{i=1}^{M} p_i [L(c,v,r_i) + I(f(c,v,r_i),a+1)] \quad ,$$

where $p_i = \text{Prob}\ (r=r_i)$. We may even allow the $\{p_i\}$ to depend upon the control and/or state, rather than being fixed constants, i.e.,

$$p_i = p_i(c,v) \ .$$

Note that the only computational change from the deterministic algorithm is that we must calculate the sum $\sum_{i=1}^{M} [\ldots]$. Hence, the computing time is increased by a factor of M but other than this additional "loop" in the computer program, all remains as before. Thus, we can deal with the foregoing type of stochastic control process with scarcely more effort than the corresponding deterministic version.

A type of problem to which we devoted some attention earlier is the control problem with quadratic criteria and linear dynamics. Let us now indicate how the results obtained there may be extended to the analogous stochastic problem.

We wish to minimize the expected value of

$$J = \sum_{k=0}^{N-1} [x^2(k) + u^2(k)] + x^2(N) ,$$

where

$$x(k+1) = ax(k) + u(k) + r(k) , \quad x(0) = c .$$

For simplicity, we assume x and u are scalars with the $\{r(k)\}$ independent random variables with common distribution function $dG(r)$. Arguing as before, we write

$$I(c,a) = \min_u E_r [J] ,$$

$$= \min_v E_r [c^2 + x^2(0) + I(ac+v+r,a+1)] , \quad a<N , \tag{3.9}$$

with

$$I(c,N) = c^2 . \tag{3.10}$$

To illustrate the great simplification that results when we assume, as above, that after any decision the stochastic state of the system may be measured, we first examine a process in which, at any particular stage, the state of the system may not be known to the decision maker.

Since there may be several successive stages for which the state of the system is unknown, we specify the state of the system by three quantities — m, n, and c. Here n is the number of stages remaining in the process, m is the number of stages in the past at which the controller last had perfect information, and c is the physical state of the system at that time. For simplicity, assume we wish to minimize the expected value of some function φ of the terminal state of the process, i.e.,

$$J = \underset{\{r(i)\}}{E} [\varphi(x_N)] ,$$

the system dynamics are

$$x(k+1) = f[x(k),u(k),r(k)] , \quad x(0) = c ,$$

and the $\{r(i)\}$ constitute a set of independent random vectors with common distribution function. Furthermore, we assume that at any stage there is a fixed probability p that we will not be able to observe the actual physical state of the system.

Let

$z_m(c)$ = the random variable designating the observed physical state of the system after m stages have passed without observation of the system, given that the state m stages back was c.

Note that $z_m(c)$ will depend upon the random vectors $r(0), r(1), \ldots, r(m-1)$ and the control vectors $u(0), u(1), \ldots, u(m-1)$.

We note that with n stages remaining, with probability (1-p) we may either be able to observe the physical state, thereby obtaining a physical state denoted by the quantity $z_{m+1}(c)$, conditioned on the random elements $r(0),\ldots,r(m+1)$, the number of remaining stages becomes n-1 and the number of stages in the past at which we last had perfect information becomes 0 or, if we cannot observe the system, then the number of remaining stages still becomes n-1, m is transformed to m+1, and the last perfect observation is unchanged. Putting these remarks together with the principle of optimality and forward dynamic programming, we obtain

$$I(m,n,c) = \min_u \left\{ p[I(m-1,n-1,c)] + (1-p) \mathop{E}_r [I(0,n-1,z_{m+1}(c))] \right\} . \tag{3.11}$$

The minimizing u is a function of m, n, and c, i.e., $u_{min} = u(m,n,c)$.

To make use of Eq. (3.11) we must determine the conditional expected value of $I(0,n-1,z_{m+1}(c))$, given the information that for $(m+1)$ successive stages the actual state of the system could not be observed and that the last known state was c. At the first stage, the control vector $u(0,m+n,c)$ (even though $m=0$ here, it is convenient to adopt this notation for a moment) was used, resulting in the new state

$$x(1) = f[c,u(0,m+n,c),r(0)] .$$

At this stage, with no information on the actual state the control $u(1,m+n-1,c)$ was used. This led to the stochastic state

$$x(2) = f[x(1),u(1,m+n-1,c),r(1)] .$$

We see that the random variable $z_{m+1}(c)$ depends upon the control vectors $u(k,m+n-k,c)$, for $k=0,1,\ldots,m$. Thus, Eq. (3.11) determines functions $I(m,n,c)$, $u(k,m+1-k,c)$ for $0 \leq m+n+1 \leq N$, $k=0,1,\ldots,m+1+n$. It is an implicit system with no straightforward way of determining the functions one at a time. For computational purposes, a method of successive approximations would most likely be used.

MARKOVIAN DECISION PROCESSES

Thus far, we have supposed that the random variables influencing our control processes have a probability distribution which is determined by the particular physical situation under consideration. Our objective has been to attempt to select our decisions so as to optimize, in a stochastic sense, with respect to this given probability distribution function.

In this section we will consider a situation in which at each stage of the process we have a choice as to which of several possible probability laws is used, i.e., the set of admissible decisions is a space whose elements are probability distributions. To obtain interesting results, we shall further particularize our treatment and consider only processes which possess a certain linear structure, the so-called Markovian decision processes. We will see that this structure is broad enough to remain analytically and computationally tractable.

Let us first consider what we mean by a Markov process (technically, a discrete Markov process). Suppose we consider a system which is capable of existing in any of N possible states i, $i = 1,2,\ldots,N$. At each time $t = 1,2,\ldots,$ there is a fixed probability p_{ij} that the system in physical state j at time t will move to physical state i at time $t+1$. Define the N-dimensional vector $x(t)$ whose i<u>th</u> component $x_i(t)$ is given by

$$x_i(t) = \text{probability the system is in physical state } i \text{ at time } t\ ,\quad i = 1,2,\ldots,N.$$

Since the system must be in some state at time $t+1$, the laws of probability and matrix multiplication give

$$x(t+1) = Px(t)\ ,\qquad x(0) = c\ , \tag{3.12}$$

where $P = [p_{ij}]$, and c is the initial state. Note that by the definitions of p_{ij} and $x_i(t)$ we must have

$$c_i , p_{ij} \geq 0 ,$$

$$\sum_{i=1}^{N} p_{ij} = 1 , \quad j = 1,2,\ldots,N ,$$

$$\sum_{i=1}^{N} c_i = 1 .$$

We shall call a process governed by the probability law (3.12) a Markov process. A central issue in the study of such processes is the limiting behavior of $x(t)$ as $t \to \infty$.

We now extend the notion of a Markov process by the introduction of a decision-making operation at each stage. We assume that at each stage the transition matrix P may be chosen from a set of such matrices. Letting q denote the decision variable, we let $P(q)$ denote the transition matrix singled out by the decision q.

In addition, to offer some rationale for choosing one particular transition matrix over another, we assume that there is a return $R(q) = [r_{ij}(q)]$ associated with the decision q at each stage. Our basic problem is to determine a sequence of decisions which maximizes the expected return from an M-stage process, given the initial state. Such a process is termed a Markovian decision process.

Let us now see how we can use functional equations to determine the optimal decision rule for the above problem. Let

$I(i,M)$ = the expected return obtained from a M-stage process, starting in physical state i and using an optimal policy, $i = 1,2,\ldots,N$, $M = 0,1,\ldots$.

Note that we now have adopted a slightly different notion of state than that given in the last paragraph. The states are now taken to be the physical states themselves, rather than the probability distribution mentioned above. The following analysis could be carried out in either setting, but for present purposes using the physical states seems to be preferable. From Eq. (3.13) below, the reader will readily observe that the identification of physical state i with abstract state x_i allows us to obtain the same fundamental results and algorithms in either setting. This point is worth emphasizing since it illustrates an earlier remark to the effect that, in any given problem, there is no such thing as "natural" state variables. One has considerable flexibility in defining what is meant by the state of the system, and often the success or failure of the problem solution critically hinges upon this definition.

Using forward dynamic programming, the principle of optimality immediately yields

$$I(i,M) = \max_q \left\{ \sum_{j=1}^{N} p_{ij}(q) \ [r_{ij}(q) + I(j,M-1)] \right\} ,$$

with

$$I(i,0) = 0 .$$

The question of primary interest now is the steady-state behavior of $I(i,M)$ as $M \to \infty$.

EXAMPLE

3.7 Let us show that if $p_{ij}(q) > 0$, the function I has the asymptotic behavior

$$I(M,i) \sim aM + g(i) , \quad i = 1,2,\ldots,N ,$$

where a is a scalar, independent of the initial state, and $g(i)$ depends upon the initial state i.

Let $b_i(q) = \sum_{j=1}^{N} p_{ij}(q)\, r_{ij}(q)$, and let $b(q) = [b_i(q)]$, $P(q) = [p_{ij}(a)]$, $I(M) = [I(M,i)]$, i.e., $P(q)$ is an $N \times N$ matrix, while b and I are N-dimensional vectors. Thus, the above equation may be written

$$I(M) = \max_{q} \, [b(q) + P(q)\, I(M-1)] \, . \tag{3.13}$$

Here q is a vector selected so that q_1 is chosen in the equation for $I(M,1)$ and so forth. Pick a particular policy and iterate it to obtain the M stage return,

$$I_q(M) = b(q) + P(q)b(q) + (P^2)(q)b(q) + \ldots + (P^{M-1})(q)b(q) \, .$$

Let us show that the vector

$$r(q) = \lim_{M\to\infty} \frac{I_q(M)}{M}$$

exists and

$$r(q) = a(q) \begin{bmatrix} 1 \\ 1 \\ 1 \\ \vdots \\ 1 \end{bmatrix} ,$$

where $a(q)$ is a scalar. Assume for simplicity that P has distinct characteristic roots. Then there exists a matrix T which diagonalizes P, i.e.,

$$P = T \begin{bmatrix} \lambda_1 & & & 0 \\ & \lambda_2 & & \\ & & \ddots & \\ 0 & & & \lambda_N \end{bmatrix} T^{-1} ,$$

where $\lambda_1, \lambda_2, \ldots, \lambda_N$ are the characteristic roots of P. Then

$$\frac{1}{n}\sum_{k=1}^{n} P^k = T \begin{bmatrix} \frac{1}{n}\sum_{k=1}^{n}\lambda_1^k & & & \\ & \frac{1}{n}\sum_{k=1}^{n}\lambda_2^k & & \\ & & \ddots & \\ & & & \frac{1}{n}\sum_{k=1}^{n}\lambda_N^k \end{bmatrix} T^{-1} .$$

Since P is a Markov matrix, $|\lambda_j| \le 1$, $j = 1,\ldots,N$. Hence,

$$\lim_{n\to\infty} \frac{1}{n}\sum_{k=1}^{n} \lambda_j^k = \alpha_j$$

exists. Actually, if $\lambda_j = e^{i\theta}$, then if $2\pi\theta$ is rational $\alpha_j \neq 0$, while if $2\pi\theta$ is irrational, then $\alpha_j = 0$. In either case

$$\lim_{n\to\infty} \frac{1}{n}\sum_{k=0}^{n} P^k$$

exists, implying the result we wished to establish.

Thus, the policy yielding the optimal steady-state policy is the policy which maximizes $a(q)$. Taking the limit as $M\to\infty$ on both sides of (3.13) and employing the optimal policy which maximizes $a(q)$ gives the desired representation.

We now use the asymptotic result (3.14) to obtain an algorithm for obtaining the optimal steady-state decision rule. This algorithm, called the "Value Determination - Policy Improvement Routine" by R. Howard, is actually a special case of approximation in policy space, which we discussed in Part 1.

From the asymptotic result

$$I(M,i) = Mr + g ,$$

and Eq. (3.13), we have

$$r + g = b(q) + P(q)g , \tag{3.14}$$

where g is the vector $g = [g(i)]$. This equation involves determining the N components of g and the scalar a appearing in r. Since P is a Markov matrix, its column sums equal one. Thus, we may normalize the asymptotic behavior by adding an equation of the form

$$\sum_{i=1}^{N} g(i) = 1 , \tag{3.15}$$

or

$$g(N) = 0 .$$

Eqs. (3.14) and (3.15) constitute a set of N+1 linear algebraic equations for the N+1 unknowns $g(i)$ and a.

To use the above equations in an iterative fashion and apply the scheme of Howard, let $q^{(o)}$ be an initial guess at the optimal policy. Determine $g^{(o)}$ from Eqs. (3.14) and (3.15). Next, determine an improved policy $q^{(1)}$ as that q which maximizes

$$b(q) + P(q)g^{(o)} - g^{(o)} .$$

Using this new policy, determine $g^{(1)}$ and $r^{(1)}$, and so on.

MULTISTAGE GAMES

In an earlier section on stochastic control processes, we discussed systems governed by the dynamic equations

$$x(n+1) = f[x(n),u(n),r(n)] , \quad x(0) = c , \tag{3.16}$$

where $x(n)$ represented the state of the system at time n, $u(n)$ the control, and $r(n)$ was a random vector with known probability distribution.

By agreeing to settle for a stochastic measure of performance of the system, we were able to extend our previous deterministic analysis to cover processes governed by Eq. (3.16). All of this was under the assumption that the distribution function of $r(n)$ was known to the decision-maker.

In most problems of real interest, the distribution function for r is not known. In this section, we wish to present the basic idea of a "game against nature" as one way of attacking the question of what to do about the unknown distribution of r. Our answer here will be to assume the distribution is chosen by an outside agency in the worst possible way from the standpoint of the controller. Clearly, this approach suffers from the defect of conservatism in that there is no a priori reason to suspect the distribution will be realized in such a disadvantageous manner. However, as we shall see, this line of attack serves as a valuable intermediate step between our previous treatment of stochastic control processes and the discussion of advaptive control problems to be presented in a later section. In addition, the results are of some independent interest in themselves in such areas as pursuit-evasion problems, card games, and mathematical economics.

Consider a situation in which there are two "players," P and Q. P has M different decisions that he may make and Q has N different alternatives. If P makes the choice i and Q the choice j, there is a return of p_{ij} units to P and q_{ij} units to Q. The question is how to determine optimal strategies for P and Q, i.e., optimal in the sense that P and Q each attempt to maximize their respective returns.

Clearly, the determination of optimal play is dependent upon the conventions in force regulating play of P and Q. If P moves first, then Q's choice of j will be made so that the quantity q_{ij} is maximized. With j chosen this way, P will choose i to maximize $p_{i,j(i)}$, where j(i) is the function giving the maximizing index for Q. Thus, the only interesting case is when the players are forced to move simultaneously.

To overcome the difficulty inherent in the concept of simultaneous moves, in place of requiring that P make a definite move i, let us require only that he give a probability distribution for his moves, $p = [p_1, p_2, \ldots, p_N]$ and that Q furnish a similar distribution for his available moves, $q = [q_1, q_2, \ldots, q_M]$, i.e., p_i = the probability P makes move i, etc.

Since we are now dealing with probability distributions rather than definite choices, we will measure the outcome of P and Q's choices by expected values. The expected return to P is

$$E_P(p,q) = \sum_{i,j} p_{ij}\, p_i\, q_j \ ,$$

while the expected return to Q is

$$E_Q(p,q) = \sum_{i,j} q_{ij}\, p_i\, q_j \ .$$

We now make the crucial assumption that $p_{ij} = -q_{ij}$; that is, P's gain is Q's loss and conversely. Such games are termed zero-sum. In view of this hypothesis, Q's desire to maximize his expected return is equivalent to his wishing to minimize P's expected return. Thus, we may consider P as the maximizing player and Q as the minimizing player.

If P must choose his probability distribution before Q, the expected return to P is

$$\max_{[p]} \min_{[q]} E_P(p,q) \ .$$

Similarly, if Q makes his choice of q first, the expected return to P is

$$\min_{[q]} \max_{[p]} E_P(p,q) \ .$$

The fundamental result of two-person game theory is the celebrated Min-Max Theorem of von Neumann which asserts

$$\min_{[q]} \max_{[p]} E_P(p,q) = \max_{[p]} \min_{[q]} E_P(p,q) .$$

The common value of these two expressions is called the Value of the game.

The interpretation of the Min-Max Theorem is that either player may announce his probability distribution in advance without giving the other player any advantage whatsoever.

Let us now consider a type of game which is a variant of the classical "gambler's ruin" problem. Suppose P possesses a quantity of resource u and Q possesses v. Decisions are made by each player until one or the other has no resource remaining. Such a process is called a "game of survival."

To treat a process such as this, let

f(u,v) = the probability that P will survive Q when P has u units of resource, Q has v units, and both players pursue optimal policies.

We readily obtain the functional equation

$$\begin{aligned} I(u,v) &= \max_p \min_q \left[\sum_{i,j} p_i \, q_j \, I(u+p_{ij}, v-p_{ij}) \right] \\ &= \min \max \left[\quad \cdots\cdot \quad \right], \quad u,v>0 . \end{aligned}$$

The boundary conditions are

$$I(u,v) = 0 , \quad u \leq 0 , \quad v > 0 ,$$

$$I(u,v) = 1 , \quad u > 0 , \quad v \leq 0 .$$

Returning now to the consideration of a control process subject to random effects as a "game against nature," let us examine the recurrence relation

$$x(k+1) = f[x(k),u(k),r(k)] , \quad x(0) = c ,$$

where $u(k)$ is a vector at our disposal, and $r(k)$ is a vector at the disposal of our opponent.

Assume that we wish to choose $\{u(k)\}$ to minimize the expected value of some function of the terminal position, while our adversary wishes to choose $r(k)$ to maximize this expected value. Say we have M choices for $u(0)$, and our opponent has N different choices for $r(0)$. That is, at each stage both we and our opponent choose a probability distribution for the controls $u(k)$ and $r(k)$.

Let $h[x(K)]$ measure the values of the final state at time K and let $p = [p_1,\ldots,p_M]$ denote the probability distribution for $u(0)$, while $q = [q_1,\ldots,q_N]$ is the distribution for $r(0)$. Define

$$I(c,N) = \min_p \max_q E[h(x(K))] ,$$

where the min and max are taken over the distributions chosen at each stage by us and our opponent. Forward dynamic programming plus the principle of optimality, in conjunction with the Min-Max Theorem, gives

$$\begin{aligned} I(c,N) &= \min_p \max_q \sum_{i,j} p_i\, q_j\, I[f(c,u^{(i)}(0),r^{(j)}(0)),K-1] , \\ &= \max_q \min_p \sum_{i,j} p_i\, q_j\, I[f(c,u^{(i)}(0),r^{(j)}(0)),K-1] , \\ &\qquad K \geq 1 , \end{aligned}$$

where $u^{(i)}(0)$ is the ith allowable choice for $u(0)$ and $r^{(j)}(0)$ is the jth allowable choice of $r(0)$, $1 \leq i \leq M$, $1 \leq j \leq N$. Clearly,

$$I(c,0) = h(c) .$$

OPTIMUM ESTIMATION

As an intermediate step toward the treatment of stochastic problems in which the probability densities are unknown and must be estimated as information becomes available (adaptive processes), we now consider the question of estimation of a process. In this problem, it is desired to find the best estimate of the state of a system on the basis of a number of noisy observations. In this section a computational procedure based on dynamic programming for solving a general class of optimum estimation problems is presented.

The optimum estimation problem is formulated as follows. The system equation for the discrete case is

$$x(k+1) = g[x(k),w(k),k] , \tag{3.17}$$

where x = n-dimensional state vector,
w = r-dimensional random forcing function,
g = n-dimensional vector function.

The equations for the measurement system are

$$z(k) = h[x(k),v(k),k] , \tag{3.18}$$

where z = s-dimensional measurement vector, where in general $s \leq n$,
h = s-dimensional vector function,
v = d-dimensional vector of random noise.

The probability density functions $p[w(k)]$ and $p[v(k)]$ are assumed to be known and independent from sample to sample. The probability density function of the initial state before any measurements are received, denoted as $p[x(0/-1)]$, is also assumed to be known.

The problem which is solved in this section can be stated as follows: given the system described by Eq. (3.17), the measurement system described by Eq. (3.18), the probability functions $p[w(j)]$ and $p[v(j)]$, $j = 1,2,\dots,k$, the a priori probability density $p[x(0/-1)]$, and the noisy measurements $z(1),\dots,z(k)$, find the maximum likelihood estimate of the entire trajectory, $x(0)$, $x(1)$, $\dots$, $x(k)$. This estimate of the trajectory is denoted as $X(k/k)$.

This problem is somewhat different from the one customarily solved by using Bayes' theorem. In the latter problem, the maximum likelihood estimate of $x(k)$, the present state, is found rather than the maximum likelihood estimate for the entire trajectory. The difference between these estimates is discussed in more detail later in this section.

An iterative relation for the maximum likelihood trajectory estimate is derived by applying Bayes' rule. The notation will be such that it is easy to see the similarity between this iterative relation and the dynamic programming functional equation. First, in order to simplify the algebra the following definitions are made:

$$\begin{aligned} X(k) &= \{x(k),\ x(k-1),\ \dots,\ x(0)\}\ , \\ Z(k) &= \{z(k),\ z(k-1),\ \dots,\ z(1)\}\ . \end{aligned} \tag{3.19}$$

The function $I[x(k),k]$ is then defined as

$$\begin{aligned} I[x(k),k] &= \max_{x(0),x(1),\dots,x(k-1)} \{p[x(k),x(k-1),\dots \\ &\qquad \dots,x(0)/z(k),z(k-1),\dots,z(1)]\}\ , \\ &= \max_{x(0),x(1),\dots,x(k-1)} \{p[X(k)/Z(k)]\}\ . \end{aligned} \tag{3.20}$$

The desired relation is obtained by deriving $I[x(k+1),k+1]$ in terms of $I[x(k),k]$.

Bayes' rule can be written as

$$p[X(k+1)/Z(k+1)] = \frac{p[z(k+1)/X(k+1),Z(k)]p[X(k+1)/Z(k)]}{p[z(k+1)/Z(k)]} . \tag{3.21}$$

Since, by Eq. (3.18), $z(k+1)$ depends only on $x(k+1)$, Eq. (3.21) can be simplified as

$$p[X(k+1)/Z(k+1)] = \frac{p[z(k+1)/x(k+1)]p[X(k+1)/Z(k)]}{p[z(k+1)/Z(k)]} . \tag{3.22}$$

The term $p[X(k+1)/Z(k)]$ can be written as

$$p[X(k+1)/Z(k)] = p[x(k+1)/x(k)]p[X(k)/Z(k)] . \tag{3.23}$$

Substituting this relation into Eq. (3.21),

$$\begin{aligned} I[x(k+1),k+1] &= \max_{x(0),x(1),\ldots,x(k)} p[X(k+1)/Z(k+1)] , \\ &= \max_{x(k)} \left\{ \frac{p[z(k+1)/x(k+1)]p[x(k+1)/x(k)]}{p[z(k+1)/Z(k)]} \cdot \right. \\ &\qquad \left. \cdot I[x(k),k] \right\} . \end{aligned} \tag{3.24}$$

This equation yields the desired iterative relation; the maximization operation is over the single value $x(k)$, rather than over the entire set of past states $x(0),x(1),\ldots,x(k)$.

The quantity $p[z(k+1)/x(k+1)]$ is determined by Eq. (3.18) and knowledge of $p[v(k+1)]$, while $p[x(k+1)/x(k)]$ is determined by Eq. (3.17) and $p[w(k)]$. The function $p[z(k+1)/Z(k)]$ serves only as a normalization factor, and the maximum likelihood estimate can be determined without explicitly computing it. The iterative relation can then be rewritten as

$$I^*[x(k+1),k+1] = \max_{x(k)} \{p[z(k+1)/x(k+1)]p[X(k+1)/x(k)] \cdot I^*[x(k),k]\} , \tag{3.25}$$

where the function $I^*[x(k),k]$ is proportional to, but not equal to, $I[x(k),k]$.

If the system equation, Eq. (3.17), is linear in $x(k)$ and $w(k)$, if the measurement system, Eq. (3.18), is linear in $x(k)$ and $v(k)$, and if the probability density functions $p[x(0/-1)]$, $p[w(k)]$, and $p[v(k)]$ are all Gaussian, then the Kalman-Bucy filter can be obtained directly from Eq. (3.20).

For the general case, a computational procedure analogous to forward dynamic programming can be implemented. The state vector $x(k)$ is quantized. The a priori probability density function is used as an initial condition

$$I^*[x(0),0] = p[x(0/-1)] . \tag{3.26}$$

At each quantized value of $x(1)$, the quantity inside the brackets in Eq. (3.25) is evaluated for every quantized value of $x(0)$. The maximum value is selected as $I^*[x(1),1]$, and the corresponding value $x(0)$ is stored along with it. This procedure, in which $I^*[x(j),j]$ is computed on the basis of $I[x(j-1),j-1]$ and the corresponding value of $x(j-1)$ is carried along, is repeated until $I^*[x(k),k]$ is reached. The maximum likelihood trajectory is then determined by computing the value of $x(k)$ which maximizes $I^*[x(k),k]$, retrieving the $x(k-1)$ that corresponds to this value, finding the $x(k-2)$ corresponding to this value of $I[x(k-1),k-1]$, and repeating until all values of $x(j)$ along the trajectory have been found.

EXAMPLE

3.8 We now examine the relation between the estimates obtained from this procedure and from the Bayesian estimation procedure for a two-stage problem (see Fig. 3.1). The state of the system is assumed to be one of two values, 0 and 1. The a priori probabilities of the two states are $p[x(0)=0] = 0.4$ and $p[x(0)=1] = 0.6$.

The system equations are such that $p[x(1)/x(0)]$ is given in Table 3.1. The observation at $k=1$ is also assumed to have two values, 0 and 1. The observation equations specify $p[z(1)/x(1)]$ as in Table 3.2.

If the observation $z(1)=1$ is received, the conditional probabilities of the four possible trajectories are as shown in

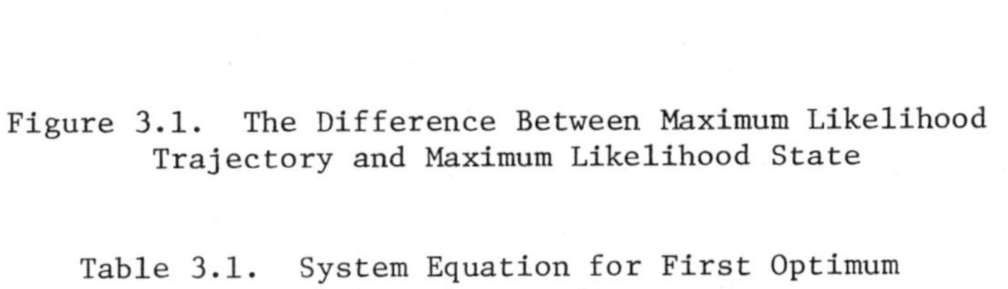

Figure 3.1. The Difference Between Maximum Likelihood Trajectory and Maximum Likelihood State

Table 3.1. System Equation for First Optimum Estimation Problem

x(0)	x(1)	p[x(1)/x(0)]
0	0	0.9
0	1	0.1
1	0	0.3
1	1	0.7

Table 3.2. Measurement Equation for First Optimum Estimation Problem

x(1)	z(1)	p[z(1)/x(1)]
0	0	0.5
0	1	0.5
1	0	0.5
1	1	0.5

Figure 3.1 (the same results are achieved if $z(1) = 0$). The maximum likelihood trajectory is found to be $x(0) = 1$, $x(1) = 1$, which has probability 0.42. On the other hand, the probability that $x(1) = 1$, which is obtained by adding the probabilities of trajectories $x(0) = 1$, $x(1) = 1$, and $x(0) = 0$, $x(1) = 1$, is 0.46, while the probability that $x(1) = 0$ is 0.54. Thus, the maximum likelihood present state is $x(1) = 0$. In this simple example, the final state along the maximum likelihood trajectory differs from the maximum likelihood state.

ADAPTIVE CONTROL PROCESSES

In place of assuming that unknown stochastic effects influencing a process are chosen by an adversary in as disadvantageous a fashion as possible, it is much more reasonable to attempt to ascertain the underlying probability law as the process unfolds and observations are made upon the behavior of the system. This idea leads us into the realm of adaptive control processes, where we admit ignorance of the probability distributions governing the process of interest, but assume that we have some mechanism for transforming the observed behavior of the system into improved estimates of the unknown distributions.

Since this is a problem of great complexity and the topic of much current research, we shall have to content ourselves in an introductory volume of this nature to consideration of simple, illustrative cases rather than attempt any sort of comprehensive treatment. As we shall see, even the treatment of simple cases involves a measure of analytic and computational complexity, the extension of which is best left to research treatises devoted entirely to the subject of adaptive control. The devoted reader will find reference to several such works at the end of the book.

To illustrate the general idea, consider the stochastic transformation

$$x(k+1) = x(k) + u(k) + r(k) \ , \quad x(0) = c \ ,$$

where $\{r(k)\}$ is a sequence of independent random variables assuming only the values 1 or 0, with probabilities p and $1-p$, respectively. Let us assume we wish to minimize the expected value of the quantity

$$J = |x(K)| + b \sum_{k=0}^{K-1} x^2(k) \ , \tag{3.27}$$

where b is some positive constant. If the probability p is known, we have a simple stochastic control process of the type already discussed. Calling the minimum expected value $I(c,N)$, we readily obtain

$$I(c,K) = \min_v \begin{Bmatrix} p[b(ac+v+1)^2 + I(ac+v+1, K-1) \\ +(1-p)[b(ac+v)^2 + I(ac+v, K-1)) \end{Bmatrix} \ , \quad K \geq 2 \ ,$$

$$I(c,1) = \min_v \begin{Bmatrix} p[|ac+v+1| + b(ac+v+1)^2] \\ +(1-p)[|ac+v| + b(ac+v)^2] \end{Bmatrix} \ .$$

Now suppose that we do not know the value of p but that we do possess an a priori distribution for it, $dG(p)$ and that we have a definite scheme for modifying the a priori estimate on the basis of observations made upon the process over time. For example, employing a Bayesian approach, if over the first $m+n$ stages we observe that the sequence $\{r(i)\}$ has taken the value 1 m times and 0 n times, and if the a priori distribution function was $dG(p)$, then we take as our new distribution function

$$dG_{m,n}(p) = \frac{p^m(1-p)^n dG(p)}{\int_0^1 p^m(1-p)^n dG(p)} .$$

The density function,

$$f(p) = \alpha p^m(1-p)^n ,$$

is called the beta density function because of its relation to the well known "beta function." The beta function $\beta(m,n)$ is defined for two real numbers, m and n as

$$\beta(m,n) = \int_0^1 p^m(1-p)^n \, dp .$$

Note that $\beta(m,n)$ is exactly the reciprocal of the normalization constant α required for $f(p)$. For integer values of m and n, it can be shown that

$$\beta(m,n) = \frac{m!\, n!}{(m+n+1)!} .$$

This equation makes it easy to calculate moments of the density function. For example, the mean value of p is

$$E\{p\} = \frac{\int_0^1 p\, p^m(1-p)^n \, dp}{\int_0^1 p^m(1-p)^n \, dp} ,$$

where now the normalization constant β has been written out in the denominator. It then follows that

$$E\{p\} = \frac{\beta(m+1,n)}{\beta(m,n)} = \frac{(m+1)!\ n!\ (m+n+1)!}{(m+n+2)!\ m!\ n!} .$$

The variance of p can be deduced by

$$\text{var}\{p\} = E\{(p-E\{p\})^2\} = [E\{p^2\} - (E\{p\})^2]$$

$$= \frac{\int_0^1 p^2\ p^m(1-p)^n\ dp}{\int_0^1 p^m(1-p)^n\ dp} - \frac{(m+1)^2}{(m+n+2)^2}$$

$$= \frac{(m+1)(n+1)}{(m+n+3)(m+n+2)^2} .$$

These functions are quite smooth and have many desirable properties. For example, they are unimodal and take on their maximum value at

$$\bar{p} = \frac{m}{m+n} .$$

The mode and the mean are thus not equal, but for large m and n the difference is small. Also, for a fixed ratio of $\frac{(m+1)}{(m+n+2)}$, the variance decreases with increasing $(m+n+2)$. It is thus possible to pick values of m and n that closely approximate the shape of any unimodal curve with arbitrary mean and variance. As an example, plots are shown of the density functions for $m=1$, $n=1$, and for $m=5$, $n=5$ in Figures 3.2a and 3.2b, respectively. Note that in both cases the mean is 1/2, but $\sigma^2_{1,1} = 1/20$, while $\sigma^2_{5,5} = 1/52$. The normalization constants are $\alpha_{1,1} = \frac{1}{\beta(1,1)} = 6$, and $\alpha_{5,5} = 2772$.

It should be pointed out that m and n do <u>not</u> need to be integers. Values of $\beta(m,n)$ for <u>real</u> m and n are tabulated

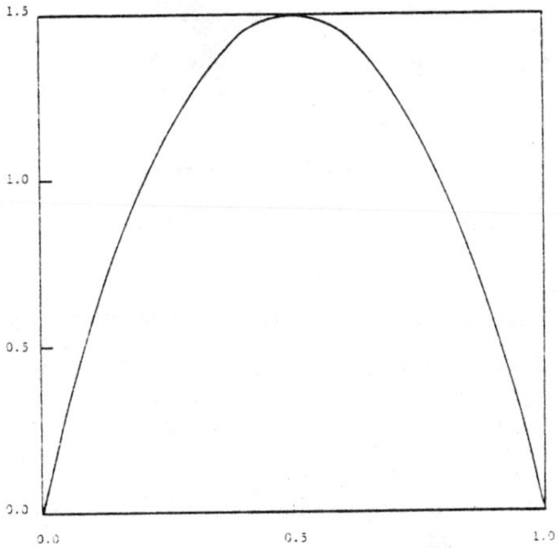

Figure 3.2a Plot of $f_1(p) = 6p(1-p)$

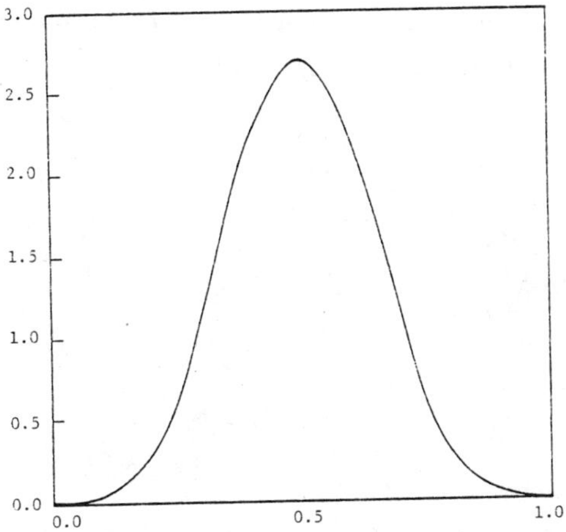

Figure 3.2b Plot of $f_2(p) = 2772\ p^5(1-p)^5$

in many places. These tables allow one to calculate means and variances in a straightforward fashion. This flexibility allows one to approximate an even greater range of functions using this class of function.

As a result of the preceding analysis, once we have specified $dG(p)$, the a priori distribution at any stage of the process depends only upon m and n. Consequently, the state of the process at any stage is specified by the physical state c and the two numbers m and n.

To employ our functional equation approach, introduce the sequence of functionals $I(c,m,n,K)$ defined as the minimum of the expected value of the function J of (3.27), starting with the information that m ones and n zeros have occurred, the system is in state c, and N stages remain.

For $K = 1$, we have

$$I(c,m,n,1) = \min_v \left\{ \begin{array}{l} p_{m,n}[b(ac+v+1)^2 + |ac+v+1|] \\ + (1-p_{m,n})[b(ac+v)^2 + |ac+v|] \end{array} \right\}, \quad (3.28)$$

where $p_{m,n}$ is the expected probability of p obtained from the distribution $dG_{m,n}(p)$, i.e.,

$$p_{m,n} = \frac{\int_0^1 p^{m+1}(1-p)^n \, dG(p)}{\int_0^1 p^m(1-p)^n \, dG(p)} . \quad (3.29)$$

For $K \geq 2$, the usual arguments yield

$$I(c,m,n,K) = \min_v \left\{ \begin{array}{l} p_{m,n}[b(ac+v+1)^2 + I(ac+v+1,m+1,n,K-1)] \\ +(1-p_{m,n})[b(ac+v)^2 + I(ac+v,m,n+1,K-1)] \end{array} \right\}. \quad (3.30)$$

Notice that (3.28)-(3.30) define a feasible computational algorithm although not an entirely straightforward one, since there is an "expanding grid" phenomena present due to the fact that $I(c,m,n,K)$ depends upon the values $I(\cdot, m+1, n, K-1)$ and $I(\cdot, m, n+1, K-1)$. Hence, various successive approximation schemes may have to be employed to satisfactorily resolve the problem, since only $I(\cdot, m, n, K-1)$ has been calculated and we actually need $I(\cdot, m+1, n, K-1)$, etc.

A more complex type of adaptive control process and one that is of considerable importance as a prototype problem in the theory of experiments is the following. Suppose a controller is called upon to estimate the value of an unknown probability p. Assume that p is the probability that a certain random variable takes the value one and $1-p$ is the probability it takes the value zero. The controller is to conduct a series of experiments, record the outcomes, and make an estimate of p on the basis of this experiment plus any a priori information available. There are costs which are associated with making each experiment and with making wrong estimates of p. We wish to determine when the experiments should be stopped and what estimate should be made by the controller.

Let us now specify the problem in more detail. Assume that before the process begins, the controller possesses on theoretical grounds the a priori information that n ones out of s trials will be observed. Insofar as observation of the outcomes of the process itself are concerned, we assume that m ones have actually been observed in r trials. Since p is unknown, we regard it as a random variable with a distribution function which changes during the course of the process. Solely on the basis of the a priori information, we consider this distribution function to be given by

$$dG(p) = \frac{p^{n-1}(1-p)^{s-n-1}}{B(n,s-n)}\, dp \ ,$$

where B is the beta function $B(m,n) = \frac{m!\ n!}{(m+n+1)!}$. After m ones have been observed in r additional trials we modify dG, as before, by the formula

$$dG_{r,m}(p) = \frac{p^{m}(1-p)^{r-m}dG(p)}{\int_0^1 p^{m}(1-p)^{r-m}dG(p)} \ .$$

Let $c_{r,m}$ represent the expected mean-square error of incorrect estimation after r additional trials have resulted in m ones and set

$$c_{r,m} = \int_0^1 (p_{r,m} - p)^2 \, dG_{r,m}(p) ,$$

where $p_{r,m}$ is the estimate which maximizes $c_{r,m}$. The value $p_{r,m}$ is given by the formula

$$p_{r,m} = \int_0^1 p \, dG_{r,m}(p)$$

$$= \frac{m+n}{r+s} .$$

A straightforward calculation then yields

$$c_{r,m} = \alpha\left(\frac{m+n}{r+s}\right)\left[\frac{m+n+1}{r+s+1} - \frac{m+n}{r+s}\right] .$$

Also, we shall assume that if r experiments have been performed, the cost of the next experiment is $k(r)$. Thus, the cost per experiment is permitted to vary during the process. In addition, we shall assume that in the absence of additional information estimated probabilities are to be regarded as true probabilities. Lastly, we require that no more than R experiments be performed, so that the process must be truncated should this point be reached. Let us now determine the optimal control policy.

Define the function $I(m,r)$ as

$I(m,r)$ = the expected cost of a process beginning with m ones in r experiments having been observed, and using an optimal sequence of decisions.

Applying the principle of optimality yields the functional equation

$$I(m,r) = \min_{C,S} \begin{cases} C: & k(r) + p_{r,m} I(m+1,r+1) + (1-p_{r,m}) I(m,r+1) \\ S: & \alpha\left(\frac{m+n}{r+s}\right)\left[\frac{m+n+1}{r+s+1} - \frac{m+n}{r+s}\right], \quad m = 0,1,\ldots,r\ , \\ & \qquad r = R-1,R-2,\ldots,0\ . \end{cases} \tag{3.31}$$

In (3.31) (C) denotes the decision to continue and (S) to stop the process. In view of the truncation assumption,

$$I(m,R) = \alpha\left(\frac{m+n}{R+s}\right)\left[\frac{m+n+1}{R+s+1} - \frac{m+1}{R+s}\right]\ . \tag{3.32}$$

Relations (3.31)-(3.32) enable us to calculate the sequence of functions $\{I(m,r)\}$ and to determine the optimal stopping rule as well as what estimate to make of p in the event the process is terminated.

The above functional equations were investigated computationally for a wide range of values for the parameters α and R, and for several cost functions $k(r)$. The results were that when the cost of experimentation was constant, or when it increased with r, and when one out of two ones had been observed a priori, i.e.,

$$dG(p) = \begin{cases} 0 & -\infty \le p < 0 \\ \frac{1}{2}\ , & 0 \le p < 1\ , \\ 1\ , & 1 \le p \le \infty \end{cases}$$

the optimal policy consisted essentially in:

(1) Continuing the experiments if r was small (not enough information present upon which to base an estimate);

(2) Stopping the experiments if r was sufficiently large;

(3) Continuing the experiments for intermediate values of r, unless extreme runs of either zeros or ones occurred.

In the case of a decreasing cost of experiment for each new experiment,

$$k(r) = \exp\left(1 - \frac{r}{15}\right) ,$$

the optimal policy is more complex. We illustrate with the case $n = 1$, $s = 2$, $\alpha = 1500$, and $R = 40$. When $r \leq 19$, generally speaking, the rule is to do at least one more experiment. When $20 \leq r \leq 33$, the rule is to stop immediately and make the appropriate estimate. If $34 \leq r \leq 39$, and approximately half the observations have been ones, then again the rule is to do another experiment. The reason for this behavior, of course, is that the cost of experimentation has dropped so low that it is profitable to do at least one more experiment before making the estimate.

COMBINED OPTIMUM CONTROL AND ESTIMATION

The general problem of finding optimal control for a system on the basis of noisy measurements has been solved by dynamic programming. This problem covers a great many cases of both practical and theoretical interest, including the general problem of optimal synthesis of adaptive and learning systems. It is called the "combined" optimum control and estimation problem because the solution for certain special cases is first to find a best estimate of the state of the system from the measurements and then to apply the control that is optimal if the state of the system is equal to this estimate. In this section, the general solution of the combined optimization problems is presented.

The problem in the discrete-time case can be visualized as in Figure 3.3. The system that is to be controlled is described by the difference equation

$$x(k+1) = g[x(k),u(k),w(k),k] , \tag{3.33}$$

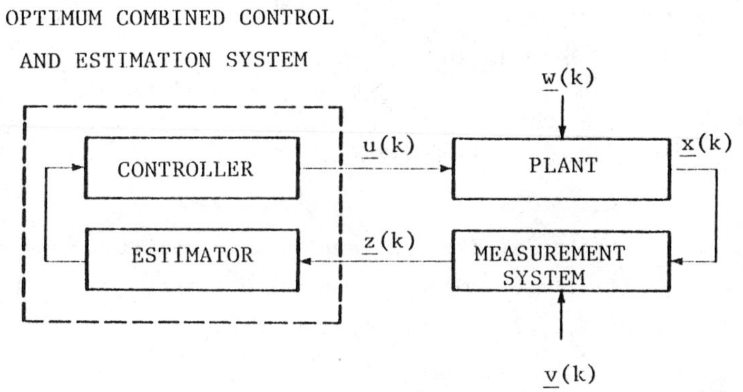

Figure 3.3 Combined Optimum Control and Estimation

where x(k) = state vector of the system,

k = index of the time instants,

u(k) = control vector,

w(k) = random forcing vector, which is statistically independent from time period to time period,

g = vector functional.

The measurement system is described by

$$z(k) = h[x(k),v(k),k] \quad , \tag{3.34}$$

where z(k) = measurement vector, which generally has dimension less than that of x(k),

v(k) = measurement noise vector, which is statistically independent from time period to time period,

h = vector function.

The performance criterion, which is to be minimized, is the expected value of a variational performance measure,

$$J = E \sum_{k=0}^{K} \ell[x(k),u(k),w(k),k] , \tag{3.35}$$

where the expectation is over the a priori probability density for $x(0)$, $p[x(0/-1)]$, and the probability densities $p[w(k-1)]$ and $p[v(k)]$, $k = 1,2,\ldots,K$. Constraints are present in the form

$$x \varepsilon X(k) , \tag{3.36}$$

$$u \varepsilon U(x,k) . \tag{3.37}$$

The combined optimization problem can be stated as follows: given the system described by Eq. (3.33), the measurement system described by Eq. (3.34), and the probability density functions $p[x(0/-1)]$, $p[w(k-1)]$, and $p[v(k)]$, $k = 1,2,\ldots,K$, find the control sequence $u(k)$, $k = 0,1,2,\ldots,K$ which minimizes J in Eq. (3.35), subject to the constraints in Eq. (3.36) and (3.37).

As an indication of the generality of this problem, the deterministic control problem, the stochastic control problem, and the optimum estimation problem are all special cases. The stochastic control problem is obtained if $z(k) \equiv x(k)$; the deterministic control problem is obtained if, in addition, $w(k) \equiv 0$; while the optimum estimation problem is obtained if g is independent of $u(k)$, and $u(k)$ then represents the best estimate. Furthermore, the control of adaptive and learning systems, in which system parameters are not known, can be treated by including the unknown parameters as additional state variables. The results of this section thus provide a general solution to the optimal control problem for such systems.

In the derivation of the solution to this problem, the algebra is simplified if the following notation is used:

$$Z(k) = \{z(k), z(k-1), \ldots, z(1)\} ,$$
$$U(k) = \{u(k), u(k-1), \ldots, u(1), u(0)\} . \tag{3.38}$$

Because the state of the system is not known but is instead a probabilistic function, it is no longer appropriate to write an iterative functional equation directly in terms of the state. Instead, it becomes appropriate to speak of an "information state," which summarizes all the information known about the state. In the general case, the information state is the set of all measurements up to the present, $z(1), \ldots, z(k)$, and all past controls, $u(0), u(1), \ldots, u(k-1)$. The minimum expected cost function can then be defined as the minimum value of the expected cost based on the past and present measurements and past controls. Using Eq. (3.38), this function can be written as

$$I^*[Z(k), U(k-1), k] = \min_{u(k), \ldots, u(K)} \left\{ \underset{x(k), w(k), \ldots, w(K), v(k+1), \ldots, v(K)}{E} \sum_{j=k}^{K} I^*[X(j), u(j), w(j), j] / Z(k), U(k-1) \right\} , \tag{3.39}$$

where the expectation is conditioned on $Z(k)$ and $U(k-1)$.

The iterative relation is found by a process similar to that used by the stochastic and deterministic control cases. The iterative relation is

$$I^*[Z(k), U(k-1), k] = \min_{u(k)} \left\{ \underset{x(k), w(k), v(k+1)}{E} [\ell[x(k), u(k), w(k), k] + I^*(h[g[x(k), u(k), w(k)], v(k+1), k+1], Z(k), u(k), U(k-1)) / Z(k), U(k-1)] \right\} . \tag{3.40}$$

The initial condition for the iterative procedure is obtained as

$$I^*[Z(K),U(K-1),K] = \min_{u(K)} \left\{ \underset{x(K),w(K)}{E} [\ell[x(K),u(K),w(K),K]/ Z(K),U(K-1)] \right\}. \tag{3.41}$$

In order to carry out the computations, it is necessary to have the conditional probability distribution $p[x(k)/Z(k),U(k-1)]$. This may be calculated iteratively using Bayes' rule. The iterative relation is

$$p[x(k+1)/Z(k+1),U(k)] = \frac{p[z(k+1)/x(k+1)] \int_{x(k)} p[x(k+1)/x(k),u(k)] \cdot p[x(k)/Z(k),U(k-1)]dx(k)}{\int_{x(k+1)} p[z(k+1)/x(k+1)] \int_{x(k)} p[x(k+1)/x(k),u(k)]p[x(k)/Z(k),U(k-1)]dx(k)dx(k+1)}, \tag{3.42}$$

where

$$\int_{x(k+1)} \ldots dx(k+1) \quad \text{and} \quad \int_{x(k)} \ldots dx(k)$$

denote integration over all possible values of $x(k+1)$ and $x(k)$, respectively. The probability density $p[z(k+1)/x(k+1)]$ is determined by Eq. (3.34) and $p[v(k+1)]$, while $p[x(k+1)/x(k),u(k)]$ is known from Eq. (3.35) and $p[w(k)]$. The iterations begin with

$$p[x(1)/z(1),u(0)] = \frac{p[z(1)/x(1)] \int_{x(0/-1)} p[x(1)/x(0/-1), u(0)\ p[x(0/-1)]dx(0/-1)}{\int_{x(1)} p[z(1)/x(1)] \int_{x(0-1)} p[x(1)/x(0/-1), u(0)\ p[x(0/-1)]dx(0/-1)dx(1)},$$

where $p[x(0/-1)]$ is the a priori probability density of $x(0)$.

The difficulty in applying the iterative relation in Eq. (3.40) is that the information state [Z(k),U(k-1)] grows with time. As more past observations and controls accumulate, the dimension of this state grows linearly; if the dimension of z(k) is d and the dimension of u(k) is q, then the dimension of the information state at time k is k(d+q). Clearly, even if d and q are reasonably small, after a few time periods the dimension of the information state will become excessive.

One method of keeping the dimensionality of the system within bounds is to use an alternate description of the information state. One alternative is to use the moments of the conditional probability density function p[x(k)/Z(k),U(k-1)]. It has been shown that if the system equations (3.33) are linear in x(k), u(k), and w(k); if the measurement equation (3.34) are linear in x(k) and v(k); if the performance criterion (3.35) is a quadratic function of x(k), u(k), and w(k); and if the probability density functions p[x(0/-1)], p[w(k)], and p[v(k)] are all Gaussian, then this conditional probability density function is also Gaussian. Consequently, it is completely specified by its mean and covariance. Thus, the information state has dimension (n+n(n+1)/2), where n is the dimension of x(k); n of the quantities are the components of vector means while the other n(n+1)/2 are terms in the covariance matrix. In this special case, the optimum combined system consists of finding the maximum likelihood estimate of x(k) and applying the control that would be optimal if this estimate were actually the state.

Even if the mean and covariance are not a set of sufficient statistics, it is often convenient to approximate the optimal system by assuming that they are. Another approximation technique is to assume that I* depends on only the last few measurements and controls, rather than on all past values.

SOLVED PROBLEMS

3.1 Consider the Van der Pol equation

$$\ddot{x} + \mu(x^2 - 1)\dot{x} + x = w(t) ,$$

where $w(t)$ is a random forcing term taking the values ± 1 with equal probability, μ a non-zero constant. Introduce a discrete approximation to this equation and, using it, derive a functional equation describing the expected value of $x + \dot{x}$.

Let us use the discrete approximation

$$x_1(n+1) - x_1(n) = x_2(n)\Delta , \qquad x_1(0) = c_1 ,$$

$$x_2(n+1) - x_2(n) = [w(n) - x_1(n) - \mu(x_1^2(n)-1)x_2(n)]\Delta ,$$

$$x_2(0) = c_2 ,$$

where $x_1(n) \equiv x_1(n\Delta)$, $x_2(n) = x_2(n\Delta)$ and $\Delta \ll 1$. Thus, $x_1(n) \cong x(n\Delta)$ and $x_2(n) \cong \dot{x}(n\Delta)$.

Since the expected value of $x + \dot{x}$ depends upon the initial conditions c_1 and c_2, as well as upon the time variable, we write

$$I_n(c_1, c_2) = E\{x_1(n) + x_2(n)\} ,$$

$$n = 1, 2, \ldots, -\infty < c_1, c_2 < \infty .$$

After one unit of time, the state variables c_1 and c_2 undergo the following transformations

$$c_1 \to c_1 + c_2\Delta ,$$

$$c_2 \to c_2 + \{w(1) - c_1 - (c_1^2 - 1)c_2\}\Delta .$$

Thus,

$$I_n(c_1,c_2) = \underset{w(1)}{E}\,[I_{n-1}(c_1 + c_2\Delta,\ c_2 + \{w(1) - c_1 - (c_1^2 - 1)c_2\}\Delta] .$$

Since $w(1) = \pm 1$ with equal probability, this becomes

$$I_n(c_1,c_2) = \frac{1}{2}[I_{n-1}(c_1+c_2\Delta,\ c_2+\{1-c_1- (c_1^2-1)c_2\}\Delta]$$
$$+\frac{1}{2}[I_{n-1}(c_1+c_2\Delta,\ c_2+\{-1-c_1-\mu(c_1^2-1)c_2\}\Delta] ,$$
$$n \geq 1 ,$$

with $I_0(c_1,c_2) = c_1 + c_2$.

3.2 At each state of a sequence of actions, we are allowed a choice of one of two decisions. The first has associated with it a probability p_1 of gaining one unit, a probability p_2 of gaining two units, and a probability p_0 of gaining no units and terminating the process. The second decision has the same set of returns with probabilities p_1', p_2', p_0'. Determine a functional equation for the sequence of choices which maximizes the probability of attaining at least n units before the process is terminated.

Let

$I(n)$ = probability of attaining at least n units before the process terminates, when following an optimal policy, $n = 1,2,\ldots$.

From the principle of optimality, we see that

$$I(n) = \max_{I,II} \begin{cases} I: & p_1 I(n-1) + p_2 I(n-2) \\ II: & p_1' I(n-1) + p_2' I(n-2) \end{cases}, \quad n=2,3,4,\ldots .$$

The initial conditions are

$$I(0) = 1 ,$$

$$I(1) = \max_{I,II} \begin{cases} I: & p_1 \\ II: & p_1' . \end{cases}$$

3.3 Consider a serial electronic system whose reliability may be taken to be the product of the reliabilities of the individual components.

To improve reliability of a particular stage, we may put several units in parallel. Let $p_k(x_k)$ be the reliability of the <u>kth</u> stage when x_k units are put in parallel at stage k; let $g_k(x_k)$ be the cost of having x_k units in parallel and let $h_k(x_k)$ be the weight of the <u>kth</u> stage when there are x_k units in parallel.

Determine a functional equation for maximizing the total reliability of the system when the total cost and weight are each subject to fixed upper bounds.

The total reliability of the system is given by

$$P_N = \prod_{k=1}^{N} p_k(x_k) .$$

We wish to maximize P_N over the x_k subject to the constraints

(a) $x_k = 1,2,3,\ldots$,

(b) $\sum_{k=1}^{N} g_k(x_k) \leq c$,

(c) $\sum_{k=1}^{N} h_k(x_k) \leq w$,

where c and w are the upper bounds on total cost and weight, respectively. Since the maximum value of P_N depends on the parameters c, w, as well as the total number of stages, let

$I_N(c,w)$ = maximum value of P_N when c is the total cost allowed, w is the total weight allowed, and an optimal policy is used, $N = 1,2,\ldots$, $c,w \geq 0$.

At stage N, an allocation of x_N units results in a cost of $g_N(x_N)$ and a weight $h_N(x_N)$. Thus, there is then $c-g_N(x_N)$ dollars left for the remaining N-1 stages and the maximum weight becomes $w-h_N(x_N)$. Thus,

$$I_N(c,w) = \max_{x_N} [p_N(x_N) I_{N-1}(c-g_N(x_N), w-h_N(x_N))] , \quad N > 1 ,$$

where the maximum is taken over the region

(a) $x_N = 1,2,3,\ldots$,

(b) $g_N(x_N) \leq c$,

(c) $h_N(x_N) \leq w$.

Obviously,

$$I_1(c,w) = \max_{x_1} [p_1(x_1)] ,$$

where the max over x_1 is over the region (a)-(c).

3.4 In the noisy communication channel problem, determine a functional equation for the case when the probability of correct transmission of the kth signal depends upon whether or not the (k-1)st signal was transmitted correctly.

Let

$f_k(x)$ = the expected value of the logarithm of the final capital obtained from the last k stages of the original N-stage process, starting with an initial capital x and the information that the (k-1)st signal was transmitted correctly, and an optimal policy is used.

$g_k(x)$ = the corresponding function when the (k-1)st signal was transmitted incorrectly.

Also, let

p_k = probability of correct transmission of the kth signal when the (k-1)st signal was transmitted correctly.

q_k = probability of correct transmission of the kth signal if the (k-1)st signal was transmitted incorrectly.

Taking into account all possibilities, the functions $\{f_k(x)\}$, $\{g_k(x)\}$ satisfy

$$f_k(x) = \max_{0 \le y \le x} \{p_{N-k+1} f_{k-1}(x+y) + (1-p_{N-k+1}) g_{k-1}(x-y)\},$$

$$g_k(x) = \max_{0 \le y \le x} \{q_{N-k+1} f_{k-1}(x+y) + (1-q_{N-k+1}) g_{k-1}(x-y)\},$$

$$k \ge 2.$$

3.5 Consider the situation in which the channel transmits any of M different symbols. Upon receiving a signal, the gambler must make a bet based upon what he believes the transmitted signal actually was. Assume the gambler knows

p_{ij} = the conditional probability that signal j was sent if signal i is received,

q_i = the probability of receiving signal i,

r_j = the return from a single unit winning bet on signal j.

Assume that the gambler is free to bet an amount z_i on the ith signal, subject to $\sum_{i=1}^{M} z_i \le x$. Determine a functional equation for the optimal return and policy.

Let $I_N(x)$ be defined as in the text. For any bet $(z_1, \ldots, z_M)$, if the gambler receives signal i his return is

$$q_i = \sum_{j=1}^{M} p_{ij} I_{N-1}\left(x + r_j z_j - \sum_{s=1}^{M} z_s\right).$$

Thus, his expected return is obtained by summing over all signals i and the optimal return is obtained by then taking the maximum over all possible decisions. Putting these remarks together, we have

$$I_N(x) = \max_Z \left\{ \sum_{i=1}^{M} q_i \sum_{j=1}^{M} p_{ij} I_{N-1}\left(x + r_j z_j - \sum_{\substack{s=1 \\ s \neq j}}^{M} z_s \right) \right\} , \quad N \geq 2 ,$$

where Z is the region

$$Z = (z_1, z_2, \ldots, z_M) : z_i \geq 0 \quad \text{and} \quad \sum_{s=1}^{M} z_s \leq x \quad .$$

The initial function $I_1(x)$ is given by

$$I_1(x) = \max_Z \left\{ \sum_{i=1}^{M} q_i \sum_{j=1}^{M} p_{ij} \log \left(x + r_j z_j - \sum_{\substack{s=1 \\ s \neq j}}^{M} z_s \right) \right\} .$$

3.6 Consider a scalar system whose dynamics are

$$x(k+1) = g(x(k), u(k)) , \quad x(0) = c , \quad k = 0, 1, \ldots, N-1 ,$$

where the $u(k)$ are to be chosen to minimize

$$J = |x(N) - d| .$$

Assume that d is not known and that we are only given its probability distribution. Determine a recurrence relation suitable for this problem.

(Remark: This exercise illustrates a different type of uncertainty — uncertainty concerning the very purpose of the process. Here, the controller wishes to have the system "in the right place at the right time" but doesn't know where the right place is.)

For simplicity, assume

$$\text{Prob}\ \{d = d_1\} = p ,$$

$$\text{Prob}\ \{d = d_2\} = 1-p ,$$

where p <u>is</u> known to the controller.

We wish to choose $u(k)$ to minimize the expected value of

$$h(x(N)) = p|x(N) - d_1| + (1-p)|x(N) - d_2| .$$

Letting

$$I(c,N) = \min_u \underset{r}{E}\ [h(x(N))] ,$$

we easily obtain

$$I(c,N) = \min_u [I(g(c,u),N-1)] , \qquad N \geq 1 ,$$

$$I(c,0) = p|c - d_1| + (1-p)|c - d_2| .$$

In the event p is also unknown to the controller with only its estimated distribution function given, we face an adaptive problem.

3.7 Consider the following problem with implicit criteria. We are given the equations

$$x_1(k+1) = x_1(k) - u_1(k) - r_1(k) , \quad x_1(0) = c_1 ,$$

$$x_2(k+1) = g[x_1(k),x_2(k),u_2(k),r_2(k)] , \quad x_2(0) = c_2 ,$$

where $a_1 \leq u_1 \leq a_2$, $b_1 \leq u_2 \leq b_2$. The objective is to choose $u_1(k)$, $u_2(k)$ to minimize the expected value of

$[x_2(T) - \alpha]^2$ when T is the first time at which $x_1(T) \leq 0$. Here the $\{r(k)\}$ are independent random variables with given distribution functions and α is a given scalar. Assuming that u_1 and r_1 are constrained by the relations

$$u_1(k) + r_1(k) \geq a_3 > 0 ,$$

which insures the problem has a solution, determine a functional equation suitable for obtaining the optimal control.

The minimum of the expected value of $[x_2(T) - \alpha]^2$ depends only upon c_1 and c_2. Thus, we write

$$I(c_1, c_2) = \min_{\{u(k)\}} \ \underset{\{r(k)\}}{E} \ [(x_2(T) - \alpha)^2] .$$

Clearly,

$$I(0, c_2) = (c_2 - \alpha)^2 .$$

For $c_1 > 0$, we see that

$$I(c_1, c_2) = \min_{u_1, u_2} \ \underset{r_1, r_2}{E} \ [I(c_1 - u_1 - r_1, g(c_1, c_2, u_2, r_2))] .$$

To obtain a computational solution to this equation, we may use the method of successive approximations.

3.8 Let $P = [p_{ij}]$ be the Markov matrix discussed in the text. Show that if $p_{ij} > 0$, $i,j = 1,2,\ldots,N$ that

(a) $\lim_{t\to\infty} x(t) = y$ exists, where y is a probability vector,

(b) y is <u>independent</u> of c,

(c) y is a characteristic vector of the matrix P with associated characteristic root 1.

(a) Consider the inner product of $x(t)$ with an arbitrary vector b. Since $x(t) = P^t c$, we have

$$[x(t),b] = (P^t c,b) = (c,P'^t b) = (c(P')^t b) , \quad (3.43)$$

where $'$ denotes the matrix transpose operation. We wish to show that $(P')^t b$ converges as $t \to \infty$ for <u>any</u> b.

Introduce the vector

$$z(t) = (P')^t b .$$

It satisfies

$$z(t+1) = P'z(t) , \quad z(0) = b .$$

For each t, let $u(t)$ be the component of $z(t)$ of largest value and let $v(t)$ be the component of smallest value.

Since

$$z_i(t+1) = \sum_{j=1}^{N} p_{ij} z_j(t) ,$$

and since $p_{ij} > 0$, and $\sum_{j=1}^{N} p_{ij} = 1$, we see that

$$u(t+1) \leq u(t) ,$$

and

$$v(t+1) \geq v(t) .$$

Thus, the sequences $\{u(t)\}$ and $\{v(t)\}$ converge as $t \to \infty$ since $\{u(t)\}$ is a monotone decreasing sequence, bounded below by zero and $\{v(t)\}$ is a monotone increasing sequence, bounded above by 1. Let $u(t) \to u$ and $v(t) \to v$, and let d be the positive lower bound for p_{ij}. Then

$$u(t+1) \leq (1-d)u(t) + dv(t) ,$$

$$v(t+1) \geq (1-d)v(t) + du(t) .$$

Thus,

$$u(t+1) - v(t+1) \leq [(1-d)u(t) + dv(t) - (1-d)v(t) - du(t)] .$$

Iterating this relation, we see that

$$u(t) - v(t) \leq (1-2d)^t [u(0) - v(0)] .$$

Since $d \leq 1/2$ if $N \geq 2$, we see that $u(t) - v(t) \to 0$ as $t \to \infty$. Hence, $z(t)$ converges to a vector all of whose components are equal. This implies via (1) that $x(t)$ converges to a vector y since b was arbitrary.

(b) Let $\lim_{t\to\infty} z(t) = z$ and $\lim_{t\to\infty} x(t) = y$. Let α be the components of z. Then

$$(y,b) = \lim_{t\to\infty} [x(t),b] = (c,z) = \alpha[c_1 + c_2 + \ldots + c_N] = \alpha ,$$

where α depends only upon b. Thus, y is independent of c.

(c) We have

$$y = \lim_{t\to\infty} x(t) = \lim_{t\to\infty} P^{t+1} c = P \lim_{t\to\infty} P^t c = Py ,$$

i.e., y is a characteristic vector of P with associated characteristic root 1.

3.9 As an example of the "value determination operation" technique, we consider Bellman's taxicab problem.

Suppose a driver has three towns he services. In towns 1 and 3 he has the alternatives:

(a) He can cruise the streets in the hope of being flagged down;

(b) He can go to the nearest taxi stand and wait;

(c) He can stop at the curb and wait for a radio call.

In town 2, he has only the alternatives (a) and (b).

For any town, and any given alternative within the town, there is a probability that the next trip will take him to each of towns 1, 2, or 3, with corresponding known return.

Let us assume the hypothetical numbers are as in Table 3.3 for these probabilities and returns.

For this problem we have three states with three choices in states 1 and 3 and two choices in state 2. The problem of

Table 3.3. Transition Probabilities and Returns for Taxicab Problem

CITY	CHOICE	TRANSITION PROBABILITY (TO CITY)			RETURN		
		1	2	3	1	2	3
1	a	1/2	1/4	1/4	10	4	8
	b	1/16	3/4	3/16	8	2	4
	c	1/4	1/8	5/8	4	7	4
2	a	1/2	0	1/2	14	0	18
	b	1/16	7/8	1/16	8	17	8
3	a	1/4	1/4	1/2	10	2	8
	b	1/8	3/4	1/8	6	4	2
	c	3/4	1/16	3/16	4	0	8

the taxi driver is to decide upon a policy which yields the maximum expected return from an N-stage process.

To utilize dynamic programming, let

$I(i,N)$ = the expected return from an N-stage process which starts in city i when an optimal policy is employed, $i = 1,2,3$, $N \geq 0$.

From the text, we have

$$I(i,N) = \max_{q} \left\{ \sum_{j=1}^{3} p_{ij}(q)[r_{ij}(q) + I(j,N-1)] \right\}, \quad N \geq 1,$$

$$I(i,0) = 0, \quad i = 1,2,3.$$

Here $q = q_N(i)$, the decision to be made in city i when N stages remain.

As an initial guess at the optimal policy, let us choose $q^{(0)} = \begin{pmatrix} a \\ a \\ a \end{pmatrix}$, i.e., in each city the driver cruises in an attempt to pick up passengers. For this policy, the transition matrix from Table 3.3 is

$$P\left(q^{(0)}\right) = \left[p_{ji}\left(q^{(0)}\right)\right] = \begin{bmatrix} 1/2 & 1/4 & 1/4 \\ 1/2 & 0 & 1/2 \\ 1/4 & 1/4 & 1/2 \end{bmatrix},$$

with the expected return

$$\left[\sum_{j=1}^{3} p_{ij} \; r_{ij}\right] = \begin{bmatrix} 8 \\ 16 \\ 7 \end{bmatrix} = \left[b_i\left(q^{(0)}\right)\right].$$

To obtain $g^{(0)}$, we solve Eqs. (3.14)-(3.15) with

$$b\left(q^{(0)}\right) = \left[b_i\left(q^{(0)}\right)\right].$$

The equations are

$$r^{(0)} + g^{(0)}(1) = 8 + 1/2\ g^{(0)}(1) + 1/4\ g^{(0)}(2) ,$$

$$r^{(0)} + g^{(0)}(2) = 16 + 1/2\ g^{(0)}(1) ,$$

$$r^{(0)} = 7 + 1/4\ g^{(0)}(1) + 1/4\ g^{(0)}(2) ,$$

$$g^{(0)}(3) = 0 .$$

The solution is

$$g^{(0)}(1) = 1.33 ,$$

$$g^{(0)}(2) = 7.47 ,$$

$$g^{(0)}(3) = 0 ,$$

$$r^{(0)} = 9.2 .$$

Thus, using the policy $q^{(0)}$ the driver will obtain 9.2 units per trip on the average.

To improve upon the policy $q^{(0)}$, we calculate the quantity

$$b_i(q) + \sum_{j=1}^{3} p_{ij}(q)\ g^{(0)}(j) - g^{(0)}(i) ,$$

for each i and q. The q which yields the maximum for each i is the improved policy. For example, for $i = 1$, we have

$$b_1(a) = \sum_{j=1}^{3} p_{ij}(a)\ r_{ij}(a)$$

$$= (1/2)(10) + (1/4)(4) + (1/4)(8) = 8 ,$$

$$b_1(b) = (1/16)(8) + (3/4)(2) + (3/16)(4) = 2.75 ,$$

$$b_1(c) = (1/4)(4) + (1/8)(6) + (5/8)(4) = 4.25 ,$$

and

$$b_1(a) + \sum_{j=1}^{3} p_{ij}(a)\ g^{(0)}(j) - g^{(0)}(1)$$

$$= (1/2)(1.33)+(1/4)(7.47)-1.33+8 = 9.20 ,$$

$$b_1(b) + \sum_{j=1}^{3} p_{ij}(b)\ g^{(0)}(j) - g^{(0)}(1) = 7.08 ,$$

$$b_1(c) + \sum_{j=1}^{3} p_{ij}(c)\ g^{(0)}(j) - g^{(0)}(1) = -4.19 .$$

Thus, for $i = 1$, the maximizing decision is (a). In a similar fashion, for $i = 2$, we find the maximizing decision to be (b), and for $i = 3$, it is also (b). Thus, the new decision vector is

$$q^{(1)} = \begin{pmatrix} a \\ b \\ b \end{pmatrix} .$$

Using the vector $q^{(1)}$, we have the transition probability matrix

$$p_{ij}\left(q^{(1)}\right) = \begin{bmatrix} 1/2 & 1/4 & 1/4 \\ 1/16 & 7/8 & 1/16 \\ 1/8 & 3/4 & 1/8 \end{bmatrix} ,$$

with

$$b\left(q^{(1)}\right) = \begin{pmatrix} 8 \\ 15 \\ 4 \end{pmatrix} .$$

Entering these numbers into (3.14) with gain $g(3) = 0$, we obtain

$$g^{(1)}(1) = 3.88 ,$$

$$g^{(1)}(2) = 12.85 ,$$

$$g^{(1)}(3) = 0 ,$$

$$r^{(1)} = 13.15 .$$

Hence, the cab now earns 13.15 units per trip on the average, an improvement over the previous 9.2 per trip.

The policy improvement routine, using these values of $g^{(1)}(i)$ and $r^{(1)}$ now gives the improved policy

$$q^{(2)} = \begin{pmatrix} b \\ b \\ b \end{pmatrix} .$$

That is, the driver must go to the nearest taxi stand whatever town he is in.

Utilizing $q^{(2)}$ in (3.14), we obtain

$$q^{(2)}(1) = -1.18 ,$$

$$q^{(2)}(2) = 12.66 ,$$

$$q^{(2)}(3) = 0 ,$$

$$r^{(2)} = 13.34 .$$

This policy gives a small improvement over $q^{(1)}$ since $r^{(2)}$ is slightly greater than $r^{(1)}$.

The policy improvement routine with these values of $g^{(2)}(i)$ and $r^{(2)}$ gives the policy $q^{(2)}$ again, which indicates the process has converged. Hence, the optimal policy is to drive to the taxi stand in any city and this policy yields an average return of 13.34 per trip.

It is of interest to note that the policy of driving to a taxi stand is the <u>worst</u> policy in terms of immediate return. However, the bad immediate return is offset by the favorable position the driver is in to obtain desirable fares for the future.

3.10 A merchant has N items and a length of time T to dispose of them. Goods may be sold at times $t = 0,1,2,\ldots,T$. Let

$\varphi(z)$ be the probability that an item of price z is sold at any particular time. Assuming independence of sales, determine a recurrence relation for maximizing the merchant's expected return over the sales period of length T.

$I(N,T)$ = the expected return from N items over a sales period of length T when an optimal pricing policy is used.

By elementary probability theory and the assumed independence of sales, the probability of the merchant selling exactly k items of price z in the first period is equal to

$$\binom{N}{k} \varphi^k(z)[1-\varphi(z)]^{N-k} = P[N,k,\varphi(z)] \ .$$

Thus, the expected return from selling k items is

$$k\ z\ P[N,k,\varphi(z)] \ .$$

Also, if exactly k items are sold in the first period, there remain $N-k$ items to be sold in the last $T-1$ periods and, by assumption, these must be sold in an optimal fashion. Putting these remarks together, and taking account of the fact that we may sell any number of items from zero to N in the initial period, we obtain

$$I(N,T) = \max_{z \geq 0} \sum_{k=0}^{N} P[N,k,\varphi(z)][kz + I(N-k,T-1)] \ , \qquad T \geq 1 \ .$$

Clearly,

$$I(N,0) = 0 \ .$$

3.11 Consider the following stochastic control problem. The system equations are

$$\dot{x} = u + w \ , \tag{3.44}$$

where x = scalar state variable,
u = scalar control variable,
w = scalar random forcing function.

The probability density function of w is taken to be Gaussian with zero mean and variance $\frac{1}{2}$,

$$p(w) = \frac{1}{\sqrt{\pi}} e^{-w^2} . \tag{3.45}$$

The performance criterion is

$$J = E \int_0^{10} (x^2 + u^2) dt + 2.5[x(1)]^2 . \tag{3.46}$$

The constraints are

$$-3 \leq x \leq 3 , \tag{3.47}$$

$$-1 \leq u \leq 1 . \tag{3.48}$$

We wish to minimize J, subject to (3.44), (3.47) and (3.48). The state variable is quantized uniformly in increments of $\Delta x = 1$, the control is quantized in increments of $\Delta u = 1$, and the time increment Δt is taken to be 1. The probability density of the random forcing function is quantized so that

$$\begin{aligned} p(w=1) &= 1/4 , \\ p(w=0) &= 1/2 , \\ p(w=-1) &= 1/4 . \end{aligned} \tag{3.49}$$

This distribution has the same mean and variance as the original density function.

The system equation in discrete form becomes

$$x(t+\Delta t) = x(t) + [u(t)+w(t)] \Delta t . \tag{3.50}$$

The iterative equation is

$$I(x,t) = \min_{u=-1,0,1} \{(x^2+u^2)\Delta t + E[I(x+(u+w)\Delta t, t+\Delta t)]\} \; . \tag{3.51}$$

The solution is exhibited in Figure 3.4.

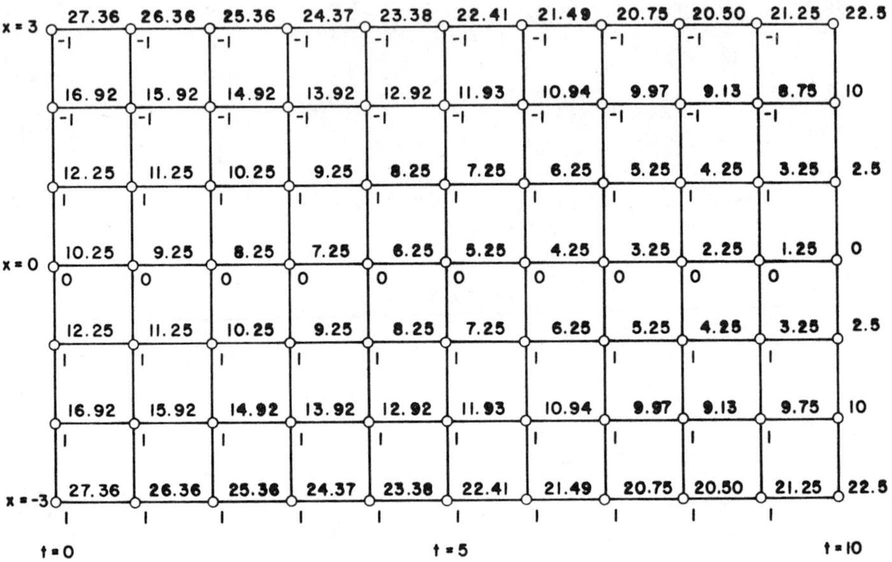

Figure 3.4. Solution to Stochastic Control Problem

3.12 For the game of survival discussed in the text, show that an equivalent functional equation may be obtained which has only one independent variable, rather than two.

Since the game is zero-sum, we may consider only the total amount of money in the game. Call this amount c. Define

$$g(u) = I(u,v) \; .$$

Then g satisfies the functional equation

$$g(u) = \max_p \min_q \Sigma\, p_i q_j\, g(u+p_{ij}) , \qquad 0 < u < c ,$$

where

$$g(u) = 0 , \quad u \leq 0 ,$$

$$g(u) = 1 , \quad u \geq c .$$

3.13 Consider the game of survival described by the payoff matrix

$$A = \begin{bmatrix} a & -1 \\ -1 & b \end{bmatrix} , \qquad a,b \text{ positive integers} .$$

Let $v_n(k)$ = the probability that P survives when the total fortune of both players is n and P possesses k units, $0 \leq k \leq n$.

Show that:

(a) $v_{n+1}(k+1) = v_n(k) + [1 - v_n(k)]v_{n+1}(1)$,

and

(b) $$v_{n+1}(1) = \frac{v_{n+1}(1+a)\, v_{n+1}(1+b)}{v_{n+1}(1+a) + v_{n+1}(1+b)} .$$

(a) It is clear by our previous remarks that $v_n(k)$ satisfies the equation

$$v_n(k) = \max_p \min_q \left[\sum_{i,j=1}^{2} p_i q_i v_n(k+a_{ij}) \right] , \quad k \geq 1 ,$$

$$v_n(k) = 0 , \quad k \leq 0 .$$

The result to be shown is clearly true for $k = 0$. Assume that

$$v_{n+1}(k) = v_n(k-1) + [1 - v_n(k-1)]v_{n+1}(1) ,$$

for some k. Then

$$v_{n+1}(k+1) = \max_p \min_q \sum_{i,j} p_i q_i v_{n+1}(k+1+a_{ij}) .$$

Then by the inductive hypothesis

$$\begin{aligned} v_{n+1}(k+1) &= \max_p \min_q \sum_{i,j} p_i q_i \{v_n(k+a_{ij}) \\ &\qquad + [1-v_n(k+a_{ij})]\, v_{n+1}(1)\} \\ &= \max_p \min_q \sum_{i,j} p_i q_i v_n(k+a_{ij}) \\ &\quad + \max_p \min_q \sum_{i,j} p_i q_j [1-v_n(k+a_{ij})]v_{n+1}(1) \\ &= v_n(k) + [1 - v_n(k)]v_{n+1}(1) . \end{aligned}$$

(b) By part (a),

$$\begin{aligned} v_{n+1}(1) &= \max_{p_1,p_2} \min_{q_1,q_2} \sum_{i=1}^{2} \sum_{j=1}^{2} p_i q_j v_{n+1}(1+a_{ij}) \\ &= \max_{p_1,p_2} \min_{q_1,q_2} [p_1 q_1 v_{n+1}(1+a) + p_1 q_2 v_{n+1}(0) \\ &\qquad + p_2 q_1 v_{n+1}(0) + p_2 q_2 v_{n+1}(1+b)] \end{aligned}$$

Taking account of the facts that $v_{n+1}(0) = 0$ and $p_1 + p_2 = q_1 + q_2 = 1$, we have

$$\begin{aligned} v_{n+1}(1) &= \max_p \min_q [p\,q\,v_{n+1}(1+a) \\ &\qquad + (1 - p - q + pq)v_{n+1}(1+b)] , \quad 0 \leq p , \\ &\qquad\qquad q \leq 1 . \end{aligned}$$

Since this expression is symmetric in p and q, the max-min occurs when $p = q$. Thus, it suffices to find the value of p which maximizes

$$p^2 v_{n+1}(1+a) + (p^2 - 2p + 1)\, v_{n+1}(1+b) \ . \tag{3.52}$$

By calculus, the maximizing p is

$$p_{max} = \frac{v(1+b)}{v(1+a) + v(1+b)} \ .$$

Substituting this value into (3.52) and simplifying, we obtain the result

$$v_{n+1}(1) = \frac{v_{n+1}(1+a)\ v_{n+1}(1+b)}{v_{n+1}(1+a) + v_{n+1}(1+b)} \ .$$

3.14 For the adaptive control problem, in Eq. (3.29):

(a) Calculate the quantities $p_{m,n}$ if we assume the initial distribution function $dG(p)$ is a beta distribution, i.e.,

$$dG(p) = \frac{p^{a-1}(1-p)^{b-1}}{B(a,b)}\, dp \ , \qquad a,b > 0 \ ,$$

where $B(a,b)$ is the beta function, $B(a,b) = \dfrac{a!\ b!}{(a+b+1)!}$.

(b) What interpretation may be attached to the numbers a and b?

(a) From the expression in (3.29), we have

$$p_{m,n} = \frac{\int_0^1 p^{m+a}(1-p)^{n+n-1} dp}{\int_0^1 p^{m+a-1}(1-p)^{b+n-1} dp}$$

$$= \frac{B(m+a+1, n+b)}{B(m+a, n+b)} \ ,$$

which, making use of elementary properties of the beta function gives

$$p_{m,n} = \frac{m+a}{(m+a)+(n+b)} .$$

(b) The numbers a and b can be interpreted as the a priori number of trials resulting in ones and zeros. If $a+b$ is large, then the first few additional observations will have little influence on the estimate $p_{m,n}$; if $a+b$ is small, the initial additional observations will have a large effect on the a priori estimate,

$$p_{0,0} = \frac{a}{a+b} ,$$

indicating little confidence in the a priori information.

3.15 Model the problem of the "efficient gambler" of the text as an adaptive control problem when the probability p of correct transmission of the signal through the communication channel is unknown.

Let $dG(p)$ be the a priori probability distribution for p and assume that if we observe m successful wagers and n unsuccessful, $dG(p)$ is transformed into

$$dG_{m,n}(p) = \frac{p^m(1-p)^n dG(p)}{\int_0^1 p^m(1-p)^n dG(p)} .$$

As before, introduce the notation

$$p_{m,n} = \frac{\int_0^1 p^{m+1}(1-p)^n dG(p)}{\int_0^1 p^m(1-p)^n dG(p)} ,$$

the expected probability of correct transmission.

Let

$I(x,m,n,N)$ = the expected value of $\varphi[z(N)]$, where $z(N)$ is the capital at the end of N states, starting with an amount x and the information that there have been m successes and n failures to date, and an optimal policy is employed.

Then we have the following recurrence relation for obtaining the optimal policy

$$I(x,m,n,N) = \max_{0 \leq y \leq x} [p_{mn} I(x+y,m+1,n,N-1) + (1-p_{mn}) I(x-y,m,n+1,N-1)] , \quad N \geq 2 ,$$

$$I(x,m,n,1) = \max_{0 \leq y \leq x} [p_{mn}\varphi(x+y) + (1-p_{mn})\, \varphi(x-y)] .$$

3.16 Show that Problem 3.15 may be solved explicitly if:

(a) $\varphi(z) = z^b$, $b > 0$,

(b) $\varphi(z) = \log z$.

(a) $$I(x,m,n,1) = \max_{0 \leq y \leq x} [p_m (x+y)^b + (1-p)\ (x-y)^b]$$

$$= x^b \max_{0 \leq y \leq 1} [p_{mn} (1+y)^b + (1-p_{mn})\ (1-y)^b] .$$

By induction on N, it is now a simple matter to show that

$$I(x,m,n,N) = S(m,n,N)\ x^b ,$$

where $\{S(m,n,N)\}$ is determined by

$$S(m,n,N) = \max_{0 \leq y \leq 1} [p_{mn} S(m+1,n,N-1)(1+y)^b + (1-p)_{mn} S(m,n+1,N-1)(1-y)^b] .$$

(b) Again by induction of N, we show that

$$I(x,m,n,N) = \log x + S(m,n,N) ,$$

where

$$S(m,n,N) = p_{mn}S(m+1,n,N-1) + (1-p_{mn})S(m,n+1,N-1)$$

$$+ \max_{0 \leq y \leq 1} [p_{mn} \log(1+y) + (1-p) \log(1-y)] .$$

Although it is difficult to find $S(m,n,N)$ explicitly, the structure of the optimal policy is clear. We see that

$$y_{opt} = \begin{cases} (2p_{mn}-1)x , & p_{mn} > 1/2 , \\ 0 , & p_{mn} \leq 1/2 . \end{cases}$$

3.17 In the case $\varphi(z) = \log z$, if the initial distribution $dG(p)$ is given by

$$dG(p) = \frac{p^{a-1}(1-p)^{b-1}}{B(a,b)} dp , \quad a,b > 0 ,$$

B the beta function, determine the fraction of capital that must be bet at each stage.

In this case,

$$p_{mn} = \frac{m + a}{(m+a) + (n+b)} \quad \text{(recall Problem 3.14a).}$$

Thus, the optimal policy when $p_{mn} > 1/2$ is to bet the fraction $(2p_{mn}-1)$, i.e.,

$$2p_{mn}-1 = \frac{(m+a) - (n+b)}{(m+a) + (n+b)} .$$

3.18 A simple problem that illustrates some of the features of the combined control and estimation problem formulation is shown pictorially in Figures 3.5 and 3.6. In this problem, an

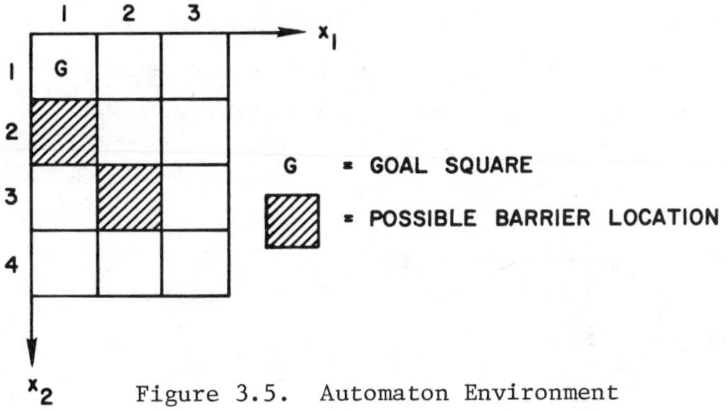

Figure 3.5. Automaton Environment

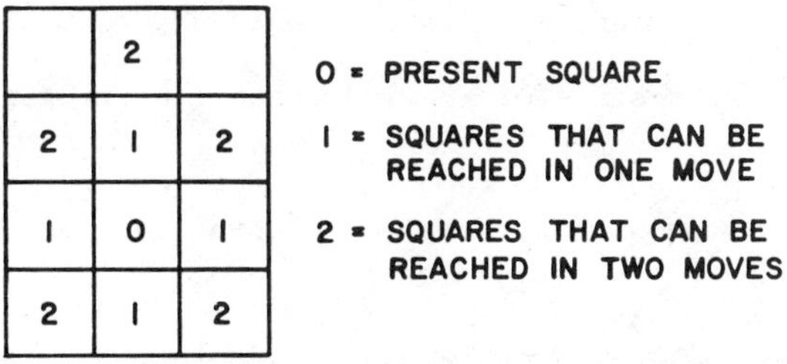

Figure 3.6. Automaton Moves

automaton is attempting to find its way through the environment shown in Figure 3.5 to a goal located at the upper left-hand square, which has coordinates ($x_1 = 1$, $x_2 = 1$). The automaton is not allowed to pass through the barriers. Initially, the automaton does not know if the barriers are present; instead it knows that there is a barrier in ($x_1 = 1$, $x_2 = 2$) with probability 0.4 and a barrier in ($x_1 = 2$, $x_2 = 3$) with probability 0.5. The robot can always "see" one move ahead — i.e., if it is within one move of a barrier location, it can find out if the barrier is there or not. For a certain price, which is expressed as a specified fraction of a move, the automaton can make an observation of all squares that are two moves away (see Figure 3.6). The objective is to find the policy for the automaton that reaches the square ($x_1 = 1$, $x_3 = 1$) from any initial square, while minimizing the sum of moves and penalties for making observations.

The problem can most easily be put into the framework of this chapter by defining the information state of the system to be the four-dimensional vector x,

$$x = \begin{bmatrix} x_1 \\ x_2 \\ x_3 \\ x_4 \end{bmatrix} , \tag{3.53}$$

where x_1 = horizontal coordinate in Figure 3.5,

x_2 = vertical coordinate in Figure 3.5,

x_3 = state of knowledge about barrier at ($x_1 = 1$, $x_2 = 2$),

x_4 = state of knowledge about barrier at ($x_1 = 2$, $x_2 = 3$).

The first two variables are quantized to the values

$$x_1 = 1,2,3 , \qquad x_2 = 1,2,3,4 . \tag{3.54}$$

The latter two variables are quite different from the usual state variables associated with dynamic systems. Each variable can take on three different values as follows:

$$x_3 = P,A,Q , \qquad x_4 = P,A,Q , \tag{3.55}$$

where P = barrier is known to be present,

A = barrier is known to be absent,

Q = absence or presence of barrier is not known.

The control vector u has three components. They are

$$u = \begin{bmatrix} u_1 \\ u_2 \\ u_3 \end{bmatrix} , \tag{3.56}$$

where u_1 = change in x_1,

u_2 = negative change in x_2*

u_3 = decision to make an observation.

The variable u_3 takes on two values

$$u_3 = L, N , \tag{3.57}$$

where L = an observation is made,

N = no observation is made.

The set of admissible controls is

$$U = \begin{bmatrix} 1 \\ 0 \\ N \end{bmatrix}, \begin{bmatrix} -1 \\ 0 \\ N \end{bmatrix}, \begin{bmatrix} 0 \\ 1 \\ N \end{bmatrix}, \begin{bmatrix} 0 \\ -1 \\ N \end{bmatrix}, \begin{bmatrix} 0 \\ 0 \\ L \end{bmatrix} , \tag{3.58}$$

corresponding to move right, move left, move up, move down, and make an observation. The first two system equations can be written as

$$\begin{aligned} x_1(k+1) &= x_1(k) + u_1(k) , \\ x_2(k+1) &= x_2(k) - u_2(k) . \end{aligned} \tag{3.59}$$

*This definition of u_2 means that a positive move is up and to the right.

The uncertainty about the barriers is taken into account by defining a random forcing function vector w with two components,

$$w = \begin{bmatrix} w_1 \\ w_2 \end{bmatrix}, \tag{3.60}$$

where w_1 = presence or absence of barrier at ($x_1 = 1$, $x_2 = 2$),

w_2 = presence or absence of barrier at ($x_1 = 2$, $x_2 = 3$).

These variables can take on the values

$$\begin{aligned} w_1 &= B, R, \\ w_2 &= B, R, \end{aligned} \tag{3.61}$$

where B = barrier is present,

R = barrier is absent.

This vector affects only the state variables x_3 and x_4. In writing the system equations for these two variables, it is useful to define two auxiliary variables, m_1 and m_2. The variable m_1 takes on the value 1 if the control is such that the presence or absence of the barrier at ($x_1 = 1$, $x_2 = 2$) will be determined; and $m_1 = 0$ otherwise. The variable $m_2 = 1$ of the control will determine the presence or absence of the barrier at ($x_1 = 2$, $x_2 = 3$), and $m_2 = 0$ otherwise.

For $u_3 = N$, $m_1 = 1$ if the move chosen causes the next square to be within one move of the barrier at ($x_1 = 1$, $x_2 = 2$). For $u_3 = L$, $m_1 = 1$ if the barrier at ($x_1 = 1$, $x_2 = 2$) is within two moves of the present square. Otherwise, $m_1 = 0$. Similar conditions can be written in the form

$$x_3(k+1) = \int_3 [x_3(k), u_3(k), m_1(k), w_1(k)], \tag{3.62}$$

where $\int_3$ is defined by Table 3.4, and where

$$\begin{aligned} p(w_1 = B) &= 0.4 , \\ p(w_1 = R) &= 0.6 , \\ p(w_2 = B) &= 0.5 , \\ p(w_2 = R) &= 0.5 . \end{aligned} \tag{3.63}$$

The equation for x_4 is

$$x_4(k+1) = \int_4 [x_4(k), u_4(k), m_2(k), w_2(k)] , \tag{3.64}$$

where $\int_4$ is specified by a table similar to that in Table 3.4.

The performance criterion, which is to be minimized, is the sum of moves and penalties for observations. This criterion can be written as

$$J = \sum_{k=0}^{\infty} \ell[x(k), u(k)] , \tag{3.65}$$

where $\ell[x(k), u(k)]$ is specified as in Table 3.5. In this table, it is assumed that the penalty for an observation is 0.3 moves.

Table 3.4. Next Value of x_3 for Automaton

$x_3(k)$	$u_3(k)$	$m_1(k)$	$w_1(k)$	$x_3(k+1) = \int_3$
P	—	—	—	P
A	—	—	—	A
Q	N	0	—	Q
Q	N	1	B	P
Q	N	1	R	A
Q	L	0	—	Q
Q	L	1	B	P
Q	L	1	R	A

— = value of $x_3(k+1)$ is the same for all values of this variable

Table 3.5. Value of $\ell[x(k),u(k)]$ for Automaton

$x_1(k)$	$x_2(k)$	$u_3(k)$	$\ell[x(k),u(k)]$
1	1	—	0
1	$\neq 1$	N	1.0
1	$\neq 1$	L	0.3
$\neq 1$	1	N	1.0
$\neq 1$	1	L	0.3
$\neq 1$	$\neq 1$	N	1.0
$\neq 1$	$\neq 1$	L	0.3

The constraints are that the set of admissible controls and the set of admissible states are defined above. Also, if a barrier is present a move to that square is forbidden.

Because the system equations and the constraints are time-invariant and because the performance criterion is the sum of time-invariant terms over an infinite number of stages, a modified version of approximation in policy space can be used. The details of this procedure were discussed earlier. The solution is shown in Table 3.6.

A number of interesting effects can be observed in Table 3.6. The first is Feldbaum's dual control effect; this effect is said to occur whenever the optimal control is used to gain more information about the system, rather than to optimize performance directly. This effect can be observed in this example whenever the optimal control is to make an observation. By imposing a penalty for the observation, the system is able to make a decision whether to gather more information or to continue moving on the basis of the information gathered so far. This occurs at $(x_1,x_2,x_3,x_4) = (1,4,Q,P)$ and $(x_1,x_2,x_3,x_4) = (1,4,Q,Q,)$.

Another effect that can be observed is Howard's value of information. This effect is related to the cost that should

Table 3.6. Solution to Automaton Problem

PRESENT STATE				OPTIMAL CONTROL			MINIMUM COST
x_1	x_2	x_3	x_4	u_1	u_2	u_3	1
1	1	—	—	—	—	—	0
2	1	—	—	-1	0	N	1
3	1	—	—	-1	0	N	2
1	2	—	—	0	1	N	1
2	2	—	—	0	1	N	2
3	2	—	—	0	1	N	3
1	3	A	—	0	1	N	2
1	3	P	A	1	0	N	4
1	3	P	P	0	-1	N	8
2	3	—	—	0	1	N	3
3	3	—	—	0	1	N	4
1	4	A	—	0	1	N	3
1	4	P	A	0	1	N	5
1	4	P	P	1	0	N	7
1	4	P	Q	1	0	N	6
1	4	Q	A	0	1	N	3.8
1	4	Q	P	0	0	L	4.9
1	4	Q	Q	0	0	L	4.5
2	4	—	A	0	1	N	4
2	4	A	P	-1	0	N	4
2	4	P	P	1	0	N	6
2	4	Q	P	-1	0	N	5.9
3	4	—	—	0	1	N	5

— = Optimal control and minimum cost are the same for all values of this variable.

be paid in order to gain information. Again, this is illustrated here in the decision of whether or not to make an observation; if the cost of 0.3 moves does not pay for the increase in performance, then the observation should not be made. At the two points where the observation is made, it is interesting to note the optimal control and minimum cost for the case where the decision to observe is not allowed. It is seen from Table 3.7 that at $(x_1,x_2,x_3,x_4) = (1,4,Q,P)$ the net profit

Table 3.7. Comparison of Optimal Control and Minimum Cost With and Without the Observation Control

x_1	x_2	x_3	x_4	MINIMUM COST WITH OBSERVATION	MINIMUM COST WITHOUT OBSERVATION	WITH NO OBSERVATION		
						u_1	u_2	u_3
1	4	Q	P	4.9	5.4	0	1	N
1	4	Q	Q	4.5	4.6	0	1	N

of making the observation is 0.5 moves, while at $(x_1, x_2, x_3, x_4) = (1,4,Q,Q)$ the net profit is 0.1 moves.

In addition to these effects, many other elements of adaptive and learning control problems can be observed either in this example directly or by a slight modification.

3.19 A type of problem which is similar to those treated above and which served as motivation for much early work in adaptive control process theory is the so-called "two-arm bandit" problem. For this problem, we have two slot machines, I and II. Machine I pays a dollar with probability r (independently of previous plays and of II) and Machine II pays a dollar with probability s. Assume s is known but r is not and that the player has N free plays. Model as an adaptive control process in which the player wishes to play the machines so as to maximize his expected take during the N plays.

We take the state of the system as the collection of triples $\{(m,n,v)\}$ where the components of the state indicate that machine I has achieved m successes and n failures, while machine II has a history of v successes. An appropriate criterion function which counts the total payoffs is

$$J(m,n,v) = \sum_{i \geq j} g[x(i),x(i+1)] \ , \quad j = 0,1,\ldots,N \ ,$$

where

$$g(x,y) = m_2 - m_1 + v_2 - v_1 \ ,$$

if

$$x = (m_1, n_1, v_1) \ , \quad y = (m_2, n_2, v_2) \ .$$

Obviously, the set of allowable decisions at each stage consists of the pair $\{I, II\}$.

The statistical law of motion of the process is given by

$$\begin{aligned} &P[(n,n,v+1) \,|\, (m,n,v), \ II] = s \ , \\ &P[(m,n,v) \,|\, (m,n,v), \ II] = 1-s \ , \\ &P[(m+1,n,v) \,|\, (m,n,v), \ I] = r_{mn} \ , \\ &P[(m,n+1),v) \,|\, (m,n,v), \ I] = 1-r_{mn} \ , \end{aligned} \qquad (3.66)$$

where $P[(m,n,v+1) \,|\, (m,n,v), \ II]$ denotes the conditional probability that playing machine II will result in a win, given that the current state is (m,n,v). A similar interpretation may be made for the other quantities in (3.66). The quantity r_{mn} is given by

$$r_{mn} = \frac{\displaystyle\int_0^1 r^{m+1}(1-r)^n \ dF(r)}{\displaystyle\int_0^1 r^m(1-r)^n \ dF(r)} \ ,$$

where $dF(r)$ is the a priori estimate of the distribution for r.

Our objective is to maximize the expected value of J. Define

$I[(m,n,v),k]$ = the expected value of J when the system is in state (m,n,v), k plays remain, and an optimal policy is followed.

Then

$$I[(m,n,v),k] = \max \begin{cases} \text{I:} & r_{mn}\{1+I[(m+1,n,v),k-1]\} \\ & +(1-r_{mn})\{I[(m,n+1,v),k-1]\}, \\ \text{II:} & s\{1+I[(m,n,v+1),k-1]\} \\ & +(1-s)\{I[(m,n,v),k-1]\}, \end{cases}$$

$$m+n+v \leq k \geq 2 \ . \qquad (3.67)$$

Clearly, if only one play is allowed and the past history of the machines is (m,n,v), then

$$I[(m,n,v),1] = \max \begin{cases} \text{I:} & r_{mn} \ , \\ \text{II:} & s \ . \end{cases} \qquad (3.68)$$

3.20 Consider the following type of bandit problem. In place of a finite process, consider the situation where we discount at a constant rate over time. Thus, if z(n) is the stochastic quantity gained on the nth trial, we wish to maximize the expected value of

$$\sum_{n=1}^{\infty} a^n z(n) \ , \qquad 0 < a < 1 \ .$$

Determine the appropriate recurrence relations for this process.

Define

$I(m,n)$ = the expected return obtained using an optimal policy for an unbounded process after the first machine has had m successes and n failures.

As above, we let

$$dF_{m,n}(r) = \frac{r^{m+1}(1-r)^n \, dF(r)}{\int_0^1 r^m(1-r)^n \, dF(r)} .$$

We readily obtain

$$I(m,n) = \max\begin{cases} \text{I:} & \int_0^1 \{r[1+aI(m+1,n)] \\ & \quad + (1-r)aI(m,n+1)\}\, dF_{mn}(r) , \\ \text{II:} & \frac{s}{1-a} . \end{cases}$$

3.21 In the "three-armed bandit" problem, the Defense has one interceptor and the Offense has two Objects, each of which will destroy the target with certainty _if_ it is a re-entry vehicle (RV) and _if_ it is not intercepted. Each Object i, $i = 1,2$, has a probability p_i of being an RV. _However_, p_i is not known — instead, the probability density function $f(p_i)$ is known. There is time to perform N discrimination experiments before deciding against which Object to launch the interceptor. The interceptor will destroy the Object with certainty. How should the discrimination experiments be distributed across the Objects to maximize probability of the target surviving?

Statistical Description of $f(p_i)$ and Discrimination Experiments

The problem can be solved by suitably defining the statistics on $f(p_i)$ and the discrimination experiments. One set of assumptions that leads to analytically tractable equations is the following:

(1) $f(p_i)$ is the "beta density function" $f(p_i) = \alpha p_i^m (1-p_i)^n$, where α is a normalization constant. Properties of this density function will be discussed in a moment.

(2) The discrimination experiment takes a sample of whether or not Object i is an RV according to the density function $f(p_i)$, i.e., the outcome of the experiment is either $\text{Object}_i = \text{RV}$ or $\text{Object}_i = \text{Decoy}$, while the probability that the result is $\text{Object}_i = \text{RV}$ is $E\{p_i\} \triangleq \hat{p}_i$ and the probability that the result is $\text{Object}_i = \text{Decoy}$ is $(1-\hat{p}_i)$.

The validity of this assumption is clearly questionable. However, the following points should be noted:

(1) The curve $\alpha p_i^m (1-p)^n$ is a relatively smooth curve with mean $\hat{p}_i = \frac{m+1}{m+n+2}$ and standard deviation

$$\sigma_{p_i}^2 = \frac{(m+1)(n+1)}{(m+n+3)(n+n+2)^2} .$$

It is thus relatively straightforward to approximate virtually any a priori distribution of p_i — the ratio $\hat{p}_i = \frac{m+1}{m+n+2}$ determines the mean, while the above term gives the "spread" of $\hat{p}_i$ around this value, i.e., the larger $\sigma_{p_i}^2$, the less confidence in the value of $\hat{p}_i$. As discussed earlier, m and n do <u>not</u> need to be integers.

(2) As will be seen shortly, the "a priori" statistics on $\alpha p_i^m(1-p_i)^n$ are equivalent to the results of $m+n$ discrimination experiments, in which the result was $\text{Object}_i = \text{RV}$ m times and $\text{Object}_i = \text{Decoy}$ n times. Thus, in order to relate these values to a real discrimination experiment, it is only necessary to determine the "relative" value of a single experiment as compared to the a priori knowledge about whether the Object is an RV or Decoy.

<u>Recursive Updating of $f(p_i)$</u>

Under the above assumptions about $f(p_i)$ and the discrimination experiments, it is straightforward to update $f(p_i)$ after one experiment. Define $f(p_i|\varphi)$ as the function before one experiment and let $f(p_i|\text{RV},\varphi)$ be the function after one experiment has been performed in which the result was $\text{Object}_i = \text{RV}$. Then, using Bayes' rule, the <u>joint density</u> $p(p_i,\text{RV},\varphi)$ can be written as

$$p(p_i,\text{RV},\varphi) = f(p_i|\text{RV},\varphi)\ p(\text{RV}|\varphi)\ p(\varphi)\ .$$

It can <u>also</u> be written as

$$p(p_i|\text{RV},\varphi) = p(\text{RV}|p_i,\varphi)\ f(p_i|\varphi)\ p(\varphi)\ .$$

Equating the two terms yields

$$f(p_i|\text{RV},\varphi)\ p(\text{RV}|\varphi)\ p(\varphi) = p(\text{RV}|p_i,\varphi)\ f(p_i|\varphi)\ p(\varphi)\ ,$$

or

$$f(p_i|\text{RV}\ \varphi)\ p(\text{RV}|\varphi) = p(\text{RV}|p_i,\varphi)\ f(p_i|\varphi)\ .$$

Note that, by definition,

$$p(\text{RV}|p_i,\varphi) = p_i\ .$$

Also note that

$$p(RV|\varphi) = \int_0^1 p(RV|p_i,\varphi) f(p_i|\varphi) dp_i$$

$$= \int_0^1 p_i f(p_i|\varphi) dp_i \ .$$

Therefore,

$$f(p_i|RV,\varphi) = \frac{p_i f(p_i,\varphi)}{\int_0^1 p_i f(p_i,\varphi) dp_i} \ .$$

For the assumed form $f(p_i|\varphi) = \alpha p_i^m (1-p_i)^n$,

$$f(p_i|RV,\varphi) = \alpha' p_i^{m+1} (1-p_i)^n \ ,$$

where α' is a new normalization factor. A similar analysis shows that if the outcome of a single experiment is $\text{Object}_i = \text{Decoy}$, then

$$f(p_i|\text{Decoy},\varphi) = \alpha'' p_i^m (1-p)^{n+1} \ .$$

By recursive application of these equations, it can be seen that after (a+b) experiments, where there are a results of $\text{Object}_i = \text{RV}$, and b results of $\text{Object}_i = \text{Decoy}$, $f(p_i)$ becomes

$$f(p_i|a\ RV,\ b\ Decoy,\ \varphi) = \alpha''' p_i^{m+a} (1-p_i)^{n+b} \ .$$

There is an immediate interpretation of $f(p_i|\varphi) = \alpha p_i^m (1-p_i)^n$. Assume that there is a total lack of knowledge of p_i and the density of $f(p_i)$ is taken to be uniform, which is the "worst-case" in terms of information content. Denoting $f(p_i|-)$ as this "zero knowledge" value, it is seen that

$$f(p_i|-) = 1 \ , \quad 0 \le p_i \le 1 \ .$$

This is a special case of the form $\alpha p_i^m(1-p_i)^n$, where $m = 0$, $n = 0$. Thus by going through the above recursive analysis, we see that

$$f(p_i|\varphi) = \alpha p_i^m(1-p_i)^n ,$$

is equivalent to starting at $f(p_i|-)$ and having the results of $(m+n)$ experiments, of which m are $\text{Object}_i = \text{RV}$ and n are $\text{Object}_i = \text{Decoy}$. This leads to the interpretation of $f(p_i|\varphi)$ noted above.

Solution of the Original Problem for a Specific Case

A key result for the following is that for $f(p_i) = p_i^m(1-p_i)^n$,

$$E\{p_i\} = \int_0^1 p_i \; p_i^m(1-p_i)^n dp_i = \frac{m+1}{m+n+2} .$$

Consider now the "tree" formed in Figure 3.7 by conducting three discrimination experiments. A "state" of the tree is denoted by the pair (a,b) where a is the number of RV outcomes and b is the number of Decoy outcomes. A "transition" is defined by the outcome "RV" or "Decoy," i.e., if the outcome is RV, a transition is made from (a,b) to (a+1,b), while if the outcome is Decoy the transition is from (a,b) to (a,b+1). Finally, if the a priori statistics of p_i are

$$f(p_i) = \alpha p_i^m(1-p_i)^n ,$$

then the expected value of p_i in any state (a,b) is simply

$$p_i(a,b) = \frac{m+a+1}{m+n+a+b+2} .$$

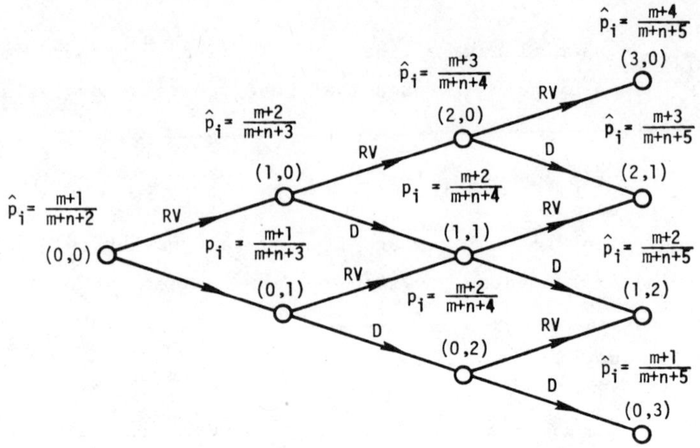

Figure 3.7. General Tree for A Priori Statistics $f(p_i) = \alpha p_i^{m+1}(1-p_i)^{n+1}$ and Three Discrimination Experiments

To be specific, let us assume that initially we have no knowledge of Object 1, so that our a priori density function for p_i corresponds to $m = n = 0$, i.e.,

$$f(p_1|\varphi) = 1 \ , \quad 0 \le p_1 \le 1 \ .$$

The tree then looks as shown in Figure 3.8.

Now, let us assume that we have a second Object, Object 2, for which $\hat{p}_2 = \frac{1}{2}$. Let us examine what we would do if we did no further experiments on Object 2. In general, at the end of three experiments we shoot at the Object most likely to be an RV. Thus, for $\hat{p}_1 = \frac{4}{5}$ and $\hat{p}_1 = \frac{3}{5}$, we shoot at Object 1, while for $\hat{p}_1 = \frac{2}{5}$ and $\hat{p}_1 = \frac{1}{5}$ we shoot at Object 2. Our probability of survival for the target is

$$p_s = (1 - \max\{\hat{p}_1, \hat{p}_2\}) \ .$$

We can summarize these outcomes by listing at the last four points of the grid the value p_s from the above equation

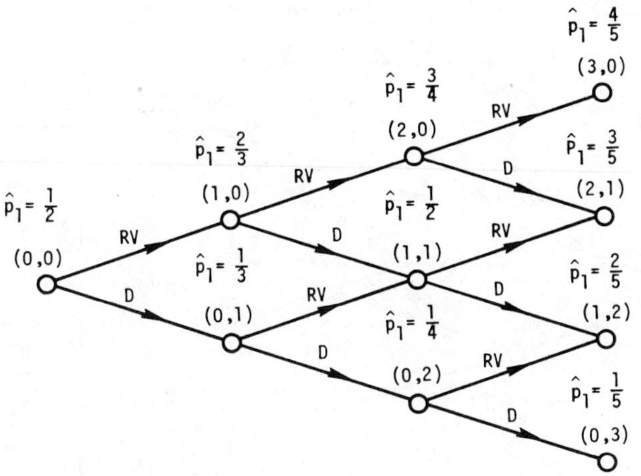

Figure 3.8. Tree for $f(p_i) = 1$ and Three Discrimination Experiments

and either $L = 1$ if we launch at Object 1 or $L = 2$ if we launch at Object 2 (see Figure 3.9).

We are now ready to formulate a dynamic programming stochastic control problem. Our "decision" can be one of three (hence, the name three-armed bandit problem): $L = 1$ (launch at Object 1), $L = 2$ (launch at Object 2), or $L = E$ (keep experimenting). We shall adopt the philosophy that if a further experiment cannot change our decision, i.e., if all states beyond the present state for which we have a non-zero probability of reaching have the same decision, and this decision is either $L = 1$ or $L = 2$, then we shall in turn make this decision at this state and discontinue further experiments. If this is not the case, we shall continue experimenting.

We shall assume that, if we keep experimenting, we shall make the transition RV with probability $\hat{p}_1$ given at that state and the transition D with probability $(1-\hat{p}_1)$. This is consistent with our assumption that p_1 does in fact obey

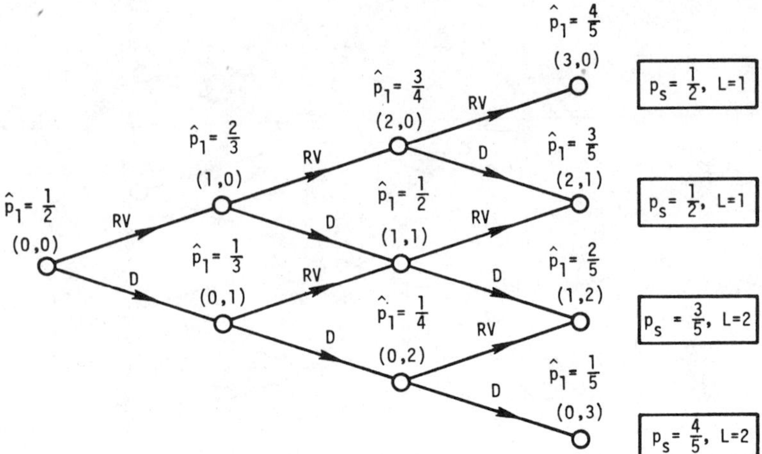

Figure 3.9. Terminal Grid

the density function $f(p_1) = \alpha p_1^a (1-p_1)^b$, where a and b are associated with the given state. Our survival probability if we keep experimenting is thus

$$p_s(a,b) = \hat{p}_1\ p_s(a+1,b) + (1-\hat{p}_1)\ p_s(a,b+1)\ .$$

The probability of survival if we adopt either $L=1$ or $L=2$ is just

$$p_s = 1 - \hat{p}_2\ , \qquad L=1\ ,$$
$$p_s = 1 - \hat{p}_1\ , \qquad L=2\ .$$

The solution to this dynamic programming problem is then obtained by performing backward one stage at a time the following steps:

(1) Check decision at states (a+1,b) and (a,b+1), denoted as L(a+1,b) and L(a,b+1), respectively.

(2) Does L(a+1,b) + L(a,b+1) = 1? If so, L(a,b) = 1 and $p_s(a,b) = 1-\hat{p}_2$. If not, continue.

(3) Does L(a+1,b) = L(a,b+1) = 2? If so, L(a,b) = 2, and $p_s(a,b) = 1-\hat{p}_1(a,b)$. If not, continue.

(4) $L(a,b) = E$ and $p_s(a,b) = \hat{p}_1(a,b)\, p_s(a+1,b) + [1 - \hat{p}_1(a,b)]\, p_s(a,b+1)$.

The solution is shown in Figure 3.10. Note that the survival probability is $\frac{6}{10}$, a significant increase over $\frac{1}{2}$, the value if no discrimination experiments are performed.

Note also that we are not really through. We must now solve the same problem with the density function for p_2. Assume that $f_2(p_2|\varphi) = 6p_2(1-p_2)$, i.e., p_2 has the a priori density function for $m=1$, $n=1$. A "quick" calculation shows that the solution is as given in Figure 3.11.

We see, then, that for this case the survival probability is $\frac{79}{140} = 0.5643$, as contrasted to $\frac{6}{10} = 0.6000$ for the former case. It is clear, then, that we should start out performing the experiments on Object 1. If we always have the optimal

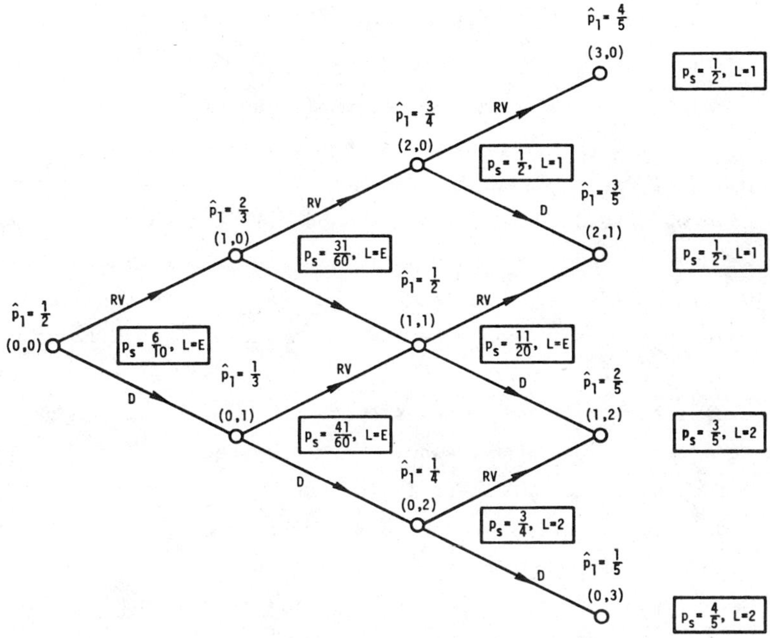

Figure 3.10. Solution to Problem With $f_1(p_1|\varphi) = 1$

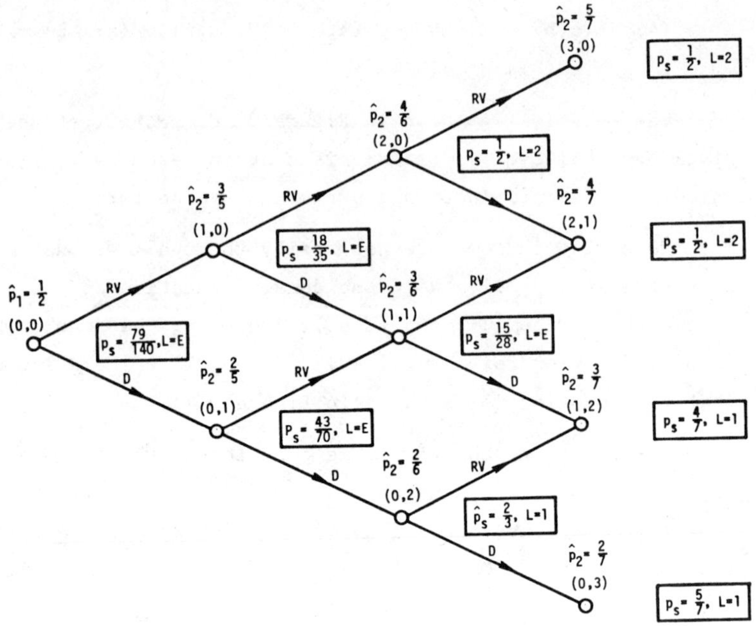

Figure 3.11. Solution for $f_2(p_2|\varphi) = 6p_2(1-p_2)$

strategy L = E, we will perform all experiments on this Object. If, however, after we perform two experiments on Object 1, we may find that either L = 1 or L = 2. In this case we have a choice of what to do with the third experiment. We have already seen that performing the experiment on Object 1 will not affect the decision either way. It is thus important to examine the value of performing a one-step experiment on Object 2. If L = 1, then $\hat{p}_1 = \frac{3}{4}$. The value of the one-step experiment on Object 2 is as shown in Figure 3.12. Here we see that our decision is not changed by the third experiment, although we will at least learn something about our survival probability. The latter occurs because we gain some information about the object we are not intercepting. For L = 2, $\hat{p}_1 = \frac{1}{4}$, the solution is as shown in Figure 3.13. Note that again the decision is not changed. Furthermore, we learn

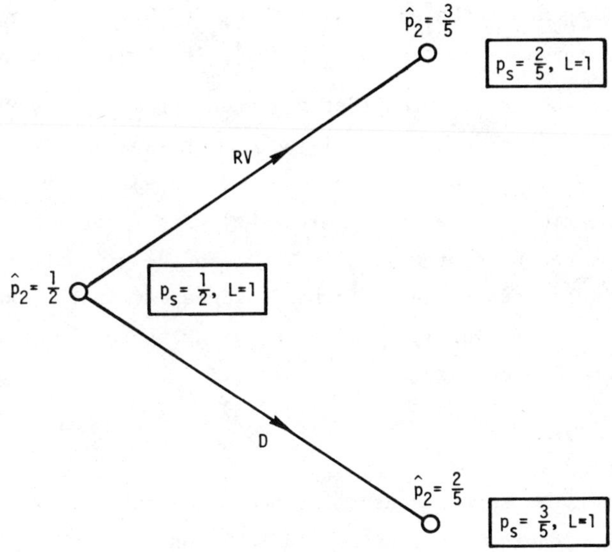

Figure 3.12. Solution to One-Step Problem for Object 2 with $\hat{p}_1 = \frac{3}{4}$

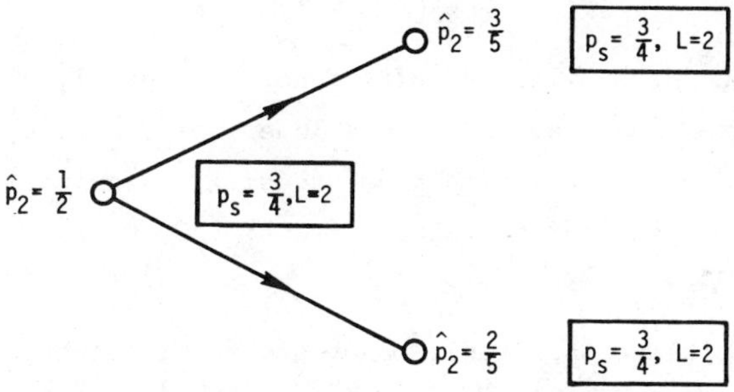

Figure 3.13. Solution to One-Step Problem for Object 2 with $p_1 = \frac{1}{4}$

nothing about our actual survival probability because the experiment is performed on the object we are intercepting. This would <u>not</u> be the case if the interceptor had a non-zero failure probability. Similarly, if we examine the solution in Figure 3.10, we see that if $L = 2$ we at least learn something about the survival probability by performing the experiment on Object 1, but we don't if $L = 1$.

It should be noted that the result that the third experiment doesn't change the decision is not generally the case. It only happens to occur for this set of values of m and n. The question of what to do with the third experiment then returns to the system level. The resource can either be used to gain the additional information inherent in the experiment or to do something more valuable to the system, such as tracking some other object.

3.22 Consider the special case of the control and estimation problem when there is no control ($u_k \equiv 0$) and

$$g(x_k, w_k, k) = \Phi_i x_k + \Gamma_k w_k \ , \tag{3.69}$$

$$h(x_k, v_k, k) = H_k x_k + v_k \ . \tag{3.70}$$

Assume the probability density functions $p(x_o)$, $p(w_k)$, and $p(v_k)$ are Gaussian, white, and independent (i.e., the random variables associated with these densities are independent) and that the means and covariances are $\hat{x}_o$, 0, 0 and P_o, Q_k, R_k, respectively.

Specialize the results above to obtain equations for the modal trajectory in this case.

It is well known that, in this case, the probability density functions $p(z_{k+1}|x_{k+1})$, $p(x_{k+1}|x_k)$ and $p(X_k|Z_k)$ are all Gaussian. Using the white assumption, $I(x_k,k)$ can be written as

$$I(x_k,k) = \max_{x_o,\ldots,x_{k-1}} c_1 \exp\left[-\frac{1}{2}\sum_{i=0}^{k}(x_i-\bar{x}_{k/k})^T P^{-1}_{i/k}(x_i-\bar{x}_{i,k})\right]$$

$$= c_1 \exp\left[-\frac{1}{2}[(x_k-\bar{x}_{k/k})^T P^{-1}_{k/k}(x_k-\bar{x}_{k/k})+c_2]\right] \ , \tag{3.71}$$

where $\bar{x}_{k/k}$ is the conditional mean, $P_{k/k}$ is the conditional covariance, c_1 is a normalization constant, c_2 is a constant independent of x_k and T denotes the transpose operation.

The probability density functions $p(z_{k+1}|x_{k+1})$ and $p(x_{k+1}|x_k)$ are characterized by

$$\begin{aligned} E(z_{k+1}|x_{k+1}) &= Hx_{k+1} , \\ \mathrm{Cov}(z_{k+1}|x_{k+1}) &= R , \\ E(x_{k+1}|x_k) &= x_k , \\ \mathrm{Cov}(x_{k+1}|x_k) &= \Gamma Q \Gamma^T . \end{aligned} \tag{3.72}$$

The recursive relation (Problem 3.18, Eq. (3.65)) then becomes

$$\begin{aligned} I(x_{k+1},k+1) = \max_{z_k} \Bigg\{ c_1 \exp\Bigg[-\frac{1}{2}[(z_{k+1}-Hx_{k+1})^T R^{-1} (z_{k+1}-Hx_{k+1}) \\ + (x_{k+1} - \Phi x_k)^T (\Gamma Q \Gamma^T)^{-1} (x_{k+1} - \Phi x_k) \\ + (x_k - \bar{x}_{k/k})^T P_{k/k}^{-1} (x_k - \bar{x}_{k/k}) + c] \Bigg] \Bigg\} . \end{aligned} \tag{3.73}$$

Maximization of an exponential is equivalent to minimization of the negative exponent. The latter minimization can be performed by differentiation with respect to x_k or by completion of the square.

The value of x_k that yields the minimum is

$$\begin{aligned} \hat{x}_k(x_{k+1},k+1) &= [P_{k/k}^{-1} + \Phi^T (\Gamma Q \Gamma^T)^{-1} \Phi]^{-1} [\Phi^T (\Gamma Q \Gamma^T)^{-1} x_{k+1} \\ &\quad + P_{k/k}^{-1} \hat{x}_{k/k}] \\ &= \hat{x}_{k/k} + P_{k/k} \Phi^T [\Phi P_{k/k} \Phi^T + \Gamma Q \Gamma^T]^{-1} \\ &\quad \cdot [x_{k+1} - \hat{x}_{k/k}] . \end{aligned} \tag{3.74}$$

When this value is substituted into (3.73), $I(x_{k+1},k+1)$ is obtained as

$$I(x_{k+1},k+1) = c_1 \exp\left[-\frac{1}{2}[x_{k+1}-\bar{x}_{k+1/k+1})^T P^{-1}_{k+1/k+1} (x_{k+1}-\bar{x}_{k+1/k+1}) + c_2]\right], \tag{3.75}$$

where

$$\begin{aligned}\bar{x}_{k+1/k+1} &= \Phi\bar{x}_{k/k} + P_{k+1/k+1}H^T R^{-1}(z_{k+1} - H\Phi\bar{x}_{k/k}), \\ P^{-1}_{k+1/k+1} &= H^T R^{-1} H + (\Phi P_{k/k}\Phi^T + \Gamma Q \Gamma^T)^{-1}.\end{aligned} \tag{3.76}$$

From (3.75) it is clear that

$$x_{k+1/k+1} = \bar{x}_{k+1/k+1}.$$

Since (3.76) represents the standard equations for the Kalman-Bucy filter, it follows that this estimate of the last state along the trajectory is the same as the Kalman-Bucy estimate.

For simplicity of notation, the argument of the functions H, Φ, Γ, R, and Q will be left out, but it is understood that they are all allowed to vary with time.

SUPPLEMENTARY PROBLEMS

3.1 In Solved Problem (3.4) show that $f_k(x) = \log x + a_k$, $g_k(x) = \log x + b_k$ and determine a_k, b_k and the structure of the optimal policy.

3.2 In Solved Problem (3.5) prove that $I_N(x) = \log x + Na_k$, determine a_k and the optimal policy. Then show that the optimal policy is independent of q_i.

3.3 For Solved Problem (3.10) consider the cases in detail when

(a) $\varphi(z) = be^{-z}, \quad b > 0$.

(b) $$\varphi(z) = \begin{cases} \frac{k-z}{k}, & 0 \leq z \leq k, \\ 0, & z \geq k. \end{cases}$$

3.4 Using the criterion function

$$J = E\left\{x^2(K) + \sum_{k=0}^{K-1} [x^2(k) + u^2(k)]\right\},$$

(a) Derive appropriate recurrence relations for minimizing J in the adaptive case (see Eq. (3.30)).

(b) Use the structure of the optimal value function $I(c,m,n,K)$ to obtain an analytic resolution of the problem.

HINT: In part (b), I has the structure

$$I(c,m,n,K) = (c,R_K(m,n)c) + (c,s_K(m,n)) + q_K(m,n),$$

where $\{R_K\}$ are $N \times N$ matrices, $\{s_N\}$ are N-dimensional vectors, and $\{q_N\}$ is a sequence of scalars.

(c) For the case when $g[x(k),u(k),r(k)] = ax(k)+u(k)+r(k)$, and

$$J = \underset{\{r(k)\}}{E}\left[x^2(K) + \sum_{n=0}^{K-1} u^2(n)\right],$$

determine explicit analytic expressions for $I(m,n,c)$ and $u(m,n,c)$.

3.5 Consider the following multi-arm bandit problem. A machine k has $\varphi(k)$ possible money prizes $\{a(k,i)\}_{i=1}^{\varphi(k)}$ which occur at each trial according to the multinomial distribution with parameters $P[a(k,i)] = r(k,i)$, $1 \leq i \leq \varphi(k)$. Assume M such machines are available. The values of $r(k,i)$ are not known to the player, but he does know that they are selected by a random mechanism with distribution function

$$F[\{r(k,i)\}], \qquad 1 \leq k \leq m .$$

If N free plays are allowed, determine a dynamic programming solution so that the player maximizes his expected take.

3.6 In the noisy communication channel problem, consider the case in which the probability of correct transmission at the k<u>th</u> stage is $p_k > 1/2$. Show that the maximum expected value of the logarithm of the capital at the end of N stages is given by

$$\log x + N \log 2 + \sum_{k=1}^{N} (p_k \log p_k + q_k \log q_k) .$$

REFERENCES

<u>Probability Theory and Stochastic Processes</u>: [B-105], [C-17], [L-25], [R-15].

<u>Markovian Decision Processes</u>: [B-15], [B-59], [D-3], [D-4], [H-23], [H-25], [M-16].

<u>Multistage Games</u>: [B-15], [B-102], [F-15], [K-12], [L-29], [V-6].

<u>Optimum Estimation</u>: [H-22], [K-5], [L-22], [R-5].

<u>Adaptive Control Processes</u>: [B-32], [C-8], [G-12], [L-16], [Y-1].

CHAPTER 4

ADVANCED COMPUTATIONAL METHODS

INTRODUCTION

In recent years a number of new computational methods based on dynamic programming have been developed. Many of these methods retain the power and generality of dynamic programming and yet have substantially reduced computational requirements over those of the standard algorithm. This chapter surveys a number of the more promising of these techniques and mentions some of their more recent applications.

The fundamental concepts of dynamic programming are first reviewed. The general deterministic variational control problem is formulated and the iterative functional equation based on Bellman's principle of optimality is then derived. Next, we describe the standard dynamic programming computational algorithm and discuss its implementation. Finally, the extension of these results to stochastic control and other problems involving uncertainty is indicated.

The new procedures are then presented and are grouped into four categories: In Group A, procedures for obtaining a complete feedback control solution are discussed. In all cases, these procedures have greatly reduced high-speed memory requirements. However, because optimal control still must be computed for every admissible state, these procedures (except where mentioned explicitly) have computing time requirements which are about the same as for the standard algorithm. In Group B, procedures for finding the optimal control sequence from a single initial state are examined. Because control does not need to be computed over much of the admissible state space, great savings in both memory and time are achieved. However, it is interesting to note that with these procedures a true feedback solution is computed in a region about the optimal trajectory. In Group C, procedures for infinite-stage problems are given. These procedures solve a multistage problem with a very large number of stages by making a single-stage calculation iteratively. Finally, in Group D, procedures that do not fit any of the other categories are discussed.

REVIEW OF DYNAMIC PROGRAMMING FUNDAMENTALS

Problem Formulation for the Deterministic Case

Most of the problems for which dynamic programming has been used to obtain numerical solutions can be formulated as deterministic discrete-time variational control problems. The general case of this problem is formulated as follows.

Given:

(1) A system described by the nonlinear difference equation

$$x(k+1) = \Phi[x(k),u(k),k] \tag{4.1}$$

where x is an n-dimensional state vector, u is an m-dimensional control vector, k is an index for the stage variable, and Φ is an n-dimensional vector function.

(2) A variational performance criterion

$$J = \sum_{k=0}^{K} L[x(k),u(k),k] \tag{4.2}$$

where J is the total cost and L the cost for a single stage.

(3) Constraints

$$x \in X(k) \tag{4.3}$$

$$u \in U(x,k) \tag{4.4}$$

where $X(k)$ is a set of admissible states at stage k, and $U(x,k)$ a set of admissible controls at stage x, stage k.

(4) An initial state

$$x(0) = c \; . \tag{4.5}$$

Find:

The control sequence $u(0),\ldots,u(k)$ such that J in (4.2) is minimized subject to the system equation (4.1), the constraint equations (4.3) and (4.4), and the initial condition (4.5).

Continuous-time variational control problems can be treated by assuming that the control is piecewise constant in time and making appropriate transformations to the discrete-time case. Extensions to problems involving uncertainty will be made later.

Derivation of the Basic Iterative Functional Equation

The dynamic programming solution to the above problem is obtained by using an iterative functional equation that determines the optimal control for any admissible state at any stage. This equation follows immediately from Bellman's principle of optimality. However, in the interest of clarity, a direct derivation will be given and the principle of optimality will then be interpreted in terms of this equation.

The first step in the derivation is to define the minimum cost function for all $x \in X$ and all k, $k = 0,1,\ldots,K$, as

$$I(x,k) = \min_{\substack{u(j) \\ j=k,k+1,\ldots,K}} \left\{ \sum_{j=k}^{K} L[x(j),u(j),j] \right\} \tag{4.6}$$

where

$$x(k) = x .$$

The summation is then split into two parts, the term evaluated for $j = k$ and the summation over $j = k+1$ to $j = K$. The minimization is similarly split into two parts. The result is

$$I(x,k) = \min_{u(k)} \min_{\substack{u(j)\\ j=k+1,\ldots,K}} \left\{ L[x,u(k),k] + \sum_{j=k+1}^{K} L[x(j),u(j),j] \right\} . \tag{4.7}$$

The first term in brackets in (4.7) is not affected by the second minimization. Thus, (4.7) becomes

$$I(x,k) = \min_{u(k)} \left\{ L[x,u(k),k] + \min_{\substack{u(j)\\ j=k+1,\ldots,K}} \left[\sum_{j=k+1}^{K} L[x(j),u(j),j] \right] \right\} . \tag{4.8}$$

The second term in brackets in (4.8) is exactly analogous to the definition in (4.6), where the argument of I is $(\Phi[x,u(k),k],k+1)$. Abbreviating $u(k)$ as u, the iterative functional equation becomes

$$I(x,k) = \min_{u} \{L(x,u,k) + I[\Phi(x,u,k),k+1]\} . \tag{4.9}$$

This equation is a mathematical statement of Bellman's principle of optimality. It states that the minimum cost for state x at stage k is found by choosing the control that minimizes the sum of the cost to be paid at the present stage and the minimum cost in going to the end from the state at stage $k+1$ which results from applying this control. The optimal control at state x and stage k, denoted by $u(x,k)$, is directly obtained as the value of u for which the minimum (4.9) is attained.

Since (4.9) determines $I(x,k)$ and $u(x,k)$ in terms of $I(x,k+1)$, it must be solved backward in k. As a terminal boundary condition

$$I(x,K) = \min_u [L(x,u,K)] . \qquad (4.10)$$

The optimization over a sequence of controls is thus reduced to a sequence of optimizations over a single control vector.

The Standard Computational Algorithm

In the standard method for solving (4.9), each state variable x_i, $i = 1,2,\ldots,n$, is quantized to N_i levels, and each control variable u_j, $j = 1,2,\ldots,m$, is quantized to M_j levels.

Initially, $I(x,K)$ is found for all quantized states $x \in X$ by evaluating $L(x,u,K)$ for each quantized control $u \in U$ and choosing the minimum value by a direct comparison. The optimal control $\hat{u}(x,K)$ is the value of u that minimizes $L(x,u,K)$. In many problems no control is applied at the final stage K; in this case $I(x,K)$ is evaluated directly as $L(x,K)$.

Next, at $k = K-1$, for each quantized state $x \in X$, each quantized control $u \in U$ is applied, and the next state $\Phi(x,u,K-1)$ is computed. The minimum cost at the next state, $I[\Phi(x,u,K-1),K]$, is found by interpolation using the values of $I(x,K)$ at quantized states. The quantity $L(x,u,K-1)$ is computed directly. The sums of these quantities for each quantized control are then compared, and the minimum value is stored as $I(x,K-1)$. The optimal control $\hat{u}(x,K-1)$ is stored as the value of u for which the minimum is attained. This procedure is illustrated for a one-dimensional problem in Figure 4.1.

The procedure continues in this manner, with $I(x,k)$ and $\hat{u}(x,k)$ being computed in terms of $I(x,k+1)$, until $k = 0$ is reached. The quantized states at which these quantities are computed in a one-dimensional example are shown in Figure 4.2.

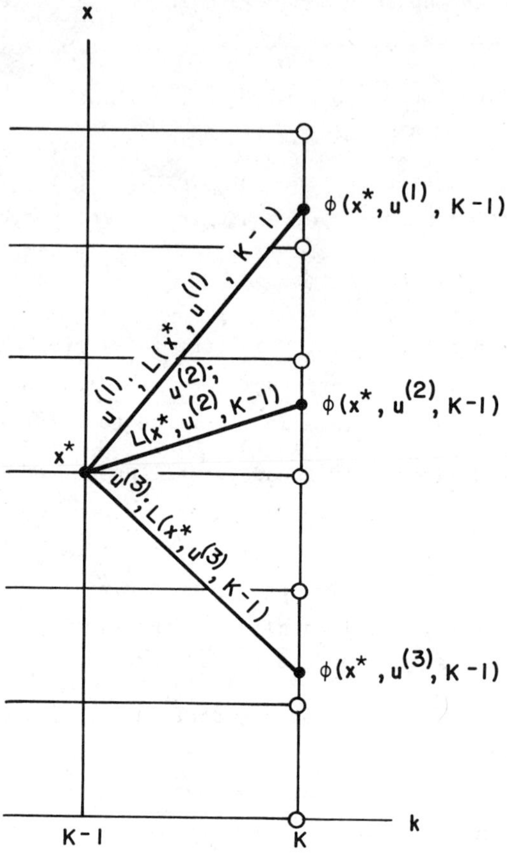

Figure 4.1 Dynamic Programming Calculation in One-Dimensional Example at State x^*, Stage $K-1$

The optimal control sequence from the given initial state c is obtained from the values of $\hat{u}(x,k)$ at quantized states. The first control is read directly as $\hat{u}(c,0)$. The next state is then computed as $\Phi[c,\hat{u}(c,0),1]$, and the corresponding control is evaluated using interpolation on values of $\hat{u}(x,1)$ at quantized x. This procedure continues until the entire control sequence is

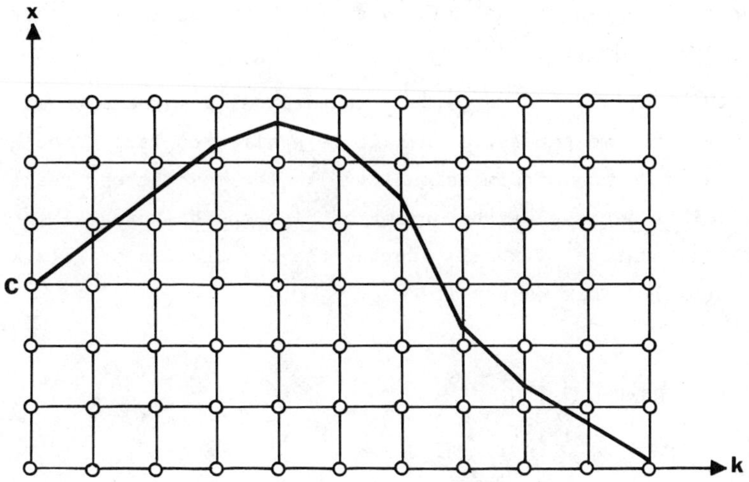

Figure 4.2 Recovery of the Optimal Trajectory from Initial State c

obtained. This is shown in Figure 4.2 for a one-dimensional example.

The same procedure can be used to compute the optimal control sequence for any initial state starting at any stage. If the initial state is not a quantized state, an additional interpolation may be necessary. Thus, the solution to many optimization problems is obtained in the same calculation as for the original problem.

This computational procedure is very appealing for a number of reasons. In the first place, thorny questions of existence and uniqueness are avoided; as long as there is at least one feasible control sequence, then the direct-search procedure guarantees that the absolute minimum cost is obtained. Furthermore, extremely general types of system equations, performance criteria, and constraints can be handled. Constraints actually reduce the computational burden by decreasing the size of the admissible sets X and U.

Finally, the optimal control is obtained as a true feedback solution in which the optimal control for any admissible state and stage is determined.

In practice, however, the computational requirements of this technique become excessive when it is applied to large problems (Bellman's "curse of dimensionality"). The most severe restriction arises because of the number of high-speed storage locations required to store $I(x,k+1)$ during the computation of $I(x,k)$ and $\hat{u}(x,k)$. This number is

$$\text{HSMR} = \prod_{i=1}^{n} N_i \ . \tag{4.11}$$

Another consideration, which is generally not as restrictive as the high-speed memory requirement (HSMR), is the total amount of computing time (CT) required. This is

$$\text{CT} = \left(\prod_{i=1}^{n} N_i\right)\left(\prod_{j=1}^{m} M_j\right) K\,\Delta t_c \ , \tag{4.12}$$

where Δt_c is the computing time required for a single evaluation of the quantity in brackets in (4.9) and a single scalar comparison.

The final factor that must be taken into account is the amount of off-line storage required to store the complete solution. The number of storage locations is

$$\text{LSMR} = \left(\prod_{i=1}^{n} N_i\right) Km \ . \tag{4.13}$$

Extension to Problems Containing Uncertainty

A number of computational methods based on dynamic programming have been developed for problems containing uncertainty. One case that we devoted some attention to in the last chapter is Bellman's optimum stochastic control problem. In this problem the state of the system is perfectly measurable, but a random forcing function $w(k)$, which has a known probability density function, enters the system equation. The problem formulation is exactly as above, except that:

(1) the system equation is affected by the random forcing function vector, so that (4.1) becomes

$$x(k+1) = \Phi[x(k),u(k),w(k),k] \; ; \qquad (4.14)$$

(2) the expected value of the performance criterion is to be minimized, where the expectation is over the sequence of random forcing functions $w(0),w(1),\ldots,w(k)$.

An iterative functional equation for this problem can be developed exactly as shown in Chapter 3. The minimum cost function is defined as

$$I(x,k) = \min_{u(k),\ldots,u(K)} \left\{ \mathop{E}_{w(k),\ldots,w(K)} \left[\sum_{j=k}^{K} L[x(j),u(j),w(j),j] \right] \right\} , \qquad (4.15)$$

where

$$x(k) = x \; .$$

Assuming that samples of $w(k)$ at different stages are uncorrelated*, the desired equation is obtained as

$$I(x,k) = \min_{u} \left\{ \mathop{E}_{w} \left[L(x,u,w,k) + I[\Phi(x,u,w,k),k+1] \right] \right\} . \qquad (4.16)$$

*This assumption can always be satisfied at the expense of defining additional state variables to account for the correlation, as was shown in Chapter 3.

As before, the optimal control $\hat{u}(x,k)$ is the control for which the minimum is attained. As a terminal boundary condition,

$$I(x,K) = \min_{u} \left\{ \underset{w}{E}[L(x,u,w,K)] \right\} . \tag{4.17}$$

A computational procedure analogous to that of the earlier section can be applied; the random forcing function w is quantized, the corresponding probability density function is converted to a discrete probability distribution, and the expectation is taken by summation.

A more general stochastic control problem, called the combined optimum control and estimation problem, is represented in Figure 4.3. Here the state of the system is not known exactly, but instead is observed through a noisy measurement system. An iterative functional equation can be derived for this problem in which the argument of the minimum cost function at a given stage is either the

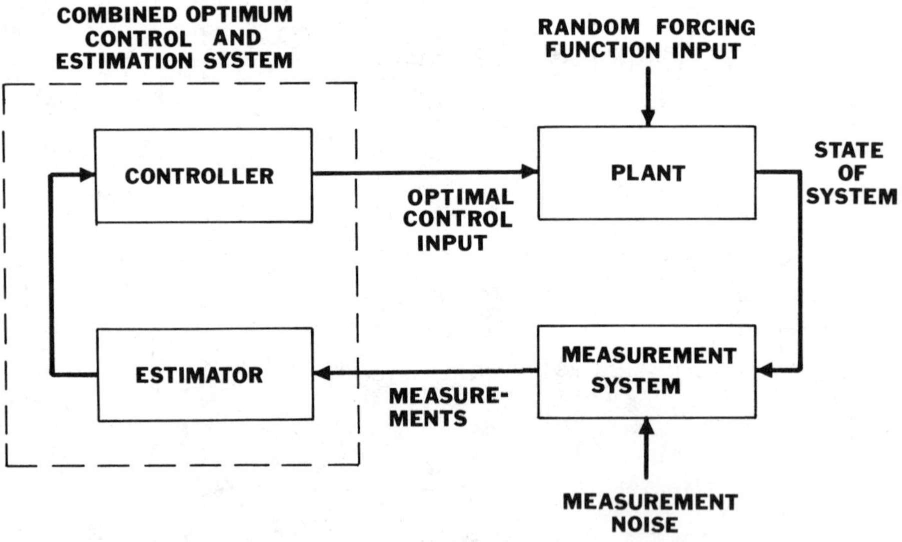

Figure 4.3 Combined Optimum Control and Estimation Problem

entire sequence of past controls and measurements or else a sufficient statistic for the a posteriori probability density function of the present state. This argument is generally called the "information state." In either case, it is necessary to calculate the latter probability density function by recursive application of Bayes' rule. Other approaches to this problem are discussed in the references.

Dynamic programming has also been applied to a number of other problem formulations involving uncertainty, notably optimum estimation and identification, optimum adaptive control, and differential games. We shall examine these areas extensively in the next chapter.

NEW COMPUTATIONAL PROCEDURES

Procedures for Obtaining a Complete Feedback Control Solution

A. *Closed-Form Solution*

The most computationally efficient method for solving (4.9) is to find a closed-form expression for $I(x,k)$. However, the number of cases where this can be done is small. The best known case is that in which the system equations are linear

$$x(k+1) = F(k)x(k) + G(k)u(k) \;; \tag{4.18}$$

the performance criterion is quadratic

$$K = \sum_{k=0}^{K} [x^T(k)Q(k)x(k) + u^T(k)R(k)u(k)] \;; \tag{4.19}$$

and there are no constraints. In this case, we have already seen that $I(x,k)$ has the structural form

$$I(x,k) = x^T P(k)x \;, \tag{4.20}$$

where $P(k)$ satisfies the well-known discrete-time Riccati equation

$$P(k) = Q(k) + F^T(k)P(k+1)F(k) - F^T(k)P(k+1)G(k) \cdot [R(k) + G^T(k)P(k+1)G(k)]^{-1} G^T(k)P(k+1)F(k) \quad . \quad (4.21)$$

The corresponding optimal control $\hat{u}(x,k)$ takes the form

$$\hat{u}(x,k) = - K(k)x \ , \quad (4.22)$$

where

$$K(k) = [R(k) + G^T(k)P(k+1)G(k)]^{-1}G^T(k)P(k+1)F(k) \quad . \quad (4.23)$$

The computations for an n-dimensional problem thus reduce to the iteration of an $n \times n$ symmetric matrix difference equation. A great saving on both storage and time is obtained.

Analogous results can be obtained for the optimum estimation problem in which the system equations and measurement equations are linear and all random variables are Gaussian, for the stochastic control problem in which the system equations are linear and the performance criterion is quadratic, and for the combined optimum control and estimation problem in which the system equations and measurement equations are linear, all random variables are Gaussian, and the performance criterion is quadratic.

B. *Approximation of the Optimal Cost Function*

In certain cases, it is possible to obtain an accurate approximation to the optimal cost function $I(x,k+1)$ by expressing it in terms of a simpler function, e.g., low-order polynomials, exponentials, spline functions, and so on. In such cases, a savings in high-speed memory is achieved by storing only the coefficients of the approximating functions and the algorithm for reconstructing values of $I(x,k+1)$ as needed. Thus, we trade the computing time required to approximate $I(x,k+1)$ and to then reconstruct its values as needed, for high-speed memory locations. Let us indicate a few examples of the technique to examine this trade-off.

Consider, at first, the case in which the state space has dimension one, i.e., dim X=1. Assume that we desire to approximate $I(x,k+1)$ by a polynomial of degree r, i.e.,

$$I(x,k+1) = a_o(k+1) + a_1(k+1)x, \ldots, a_r(k+1)x^r ,$$

where the coefficients $a_i(k+1)$, $i = 0,\ldots,r$, are to be determined. A standard procedure for obtaining the numbers $a_i(k+1)$ is Lagrangian interpolation. We evaluate the function $I(x,k+1)$ at $(r+1)$ values of x, say $X_1, X_2, \ldots, X_{r+1}$, obtaining the values $I(X_1,k+1),\ldots,I(X_{r+1},k+1)$. Then the coefficients $a_i(k+1)$ are determined by solving the linear system

$$\begin{aligned}
a_o(k+1) + a_1(k+1)X_1, \ldots, a_r(k+1)X_1^r &= I(X_1,k+1) \\
a_o(k+1) + a_1(k+1)X_2, \ldots, a_r(k+1)X_2^r &= I(X_2,k+1) \\
&\vdots \\
a_o(k+1) + a_1(k+1)X_{r+1}, \ldots, a_r(k+1)X_{r+1}^r &= I(X_r,k+1) .
\end{aligned}$$

The above "rule" for determining the a_i comes about by demanding that the approximation be exact for those values of x at which a direct evaluation of $I(x,k+1)$ is made, i.e., for $x = X_i$, $i = 1,2,\ldots,r$.

There are well known difficulties associated with using an approximation of the above type; however, if the function $I(x,k+1)$ is sufficiently smooth in x, it can be shown in many cases that the approximation will be quite good for moderately low values of r. Thus, in place of storing the values of $I(x,k+1)$ for all quantized values of x, we need only evaluate I for r values of x, and solve a linear algebraic system in $(r+1)$ variables. Having the coefficients $\{a_i\}$ at our disposal, the value of $I(x,k+1)$ may be computed as needed.

It is evident that the success or failure of the foregoing procedure will depend upon the degree of smoothness of the function I(x,k+1). If it is extremely well-behaved, then a small value of r, say $r = 2$ or 3, will suffice to obtain an accurate representation and the amount of work needed to determine the coefficients $\{a_i\}$ will be nominal. However, for badly behaved I(x,k+1), more sophisticated procedures must be invoked. Unfortunately, in many problems of practical concern, especially in the general area of operations research, it is not unusual to have the function I(x,k+1) being non-differentiable or even discontinuous. Clearly, in such instances an elementary approximation of the foregoing type will be of no particular benefit.

There are many ways to improve upon the polynomial approximation technique sketched above at very nominal analytic and computational cost. We give a few examples now to indicate some of the many possibilities:

(1) Local Approximations — In many instances, the function I(x,k+1) will not be globally smooth due to points of discontinuity; however, often the function will be differentiable (or even analytic) between such points, e.g., a function such as the following form will be poorly approximated by the above global polynomial technique due to its discontinuities at x_1, x_2, x_3:

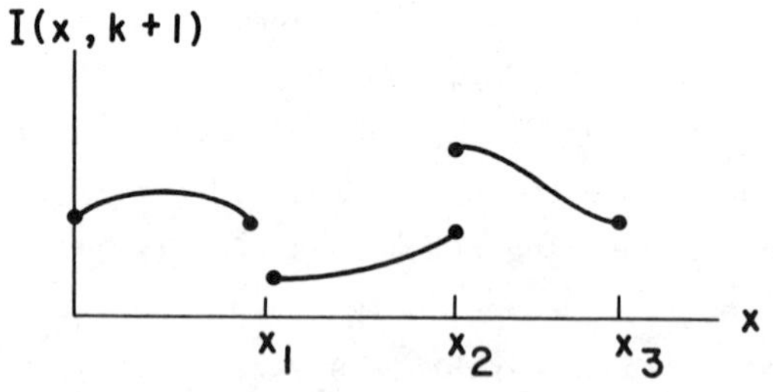

However, if we construct separate polynomial approximations on the intervals $(0,x_1)$, (x_1,x_2), (x_2,x_3), then we shall be able to accurately determine I with modest effort. The real problem in this case is to determine the points of discontinuity in advance, in order to separate the relevant regions of x-space. Fortunately, in a number of cases, the points of discontinuity will be evident, even by inspection of numerical data. We note in passing that the theory and methodology of spline approximations may be put to good use in connection with approximating I locally in an optimal manner.

(2) Regression Techniques — Even in the case when $I(x,k+1)$ is sufficiently smooth, a polynomial approximation may be materially improved by using more than (r+1) values of x to determine the coefficients $\{a_i\}$. For instance, we could evaluate $I(x,k+1)$ at $n \gg r+1$ points and then use a regression technique to obtain the values of $\{a_i\}$ such that the best least-squares fit is obtained between the approximation and the evaluated data, i.e., we determine the $\{a_i\}$ to minimize

$$\sum_{i=1}^{n} [I(X_i,k+1) - (a_o + a_1 X_i + \ldots + a_r X_i^r)]^2 .$$

(3) Orthogonal Polynomials — Assuming requisite smoothness for $I(x,k+1)$, the biggest single difficulty in employing polynomial approximation is the need to solve a linear algebraic system to determine the coefficients of the approximation. This problem may be overcome by a more judicious choice of the approximating polynomials.

Consider the case in which the state variable x is confined to the region $a \leq x \leq b$ (there is no loss of generality here, as we could have $a = -\infty$ and/or $b = +\infty$).

Let $\{\Phi_i(x)\}$ be a set of polynomials which are orthogonal over $[a,b]$, i.e.,

$$\int_a^b \Phi_i(x)\Phi_j(x)\,dx = \begin{cases} 1, & i=j \\ 0, & i \neq j . \end{cases}$$

If we now approximate $I(x,k+1)$ as

$$I(x,k+1) \cong \sum_{i=0}^{r} a_i(k+1)\ \Phi_i(x) ,$$

and employ the same chain of reasoning given above for Lagrangian interpolation, the orthogonality property of the set $\{\Phi_i\}$ will, in effect, diagonalize the resulting linear algebraic system, thereby enabling the coefficients $\{a_i\}$ to be obtained by inspection, i.e.,

$$a_i(k+1) = \int_a^b I(x,k+1)\ \Phi_i(x)\ dx .$$

Although for sake of exposition we have presented the main ideas surrounding approximation of the optimal cost function $I(x,k+1)$ in the case of x being a scalar quantity, the real situations of interest arise when x is multidimensional. Formally, there is no difficulty associated with extending the above ideas to this setting. For instance, if our process involved the two state variables x and y, we could approximate $I(x,y,k+1)$ as

$$I(x,y,k+1) = \sum_{i,j=0}^{r} a_{ij}(k+1)\ x^i y^j .$$

Here, of course, we would have to evaluate I at $(r+1)^2$ values in (x,y)-space in order to obtain data sufficient to compute the matrix of coefficients $\{a_{ij}(k+1)\}$. In general, we see that if x is an n-dimensional vector and an elementary approximation of the above type is used, then a linear algebraic system of order r^n must be solved to determine the expansion coefficients. Clearly, for large n this will be feasible only if r is small, i.e., only if the analytic structure of I is such that a low-order polynomial is sufficiently accurate for the purposes at hand.

The exponential growth in the computing requirement to obtain the approximation coefficients makes the use of orthogonal polynomials and local approximation of greater importance as the dimension of x increases.

EXAMPLE

4.1 To illustrate a few of the foregoing points, we consider the problem of maximizing

$$J = \sum_{i=1}^{3} u_i^2 ,$$

over the region $u_1 + u_2 + u_3 = x$. If we let

$$I(x,k) = \max J ,$$

an easy application of the principle of optimality yields

$$I(x,k) = \max_{0 \leq u \leq x} [u^2 + I(x-u,k-1)], \quad k = 2,3 ,$$

$$I(x,1) = \max_{u=x} (u^2) = x^2 .$$

Let us pretend, for the moment, that we do not know that $I(x,k)$ is a quadratic form in x and that we wish to approximate to I by a quadratic polynomial, i.e., we write

$$I(x,k) = a_0(k) + a_1(k)x + a_2(k)x^2 ,$$

and attempt to find the coefficients a_0, a_1, a_2 as above. Of course, since I is a pure quadratic, we must ultimately obtain $a_0(k) = a_1(k) = 0$.

Substituting the above approximation into the recurrence relation for I, we have

$$a_0(k)+a_1(k)x+a_2(k)x^2 = \max_{0\leq u\leq x} \left[\begin{array}{c} u^2+a_0(k-1)+a_1(k-1)(x-u) \\ +a_2(k-1)(x-u)^2 \end{array} \right], \quad k=2,3,$$

$$[a_0(1),a_1(1),a_2(1)] = (0,0,1) .$$

To obtain $[a_0(2),a_1(2),a_2(2)]$, we need to solve the above equation for three values of x. For example, suppose we choose $X_1 = 1$, $X_2 = 2$, $X_3 = 3$: then we have

$$\max_{0\leq u\leq 1} [u^2 + a_0(2) + a_1(2)(1-u) + a_2(2)(1-u)^2] = 1 ,$$

$$\max_{0\leq u\leq 2} [u^2 + a_0(2) + a_1(2)(2-u) + a_2(2)(2-u)^2] = 2 ,$$

$$\max_{0\leq u\leq 3} [u^2 + a_0(2) + a_1(2)(3-u) + a_2(2)(3-u)^2] = 9 .$$

Thus, we formally obtain

$$a_0(2) + a_1(2) + a_2(2) = 1 ,$$

$$a_0(2) + 2a_1(2) + 4a_2(2) = 4 ,$$

$$a_0(2) + 3a_1(2) + 9a_2(2) = 9 ,$$

which has the solution $a_0(2) = 0$, $a_1(2) = 0$, $a_2(2) = 1$, as expected. The same result is then obtained for $a_0(3)$, $a_1(3)$, $a_2(3)$.

The important point to note about the above example is that we needed only to store the three numbers $a_0(k-1)$, $a_1(k-1)$, $a_2(k-1)$, and to obtain the three values $I(1,k-1)$, $I(2,k-1)$, $I(3,k-1)$, in order to carry out the approximation of $I(x,k)$. It is evident that this represents a substantial computational saving over the determination of I by the standard procedures. This saving is possible only because I had the structure of the hypothesized form. This illustrates that the success or failure of the approximation methods will hinge critically upon our ability to determine as much as possible about the analytic structure of I so that a judicious choice of the approximating functions can be made.

C. *Search Procedures Other Than Quantizing Control*

If the interpolation formulas for $I(x,k+1)$ are the proper form, it is sometimes possible to use efficient search procedures in the minimization in (4.9), rather than quantization of control and direct comparison. Several different search procedures can be employed, depending upon the particular analytic structure of the function $L(x,u)$ and the optimal return function $I(x,k+1)$.

For example, if the single-stage cost function $L(x,u)$ is a linear form in u,

$$L(x,u) = \alpha u + \beta \ ,$$

while $I(x,k+1)$ is also linear in u, then the determination of a minimizing control reduces to the problem of minimizing a linear form, subject to linear inequality constraints. In this case, the well known simplex algorithm from linear programming provides a very efficient and effective procedure for carrying out the minimization over u. Fortunately, it is easily established that the above linearity condition on I will hold if L is linear in u and x and if u is subject to linear side constraints (possibly dependent upon x).

In other cases, we may be able to employ techniques from nonlinear programming to effect the minimization over u. If L is convex in u, then use of a steepest descent algorithm is suggested, since such a constraint on L will, under modest additional hypotheses, induce convexity upon the optimal value function I. Of course, even if the cost functions are not convex, a steepest descent algorithm may be used provided the requisite smoothness properties are present in L and I; however, in this case we run the usual risks of finding a relative, rather than global minimum, or what is far worse, a maximum or an inflection point!

Another type of search procedure that is remarkably efficient (optimal, even) for one-dimensional problems is the so-called "Fibonacci search" technique.

Let us first consider the application of this procedure to the simple problem of finding the minimum of a unimodal function $f(x)$ of one variable (recall that function $f(x)$ is unimodal in an interval $a \leq x \leq b$ if there is a point x^* $[a,b]$, such that $f(x)$ is either strictly increasing for all $x \leq x^*$ and strictly decreasing for $x > x^*$ or vice-versa). For our purposes, the most important examples of such functions are those which are convex or concave.

Let the domain of $f(x)$ be given by the interval $0 \leq x \leq k_n$, and suppose it is desired to determine the minimum of f to within a unit interval by means of n evaluations. The basic question is: what is the largest value of k_n for which this determination can be carried out? In other words, what is the largest interval length for which it is possible to locate the minimum of f to within a unit length by means of n function evaluations?

If we denote

$$f_n = \sup k_n ,$$

then it may be shown that the sequence f_n satisfies

$$f_n = f_{n-1} + f_{n-2} , \quad n \geq 2 ,$$

$$f_o = 1 , \quad f_1 = 1 .$$

The sequence of numbers $\{f_n\}$ is the Fibonacci sequence which turns up unexpectedly in many diverse areas of mathematics, physics and biology.

Although the sequence $\{f_n\}$ starts out slowly,

$$1, 1, 2, 3, 5, 8, 13, 21, \ldots,$$

we have $f_{20} > 10^4$. Thus, the position of the minimum can always be located to within 10^{-4} of the original interval length by at most 20 function evaluations.

The ratio $f_n/f_{n-1} \rightarrow (\sqrt{5}+1)/2 \cong 1.62$ as $n \rightarrow \infty$. Thus, the first two values for evaluation should be chosen at a distance of $0.62L$ from either end of an interval of length L. This is a good uniform testing policy, which is very close to the optimal policy except for the last few stages of the procedure, in which case an optimal search is of greatly reduced interest and utility.

The use of Fibonacci search in dynamic programming is now fairly evident. In carrying out the minimization of

$$L(x,u) + I[T(x,u),k+1] ,$$

over all u, for fixed x, if L is convex and u is a scalar, we choose the evaluation points of u according to the Fibonacci rule. In this way, we locate the minimizing u* utilizing a minimal number of evaluations of the expression $L(x,\cdot)+I[T(x,\cdot),k+1]$.

The major drawback to Fibonacci search is the fact that it is optimal only in the one-dimensional case. However, it should be noted that the dimensionality involved here is that of the decision variable, u, and not the state. Fortunately, there are many problems in optimal control and operations research where scalar controls appear.

Now, let us briefly consider the case where f(x) is defined only over a discrete set of points. Assume f is defined on a set of M_n points, where M_n is an integer such that the minimum value of f can always be identified by n evaluations and comparisons. Setting

$$K_n = \max M_n ,$$

we can obtain

$$K_n = -1 + f_{n+1} , \qquad n \geq 1 .$$

The proof of this basic result is by induction. To begin with, we number the points in the domain of f in some order $1,2,3,\ldots,M_n$. It is clear that

$$K_1 = 1 , \qquad K_2 = 3 , \qquad K_3 = 4 .$$

Now let $n > 3$ and assume

$$K_k = -1 + K_{k+1} , \qquad k < n .$$

We calculate f(x) at two points x_1 and x_2. It is easy to see that

$$x_1 \leq K_{n-2} + 1 , \qquad M_{n-x} \leq K_{n-1} .$$

hence,

$$M_n \leq K_{n-2} + 1 + K_{n-1} = (f_{n-1} - 1) + 1 + (f_n - 1) = f_{n+1} - 1 \ .$$

The maximum is attained if $x_1 = f_{n-1}$ and $x_2 = f_n$. This proves the result, as well as establishes the optimal search policy (choice of x_1 and x_2).

To summarize the above result, we see that the optimum value of a unimodal function defined on f_n points can be determined by evaluating the function at n points, where

$$f_n = f_{n-1} + f_{n-2} + 1 \ , \qquad n \geq 2 \ ,$$

$$f_o = f_1 = 1 \ .$$

The first two evaluations are made at the $(f_{n-2}+1)$st point and at the $(f_{n-1}+1)$st point. This reduces the region under consideration to f_{n-1} points, with an evaluation already having been made at one of the appropriate points in this interval.

Use of Analytical Results and Necessary Conditions

Analysis of a particular problem may sometimes reveal useful information about the solution. For example, in the bang-bang control problem it can be shown that the optimal control is always one of the two extreme admissible values. Conditions such as this can reduce computational requirements by decreasing the size of the admissible control set $U(x,k)$.

In another direction, we can sometimes exploit the analytic structure of a problem to reduce the number of "relevant" state variables to a manageable level. The following example illustrates the application of this general remark.

EXAMPLE

4.2 A Terminal Control Process. Let us begin by considering the following terminal control problem. We wish to minimize the quadratic functional

$$J(x,y) = [x(T),Bx(T)] + \int_0^T (y,y)dt \ , \tag{4.24}$$

where x and y are connected by the linear relation

$$\dot{x} = Ax + y \ , \quad x(0) = c \ . \tag{4.25}$$

We have already indicated in Volume I some of the computational obstacles associated with this simple analytic problem in the case where x is of large dimension. One type of difficulty is encountered using the calculus of variations; another occurs when dynamic programming is employed. Here we wish to consider the case where the matrix B has the special form

$$B = \begin{bmatrix} \begin{pmatrix} b_{11} & \cdots & b_{1k} \\ b_{21} & \cdots & b_{2k} \\ & \vdots & \\ b_{k1} & \cdots & b_{kk} \end{pmatrix} & \begin{matrix} 0 \\ \cdot \\ \cdot \\ \cdot \end{matrix} \\ 0 \quad \cdots & 0 \end{bmatrix} , \tag{4.26}$$

with $k \ll N$.

This is equivalent to the statement that in evaluating the terminal behavior of the system only the first k components of $x(T)$ are of significance, a situation which occurs frequently in applications. The other components, $x_{k+1}, x_{k+2}, \ldots, x_N$, play the role of "hidden variables." Although they are vital for a classical deterministic description of the system, we have no specific interest in their precise behavior at the end of the process.

A particular case of the foregoing general problem is that in which we wish to minimize an expression of the form

$$J(x,u) = \sum_{i=1}^{k} b_i x^{(i)}(T)^2 + \int_0^T u^2\, dt , \tag{4.27}$$

with x and u connected by the relation

$$\begin{aligned} & x^{(N)} + a_1 x^{(N-1)} + \dots + a_N x = u , \\ & x^{(i)}(0) = c_i , \qquad i = 0,1,\dots,N-1 . \end{aligned} \tag{4.28}$$

The linearity of (4.25) is the basic factor. We begin by noting the relation

$$x(t) = e^{At}c + \int_0^t e^{A(t-s)} y(s)ds ,$$

whence

$$x(T) = e^{AT}c + \int_0^t e^{A(T-s)} y(s)ds .$$

We presage our subsequent procedure by writing this in the form

$$x(T) = z + \int_0^T K(T-s)y(s)ds , \tag{4.29}$$

where z is a vector parameter and K is a given matrix. This parameter z has an extremely interesting physical significance; namely, it is the state of the system at time T if no control is exerted at all, i.e., if $y = 0$, $0 \leq t \leq T$.

The ith component of x(T) then has the form

$$x_i(T) = z_i + \int_0^T \left[\sum_{j=1}^{N} k_{ij}(T-s)\, y_j(s) \right] ds \ .$$

Hence, if B has the special form indicated in (4.26), we see that J(x,y) takes the form

$$\sum_{i,j=1}^{k} b_{ij} \left(z_i + \int_0^T [\ldots]\, ds \right) \left(z_j + \int_0^T [\ldots]\, ds \right) + \int_0^T (y,y)\,dt \ , \qquad (4.30)$$

where [...] contains terms linear in the y_j.

Starting from (4.30), we consider the z_i as new state variables and introduce the new criterion function

$$\varphi(z,T) = \min_y \left[\sum_{i,j=1}^{k} b_{ij} \left(z_i + \int_0^T [\ldots]ds \right) \left(z_j + \int_0^T [\ldots]ds \right) + \int_0^T (y,y)dt \right] ,$$

defined for $T \geq 0$, $-\infty < z_i < \infty$, $i = 1,2,\ldots,k$.

What is of basic importance in what follows is that $\varphi(z,T)$ is a function of only k variables. We have thus converted the original optimization problem involving a function of N state variables into one involving k state variables. If $k \ll N$, this new problem is much easier to treat. In particular, a dynamic programming solution is now feasible in a number of cases.

Let us turn to the case where the criterion function takes the form

$$J(x,y) = g[x(T)] + \int_0^T h(y)dt .$$

If we assume once again that g depends only upon k components of $x(T)$, there is the possibility that we can employ the general computational approach of dynamic programming. Arguing as before, we obtain a recurrence relation

$$\psi(z,T) = \min_{y(0)} [h(y(0))\Delta + \psi(z+K(T)y(0)\Delta, T-\Delta)] ,$$

with

$$\psi(z,0) = g(z),$$

where $\psi(z,k\Delta) \cong \varphi(z,k\Delta)$. If $k \leq 3$, we can employ direct methods; if $k \geq 4$, we must employ polynomial approximation to circumvent the limitations on rapid-access storage, or alternative methods, as we shall discuss below.

State Increment Dynamic Programming

In discrete-time optimal control problems that have been transformed from continuous-time problems, a substantial saving in high-speed memory requirements can be obtained by using state increment dynamic programming. The system equations for the continuous-time problem are

$$\dot{x} = f(x,u,t) , \tag{4.31}$$

where t is a continuous stage variable (usually time), f an n-dimensional vector functional, and the dot denotes d/dt. The performance criterion is

$$J = \int_{t_o}^{t_f} \ell(x,u,t)dt + \psi[x(t_f),t_f] , \tag{4.32}$$

where J is the total cost, t_o the initial time, t_f the final time, and ℓ and ψ are scalar functionals representing cost per unit time and terminal cost, respectively, with the approximations made so that

$$x(t+\delta t) \cong x(t) + f[x(t),u(t),t]\delta t \quad , \tag{4.33}$$

and

$$\int_t^{t+\delta t} \ell[x(\sigma),u(\sigma),\sigma]d\sigma \cong \ell[x(t),u(t),t]\delta t \ . \tag{4.34}$$

The iterative functional equation then becomes

$$I(x,t) = \min_u \{\ell(x,u,t)\delta t + I[x+f(x,u,t)\delta t, t+\delta t]\} \ . \tag{4.35}$$

In the state increment dynamic programming computational procedure, the state x and control u are quantized as before. The stage variable t is also quantized with an increment size Δt, and minimum cost and optimal control are computed only at quantized values of x and t. However, δt, the time increment over which a given piecewise constant control is applied, is determined as

$$\delta t = \min_{\substack{i=1,2,\ldots,n \\ u \in U(x,t)}} \left\{ \left| \frac{\Delta x_i}{f_i(x,u,t)} \right| , \Delta t \right\} , \tag{4.36}$$

where Δx_i is the quantization increment in the ith state variable and $f_i(x,u,t)$ the ith component of $f(x,u,t)$. This equation ensures that the change in any state variable x_i over the increment δt is at most Δx_i and that δt is at most Δt. Representative values of Δt for a one-dimensional example are shown in Figure 4.4.

The major consequence of this value of δt is that $I[x+f(x,u,t)\delta t, t+\delta t]$ can be determined using only values of the minimum cost function at neighboring quantized states for two or three time increments. Interpolation is in $(n-1)$ state variables and t if $\delta t < \Delta t$ and in n state variables if $\delta t = \Delta t$.

This result can be exploited by processing data in units called blocks, which cover a few increments along each state variable axis but several increments in t. A great savings in high-speed memory requirement can thus be achieved at a small cost in computing time.

We now consider the basic procedure in some detail. The sets of allowable states and admissible controls are determined exactly as in conventional dynamic programming. Constraints

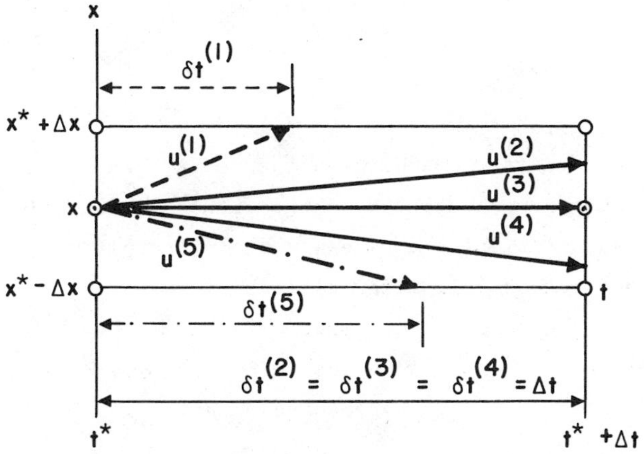

Figure 4.4. Values of Δt in State Increment Dynamic Programming for Controls $u^{(i)}$, $i = 1,2,\ldots,5$, Applied at State x^*, time t^*

in the problem are used to restrict each state variable to a finite range,

$$\beta_i^- \le x_i \le \beta_i^+ , \quad i = 1,2,\ldots,n .$$

Within this range each state variable is quantized in uniform increments x_i,

$$x_i = \beta_i^- + j_i \Delta x_i , \quad j_i = 0,1,\ldots,N_i ,$$

$$N_i \Delta x_i = \beta_i^+ - \beta_i^- , \quad i = 1,2,\ldots,n .$$

The set of allowable states X consists of vectors x, where each component x_i is of the form above. The set can lie in any convex region of the total state space and can vary with time.

The set of admissible controls can be determined by a similar quantization. It is sufficient to state that the set of admissible controls U contains a finite number of elements

$$U = \{u^{(1)}, u^{(2)}, \ldots, u^{(K)}\} .$$

The basic difference between the new procedure and conventional dynamic programming is in the method of determining the time interval over which a given control is applied. With the conventional method this interval is a fixed value Δt, $\delta t = \Delta t$. On the other hand, in the new procedure this time interval is determined as the interval necessary for at least one of the n state variables to change by one increment,

$$\delta t = \min_{i=1,2,\ldots,n} \left\{ \left| \frac{\Delta x_i}{f_i(x,u,t)} \right| \right\} .$$

The fact that control is applied until a state changes by one increment rather than for a fixed time interval is the source of the term "state increment" dynamic programming.

An important consequence of this choice of δt is that the next state always lies within a small neighborhood of the present state. Specifically, the next state lies on the surface of an n-dimensional hypercube centered at the present state and with length $2\Delta x_i$ along the x_i axis. The new procedure uses this property to reduce the high-speed memory requirements from that of the conventional method.

INITIALIZATION PROCEDURE — In order to begin the computational procedure, it is necessary to have available one value of the optimal cost function for each quantized state, $x \in X$. It is not necessary that each value be computed at the final time t_f as in conventional dynamic programming. However, if the value for a given state is computed at t', then, for $t > \Delta t'$, no computation of optimal control will be performed at that state.

An especially convenient choice of t' for each state can be made when the terminal state $x(t_f)$ is constrained to fall within C, a subset of X. In this case, $t' = t_f$ if $x \in C$, while if $x \notin C$, t' is taken to be the least time for which it is still possible to reach C at t_f from a given state using admissible control.

DEFINITION OF THE BLOCK - The first step in the reduction of the high-speed memory requirement is the partitioning of the (n+1)-dimensional $(X-t)$ space into rectangular units called blocks. Each block covers w_i increments along the x_i axis and T units along the t axis. A given block is denoted by the largest values of the coordinates that are contained in the block:

$$B(j_1, j_2, \ldots j_n, m) = \{x, t\} ,$$

such that

$$(j_i - 1) w_i \, \Delta x_i \le x_i \le j_i w_i \, \Delta x_i , \qquad (4.37)$$

$$(m-1)\Delta T \le t - t_o \le m\Delta T ,$$

where

$$j_i = 1, 2, \ldots, J_i ,$$

$$m = 1, 2, \ldots, M ,$$

and

$$J_i w_i \, \Delta x_i = \beta_i^+ - \beta_i^- ,$$

$$i = 1, 2, \ldots, n ,$$

with

$$M\Delta T = t_f - t_0 .$$

As indicated in the foregoing equation, the boundary between adjacent blocks is considered to be included in both blocks. The number w_i is taken to be a small integer, usually between 2 and 5; and ΔT is taken to be considerably larger than the average value of δt determined by (4.29).

COMPUTATIONS WITHIN A BLOCK — The next step in the reduction of the high-speed memory requirement is to constrain the next state for any control to be in the same block as the state at which the control is applied. Since by the definition of δt, the next state must always lie in a neighborhood of the present state, this constraint is easy to implement. The reduction in high-speed memory requirement occurs because only one or two values of the optimal cost function per state in the block are needed in the high-speed memory, rather than one value per state in the entire state space as in the conventional method.

In the next two sections, the detailed procedure, first for the general case and next for a special case, will be presented. The section after that will demonstrate that if specialized computations are performed near the boundaries of a block, then the optimal trajectories can have any desired degree of freedom in passing from block to block. Finally, the complete computational procedure will be discussed.

Procedure in the General Case — In the general case, the time interval ΔT is further divided into smaller increments Δt. A set of quantized times is determined

$$t_m = t_0 + (m-1)\Delta T + s\Delta t \ ,$$

where

$$s = 0,1,\ldots,S \ ,$$

$$S\Delta t = \Delta T \ .$$

Typical values of S range from 5 to 15. Optimal control is computed at each quantized $x \in X$ for each quantized t of the form above. The operations at a given x and t take place as follows.

Each admissible control $u^{(k)} \in U$ is applied. For each control the time over which the control is applied is determined as $\delta t^{(k)}$ from (4.29),

$$\delta t^{(k)} = \min_{i=1,2,\ldots,n} \left\{ \left| \frac{\Delta x_i}{f_i(x,u^{(k)},t)} \right| \right\} .$$

The corresponding next state is computed as

$$x^{(k)} = x + f(x,u^{(k)},t)\Delta t^{(k)} \ .$$

The optimal cost function at this state and time $I(x^{(k)}, t+\delta t^{(k)})$ is computed by interpolation in (n-1) state variables and time using the previously computed values at quantized states and at times $t+\Delta t$, $t+2\Delta t$, ... The state variable not used in the interpolation formula is the one for which $\delta t^{(k)}$ takes on the minimum value. If the increment Δt is chosen to be of the same order of magnitude as the average value of Δt, then linear interpolation in time based on values at $t+\Delta t$ and $t+2\Delta t$ is generally sufficient.

If a control $u^{(k)}$ is such that none of the state variables changes, i.e., if $f(x,u^{(k)},t) = 0$, then $x^{(k)} = x$ and $\delta t^{(k)}$ is set equal to Δt.

The resulting next states for a one-dimensional example are shown in Figure 4.5, where $u = \{u^{(1)}, u^{(2)}, u^{(3)}, u^{(4)}\}$. Values of the optimal cost function are known at the points indicated by small dark circles.

Once the optimal cost for each next state has been found, the cost of each control over the interval $\delta t^{(k)}$ is found by evaluating $\ell(x,u^{(k)},t)$, the cost function per unit time. The optimal

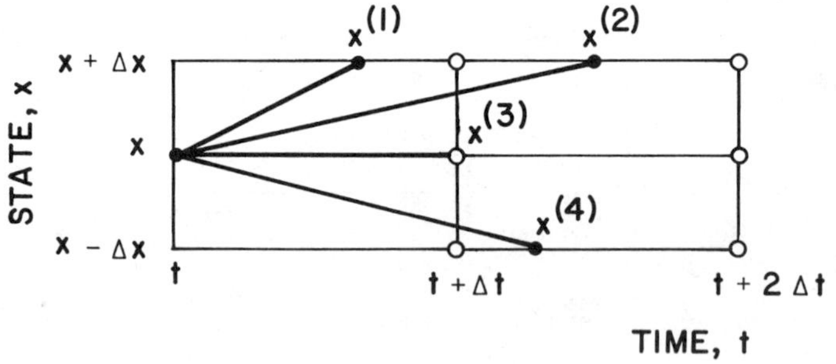

Figure 4.5. Procedure in the General Case for a One-Dimensional Example

cost and optimal control at x, t are then calculated from the principle of optimality

$$I(x,t) = \min_{k=1,2,\ldots,K} \{\ell(x,u^{(k)},t)\delta t^{(k)} + I(x^{(k)},t+\delta t^{(k)})\} ,$$

$$\hat{k} = \arg \min_{k=1,2,\ldots,K} \{\ell(x,u^{(k)},t)\delta t^{(k)} + I(x^{(k)},t+\delta t^{(k)})\} ,$$

$$\hat{u}(x,t) = u^{(k)} ,$$

where $\hat{u}(x,t)$ denotes the optimal control at (x,t).

Procedure With as Many Control Variables as State Variables — When the set of admissible controls is such that each admissible control changes at most one state variable, a modified procedure can be used. The most important case where this situation occurs is where m, the number of control variables, is greater than or equal to n, the number of state variables. When this modified procedure is followed, interpolations take place in one dimension t rather than in n dimensions. There is also a significant reduction in both the high-speed memory requirement and computing time.

An important difference between this procedure and the one just described is that optimal control is not computed at quantized times. Instead, the time at which optimal control is computed is determined as the time at which the next state occurs minus δt, the increment over which the control is applied.

In order to obtain the reduction in interpolations, the set of admissible controls is taken to be the following

$$U = \left\{u^{(-n)}, u^{(-n+1)}, \ldots, u^{(-1)}, u^{(0)}, u^{(1)}, \ldots, u^{(n)}\right\} ,$$

where for $k = 1,2,\ldots,n$, control $u^{(k)}$ changes only coordinate k in the positive direction, control $u^{(-k)}$ changes only coordinate k in the negative direction, and where control $u^{(0)}$ changes no coordinate.

If computations are performed at state x, δt, the time increment over which control $u^{(k)}$, $k \neq 0$, is applied, becomes

$$\delta t^{(k)} = \frac{\Delta x_i}{f_k(x, u^{(k)}, t^{(k)})}$$

$$\delta t^{(-k)} = \frac{-\Delta x_i}{f_k(x, u^{(-k)}, t^{(-k)})}, \qquad k = 1, 2, \ldots, n .$$

The next state for each of these controls is computed from

$$x^{(k)} = x + \epsilon_k \Delta x_k$$

$$x^{(-k)} = x - \epsilon_k \Delta x_k$$

where ϵ_k is the unit vector along the kth coordinate.

The times $t^{(k)}$ and $t^{(-k)}$ are the least values of time at which optimal control has been previously computed for states $x^{(k)}$ and $x^{(-k)}$, respectively. The values of the minimum cost function at these points are denoted as $I^{(k)}$ and $I^{(-k)}$, respectively.

The state x is determined as that state in the block where $t^{(0)}$, the least time at which optimal control has been previously computed for this state, is greater than the corresponding time for any other state in the block.

The time t is determined from

$$t = \min_{k=1,2,\ldots,n} \left\{ \bar{t}^{(k)}, \bar{t}^{(-k)} \right\},$$

where

$$\bar{t}^{(k)} = t^{(k)} - \delta t^{(k)},$$

$$\bar{t}^{(-k)} = t^{(-k)} - \delta t^{(-k)} .$$

Then, the optimal cost at state x, time t, is determined from

$$I(x,t) = \min_{u \in U} \left\{ \ell(x,u^{(k)},t)\delta t^{(k)} + \Delta J^{(k)} + I^{(k)}) , \right.$$

$$(\ell(x,y^{(-k)},\delta t^{(-k)} + \Delta J^{(-k)} + I^{(-k)}) ,$$

$$\left. (\ell(x,u^{(0)},t)(t^{(0)}-t) + I^{(0)}) \right\} ,$$

$$k = 1,2,\ldots,n ,$$

where $\Delta J^{(k)}$ and $\Delta J^{(-k)}$ represent a one-dimensional interpolation in time from t to $\bar{t}^{(k)}$ and $\bar{t}^{(-k)}$, respectively. This interpolation can be regarded as the cost of holding the state fixed at x from t to $\bar{t}^{(k)}$ or $\bar{t}^{(-k)}$. In this case

$$\Delta J^{(k)} = (x,u^{(0)},t(\bar{t}^{(k)}-t) ,$$

$$\Delta J^{(-k)} = (x,u^{(0)},t)(\bar{t}^{(-k)}-t) .$$

In order to prevent the results of interpolations from appearing as portions of optimal trajectories, the interpolation term is subtracted from I and the new optimal control is computed at $\bar{t}^{(k)}$, where $\hat{k}$ is the index of the control for which I is minimized,

$$\hat{u}(x,t^{(\hat{k})}) = \hat{u}^{(k)} ,$$

$$I(x,t^{(\hat{k})}) = I(x,t) - \Delta J^{(k)} , \qquad \hat{k} \neq 0 ,$$

$$= I(x,t) , \qquad \hat{k} = 0 .$$

It is then true that every optimal control goes exactly from one optimal point to another; thus, it is never necessary to interpolate when generating an actual optimal trajectory.

The complete procedure for a one-dimensional example is illustrated in Figure 4.6.

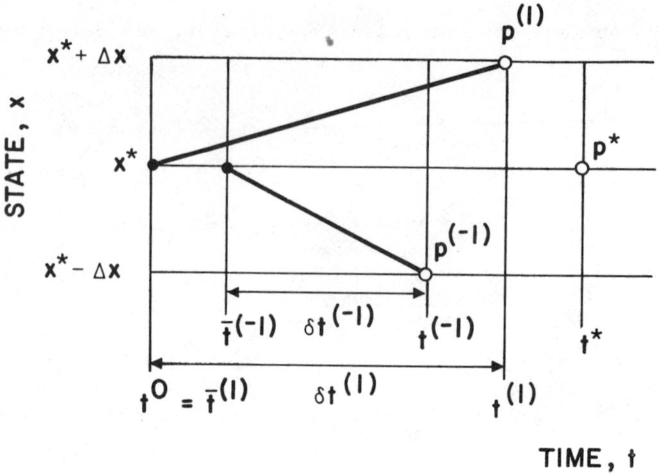

Figure 4.6. Procedure when $q \geq n$ for a One-Dimensional Example

INTERBLOCK TRANSITIONS — In the procedures described so far, it has been assumed that the next state is always in the same block as the present state. In order to allow optimal trajectories to pass from block to block, the computations near the boundaries of the block must be slightly modified. The modifications are similar for both of the previously described procedures but a different modification is used, depending on whether or not the adjacent block at the boundary has been processed.

For two adjacent blocks, one of which has been processed and the other has not, a sequence of optimal points has been computed along the boundary between these blocks. In the computation of the latter block, the optimal points are stored in the high-speed memory, and no new optimal points are computed along the boundary. Values of the optimal cost function for these points are used in evaluating the cost of those controls applied at states within one increment of the boundary for which the next state lies on this boundary. It is then true that an optimal trajectory can pass from a state in the interior of the block being computed to the boundary with a previously computed block. From this boundary the optimal trajectory can then travel further into the previously computed block. In this manner, transitions from a given block to an adjacent block that has been previously processed can take place without constraint, provided that optimal points on the boundary are available.

In most problems of physical significance, it is possible to obtain enough feeling about the system so that it is possible to determine the direction which optimal trajectories are most likely to take. This direction is called the "preferred" direction of motion. In many linear problems, it is known that the preferred direction is toward the origin. In the supersonic aircraft control problem, discussed in a later section, it is known that the aircraft should travel at higher altitudes and higher velocities as it consumes fuel.

When a preferred direction of motion can be defined, adjacent blocks in a given time interval are processed in order such that the preferred direction is from the block processed later to the one processed first. In this manner, interblock transitions in the preferred direction take place without constraint.

Although interblock transitions in the opposite direction will occur less frequently, nevertheless they must be considered. Several procedures are available for implementing these transitions. First, after two adjacent blocks in the same time interval have been processed, it is possible to compute interblock transitions in the nonpreferred direction at $t_m = (m-1)\Delta T$, the least time in the block. These transitions can be compared with the previously computed optimal points on the boundary at this time. In this manner the rate of change in a nonpreferred direction is constrained to be less than $(w_i \Delta x_i)/\Delta T$. The quantity ΔT can sometimes be chosen so that this constraint is not unduly restrictive, and yet a reasonable number of computations per state are obtained in the interval.

If more frequent interblock transitions in the non-preferred direction are desired, it is possible to use values of the optimal cost function into a not-yet-computed block. Since the extrapolation is limited to extend at most Δx_i along any coordinate x_i, the errors introduced by this technique can be bounded. The computations described in the previous paragraph can be used to provide more accurate results every ΔT seconds, and hence, limit the accumulation of errors from the extrapolation process.

Recomputation of optimal points along the boundary between blocks eliminates much of the inaccuracy introduced in these transitions. However, the extra computing time required may be excessive in terms of the improvement obtained, especially if a combination of the other two methods yields satisfactory results.

HIGH-SPEED MEMORY REQUIREMENT — For the conventional dynamic programming procedure, if there are n state variables and N_i quantization levels in the ith state variable, the high-speed memory requirement is

$$N = \prod_{i=1}^{n} N_i \ .$$

For a three-dimensional system with 100 quantization levels in each state

$$N = (100)^3 = 10^6 \ .$$

In the new procedure, however, a much lower number is required. If a block covers $w_i \, \Delta x_i$ units along the ith axis, so that there are (w_i+1) quantization levels in each block, then for the case with $m \geq n$, the high-speed memory requirement is

$$N' = \prod_{i=1}^{n} (w_i + 1) \ .$$

For the general case using linear interpolation in time, twice as many locations ($2N'$) are required. If, in addition, points on the boundaries with previously computed blocks are stored in the high-speed memory, and if there are S optimal points per state in an interval ΔT, then the high-speed memory requirement becomes

$$N'' = 2\left\{\prod_{i=1}^{n} w_i + S\prod_{i=1}^{n}(w_i+1) - \prod_{i=1}^{n} w_i\right\}.$$

In the problem where $N = 10^6$ for the conventional method, if the block size is chosen so that $(w_i+1) = 3$, $i = 1,2,3$, and $S = 5$, then $N'' = 111$. If $(w_i+1) = 4$, $i = 1,2,3$, and $S = 10$, a rather large set of values, then $N'' = 424$, still far less than 10^6 locations.

SUMMARY OF PROCEDURE — The entire state-time space over which optimal control is to be computed is partitioned into blocks according to (4.37). Within each time interval the blocks are ordered so that most interblock transitions will take place in the preferred direction of motion. The blocks at $t_f - \Delta T \leq t - t_0 \leq t_f$ are processed first. If no initial points for a block are given from either the initialization procedure or from previous processing of blocks containing these states, then no computations are made in the block. The blocks in the time interval $(m-1)\Delta T \leq t - t_0 \leq m\Delta T$ are computed on the basis of data obtained at $t = m\Delta T$ by the processing of blocks in the interval $m\Delta T \leq t - t_0 \leq (m+1)\Delta T$. The processing continues until time t_0 is reached.

EXAMPLE

4.3 CONTROL OF A SUPERSONIC AIR TRANSPORT

In this section the application of state increment dynamic programming to the computation of minimum fuel trajectories for a supersonic transport (SST) is considered. This problem is of considerable interest because the cost of fuel is estimated to be as much as 60 percent of the total cost of a flight; hence, knowledge of minimum fuel trajectories is an important factor in determining the economic feasibility of these aircraft. In order to evaluate different configurations, a computer program was developed to compute the optimal trajectory from any possible initial condition using any airframe/engine combination. In order to study a particular configuration, it is necessary only to insert tabular data for the airframe and the engine into the program. The program can be adapted to treat more conventional aircraft, such as the jets currently in use.

The equations of motion used to represent the aircraft are derived from quasi-steady-state assumptions. These assumptions have led to good results in applications very similar to this problem. Under these assumptions, the forces on an aircraft with an angle of attack α flying to an angle γ to the horizontal, as in Figure 4.7, can be resolved into components along and perpendicular to the direction of flight, as follows:

$$F = \frac{W}{G}\ \frac{dv}{dt} = T \cos \alpha - D - W \sin \gamma \ , \tag{4.38}$$

$$0 = L - T \sin \quad - W \cos \gamma \ , \tag{4.39}$$

where

W = weight of aircraft, including fuel,

g = acceleration due to gravity,

v = velocity of aircraft,

h = altitude,

T = engine thrust,

D = drag,

L = lift,

γ = flight path angle with respect to horizontal,

α = angle of attack,

r = range.

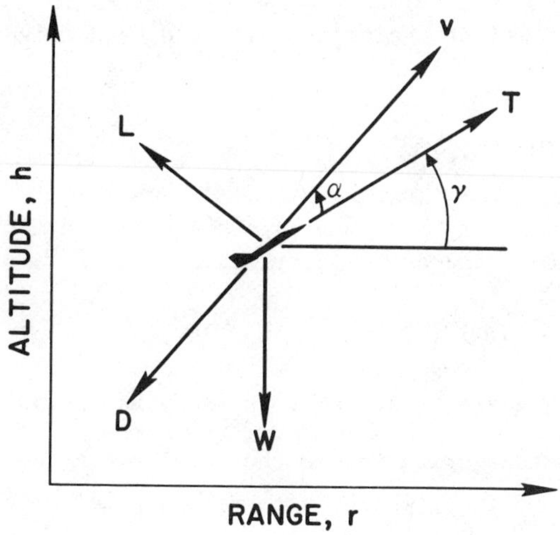

Figure 4.7. Force Diagram for Aircraft

The following differential equations can be written to describe the motion of the aircraft:

$$\frac{dv}{dt} = \frac{g}{W} (T \cos \alpha - D - W \sin \gamma) , \tag{4.40}$$

$$\frac{dh}{dt} = v \sin \gamma , \tag{4.41}$$

$$\frac{dr}{dt} = v \cos \gamma , \tag{4.42}$$

$$\frac{dW}{dt} = W_F , \tag{4.43}$$

where W_F is the fuel flow of the engines. The fuel flow is a function of thrust, velocity, altitude, and weight of the aircraft,

$$W_F = W_F(T,v,h,W) . \tag{4.44}$$

This dependence for a given engine is completely specified by the tabular data for that engine. Similarly, drag can be written as a function of weight, velocity, altitude, and flight-path angle,

$$D = D(W,v,h,\gamma) \ . \tag{4.45}$$

This equation is obtained by noting that lift is a function of velocity, altitude, and angle of attack,

$$L = L(v,h,\alpha) \ , \tag{4.46}$$

and then setting angle of attack to satisfy Eq. (4.39).

The performance criterion is the total amount of fuel consumed during the flight, which is to be minimized. This criterion can be expressed as

$$J = \int_{t_0}^{t_f} W_F(T,v,h,W)dt \ . \tag{4.47}$$

The fuel consumed, J, is to be minimized subject to a number of constraints. The first set of constraints determines the maximum and minimum allowed values of thrust and flight-path angle:

$$0 \leq \alpha_1^-(v,h) \leq T \leq \alpha_1^+(v,h) \ , \tag{4.48}$$

$$\alpha_2^-(v,h) \leq \gamma \leq \alpha_2^+(v,h) \ . \tag{4.49}$$

The functions $\alpha_1^-(v,h)$ and $\alpha_1^+(v,h)$ are specified by the engine data and the functions $\alpha_2^-(v,h)$ and $\alpha_2^+(v,h)$ are specified by the airframe data.

Another set of constraints limits the region of v-h space in which the aircraft can fly. First, there is a maximum operational altitude and maximum operational speed. Second, for a given altitude, if the speed is too low, drag will be greater than maximum thrust; and if speed is too high, aerodynamic heating of the aircraft becomes too great. These constraints can be summarized as:

$$0 \leq h \leq \beta_2^+ , \tag{4.50}$$

$$\beta_1^-(h) \leq v \leq \beta_1^+(h) . \tag{4.51}$$

The functions $\beta_1^-(h)$ and $\beta_1^+(h)$ are determined from considering both the engine and airframe data. The region of the v-h plane which satisfies both Eqs. (4.50) and (4.51) is called the flight envelope. With some airframe/engine combinations, the flight envelope varies with time, and the boundary in Eq. (4.51) must then be constantly recomputed throughout the process.

As part of the study, it was desired to insert still other constraints, such as maximum sonic boom and air traffic control restrictions. It was also desired to consider flights under variable weather conditions, including temperature and wind variations; emergency conditions, including loss of one or more engines, loss of cabin pressurization, or loss of trim drag control; and other conditions.

The aircraft is further constrained to fly to some specified set of terminal values of range, altitude, and velocity. These values can correspond to the entrance to the air traffic control zone at the destination airport, to the aircraft landed on the ground at the destination, or to any other desired condition.

The formulation of the problem according to state increment dynamic programming proceeds as follows. The system equations (4.40) through (4.43) are reduced to two equations by considering the range, r, to be the independent variable rather than time, t. However, since it is desirable to know the time to the end of the flight, as well as the distance to the end, the differential equations (4.40) and (4.41) are used as the system equations, and the range covered by a given flight is related to the time consumed by Eq. (4.42). The two-dimensional state vector thus consists of velocity and altitude

$$x = \begin{pmatrix} v \\ h \end{pmatrix} . \tag{4.52}$$

The control variables are taken to be thrust and flight path angle:

$$u = \begin{pmatrix} T \\ \gamma \end{pmatrix} . \tag{4.53}$$

The state transformation equation is taken to be as follows:

$$x(t+\delta t) = x(t) + f(x,u,t)\delta t , \tag{4.54}$$

where

$$f(x,u,t) = \begin{pmatrix} \frac{dv}{dt} \\ \\ \frac{dh}{dt} \end{pmatrix} = \begin{pmatrix} \frac{g}{W} (T \cos \alpha - D) - g \sin \gamma \\ \\ v \sin \gamma \end{pmatrix} . \tag{4.55}$$

The last of Eqs. (4.40)-(4.43) is used in the performance criterion

$$J = \int_{t_0}^{t_f} \ell(x,u,\sigma)\, d\sigma = \int_{t_0}^{t} \frac{dW}{dt}\, dt = \int_{t_0}^{t_f} W_f dt . \tag{4.56}$$

The boundaries of the allowed state space, X, are determined by taking the maximum altitude as 90,000 ft, and the maximum velocity as Mach 3.00. The increment sizes are as follows:

$$\begin{aligned} \Delta x_1 &= v = \text{Mach } 0.03 , \\ \Delta x_2 &= h = 1{,}000 \text{ ft} . \end{aligned} \tag{4.57}$$

The blocks are taken to be two increments in velocity by two increments in altitude by ten nautical miles (nm) in range,

$$w_1 \Delta x_1 = 2\Delta v = \text{Mach } 0.04 ,$$

$$w_2 \Delta x_2 = 2\Delta h = 2{,}000 \text{ ft} , \tag{4.58}$$

$$\Delta R = 10 \text{ nm} .$$

Within this range, the flight envelope is determined by Eqs. (4.50) and (4.51). A typical flight envelope is shown as the shaded area of Figure 4.8.

Since there are two control variables and two state variables, the conditions for complete state-transition control are met. Furthermore, a choice of U, the set of admissible controls, which satisfies the conditions for complete state-transition control, has been used in previous aircraft studies with good results. Consequently, this procedure will be used here.

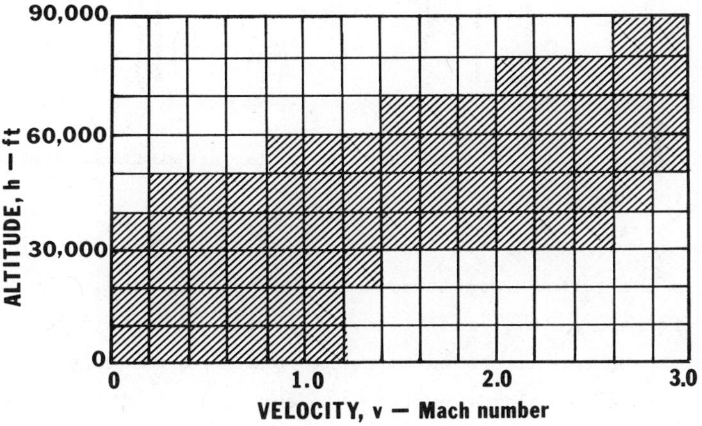

Figure 4.8. Shape of Flight Envelope for a Typical Mach 3 Aircraft

The set of admissible controls consists of the following: climb at constant velocity, dive at constant velocity, accelerate at constant altitude, decelerate at constant altitude, and cruise at constant altitude and velocity. The maximum thrust, $T^+ = \beta^+$ from Eq. (4.50), is used for the controls climb and accelerate; minimum thrust, $T^- = \bar{\beta}$ from Eq. (4.50), is used for the controls dive and decelerate; and thrust is set equal to drag for the control cruise. The flight path angle, γ, is chosen as $\gamma = 0$ for accelerate, decelerate, and cruise; and it is set at that value which keeps $dv/dt = 0$ in Eq. (4.40) for climb and dive. The five admissible control functions are then as follows:

$$U = \left[u^{(-2)}, u^{(-1)}, u^{(0)}, u^{(1)}, u^{(2)}\right]$$

$$= \left\{ \begin{bmatrix} T^- \\ \sin^{-1}\left(\dfrac{T^- \cos\gamma - D}{W}\right) \end{bmatrix}, \begin{bmatrix} T^- \\ 0 \end{bmatrix}, \begin{bmatrix} D \\ 0 \end{bmatrix}, \right. \tag{4.59}$$

$$\left. \begin{bmatrix} T^+ \\ 0 \end{bmatrix}, \begin{bmatrix} T^+ \\ \sin^{-1}\left(\dfrac{T^+ \cos\gamma - D}{W}\right) \end{bmatrix} \right\},$$

where the five controls correspond to dive, decelerate, cruise, accelerate, and climb, respectively.

The set of admissible final states contains one element which specifies the terminal altitude and terminal velocity. The weight of the aircraft at the final time, W_0, is also specified. The terminal state can be chosen to correspond to a landed position, the entrance to a holding pattern, or any other desired condition.

The initial set of optimal points uses a technique which has been extensively investigated. In this technique, if the terminal state is $v(t_f) = 0$, $h(t_f) = 0$, then the optimal trajectory of the aircraft from any state to the terminal state is calculated under the restriction that the only admissible controls are a maximum-rate dive and maximum-rate deceleration. The optimal points so generated are optimal in the sense that at the state where the optimal point in this set is generated, any other admissible control applied at $t \geq \bar{t}$, where $\bar{t}$ is the time of the optimal point, either fails to get the aircraft to the terminal state at $t = t_f$ or has a higher cost than the optimal control. A slight modification is necessary if the terminal state is not the origin.

An optimal point consists of the normal data for an optimal point, plus the time to the end of flight. The index of an optimal point is determined by numbering the points consecutively as they are generated. Since in the case of complete state-transition control the next state is always an existing optimal point, the number of the next optimal point is also listed. In addition, the total weight, W, of the aircraft plus fuel is listed in the optimal point rather than the fuel consumed. The fuel consumed can be found by subtracting from W the quantity W_0, the weight at the final time.

Thus, the information at an optimal point has the following form:

$$p(d) = (a,b,r,t,W,\hat{k},c) \, , \tag{4.60}$$

where

$v = a\Delta v$ = velocity at point,

$h = b\Delta h$ = altitude at point,

r = range at point,

t = time to the end of flight point,

W = weight of aircraft plus fuel,

= W_0 plus minimum fuel burned in reaching terminal state from point

$\hat{k}$ = index of optimal control; one of $\{-2, -1, 0, 1, 2\}$,

c = number of next optimal point,

d = number of this optimal point.

It is known from previous studies that as fuel is consumed and the weight of the aircraft plus fuel decreases, the optimal altitude and velocity at which the aircraft should fly both increase. Thus, except for the terminal descent portion of the flight, optimal trajectories tend to be in the direction of increasing altitude and increasing velocity. As a result, a preferred direction of motion for the problem is well defined.

The blocks in a given time interval are processed in an order such that interblock transitions in the preferred directions of motion take place without constraint. Specifically, all the blocks at a given altitude are processed consecutively in order of decreasing velocity, and the altitudes are taken in order from the highest to the lowest. Only blocks in the flight envelope are processed. In Figure 4.8, the top row of blocks is computed first, then successively lower rows until the row at $h = 0$ is reached. Each row is processed from right to left, and only shaded blocks are computed.

Within each block, the computations follow the procedure described above. The change in range for control k, $\delta r^{(k)}$, is related to the corresponding change in time, $\delta t^{(k)}$, by

$$\begin{aligned} \delta r^{(k)} &= v\, \delta t^{(k)}, \qquad k = 0, -2, 2, \\ \delta r^{(1)} &= (v + \tfrac{1}{2}\Delta v)\delta t^{(1)}, \\ \delta r^{(-1)} &= (v - \tfrac{1}{2}\Delta v)\delta t^{(-1)}. \end{aligned} \tag{4.61}$$

The interpolation in t is performed by computing the cost of holding altitude and velocity constant over the interval. This amounts to introducing an "incremental cruise" into the flight. This incremental cruise is used for interpolation purposes only, and it is subtracted out from the actual optimal point.

The high-speed memory requirement for a single block is, on the average, 10 locations for each state on the boundary with a previously computed block plus one location for each of the other states in the block, a total of 54 locations. Actually, since this number is so small, all the optimal points along the boundary at the top altitude of a row of blocks are stored in the high-speed memory, then the entire row of blocks can be computed without further reference to the slow memory. At the same time optimal points along the lowest altitude of the row are retained in the high-speed memory to serve as the points on the boundary at the upper altitude of the row of blocks immediately below. Thus, referral to the high-speed memory takes place only once for each row of blocks, and the transfer of relatively few points from the slow memory to the high-speed memory makes possible the computation of a large number of optimal points. The amount of high-speed memory needed to implement this procedure is only about 1,100 locations.

A computer program to carry out these computations was written. Data for a specific aircraft were used in the program, and a portion of the program output is shown in Table 4.1. Part of the optimal trajectory from a specific initial condition is shown in Table 4.2. In both of these figures, the first column is the altitude in thousands of feet; the second column is Mach number in hundredths; the third column is the distance to end of flight in nautical miles; the fourth column is time to end of flight in seconds; the fifth column is weight in pounds of aircraft plus fuel, from which the total fuel consumed can be computed by subtracting 80,000 lbs, the weight of the aircraft with no fuel on board; the sixth column is the index of the present point (not all consecutive numbers appear because this particular program assigns numbers to each possible control, not just the control that results in an optimum point); the seventh column is the index of the next point; and the last column is the optimal control that is to be performed.

On the basis of the running time for this program, a complete specification of all possible optimal trajectories would

Table 4.1

Page of Print-Out from Program with $\Delta h = 2{,}000$ ft, $M = 0.04$ M, $0 \leq h \leq 70{,}000$ ft, $0 \leq M \leq 3.00$, Weight of Empty

Alt.	Mach x 10^{-2}	Dist.	Time	Weight	PPT	NPT	Opt.Cont.
60	288	482.4	1218.7	89178.1	9104	9057	Climb
58	288	484.0	1222.3	89256.6	9108	9104	Climb
60	260	514.2	1306.2	90329.6	9112	9065	Climb
60	256	522.0	1325.4	90564.2	9116	9075	Climb
50	252	502.1	1285.2	90605.9	9122	8418	Cruise
54	252	503.0	1287.3	90471.5	9124	8958	Dive
54	256	499.4	1276.2	90328.0	9125	8950	Dive
52	260	500.2	1278.2	90343.1	9216	8950	Decel.
50	260	501.3	1280.8	90510.7	9127	8954	Decel.
68	292	486.3	1226.5	88973.9	9135	8825	Cruise
68	288	496.2	1247.6	89264.3	9158	9019	Climb
70	284	501.8	1260.0	89402.8	9162	9019	Accel.
66	280	490.3	1236.3	89339.2	9164	9022	Climb
68	276	494.3	1245.5	89445.9	9169	9022	Accel.
70	268	496.9	1256.0	89595.8	9172	8854	Accel.
68	268	502.1	1268.2	89746.0	9174	9172	Climb
64	292	492.3	1237.3	89160.4	9187	9041	Cruise
62	296	484.8	1223.0	88991.4	9193	9103	Cruise
62	284	488.0	1231.1	89313.7	9202	9061	Climb
64	280	493.1	1242.9	89441.9	9206	9164	Climb
62	280	495.5	1248.2	89535.7	9210	9206	Climb
64	276	498.0	1253.8	89611.3	9215	9206	Accel.
62	276	500.4	1259.4	89706.1	9218	9215	Climb
64	252	524.2	1331.0	90580.6	9222	9079	Climb
62	252	527.4	1339.2	90700.7	9226	9222	Climb
60	296	486.6	1227.0	89073.8	9232	9193	Climb
58	296	488.2	1228.8	89139.9	9241	8899	Cruise
60	284	489.9	1235.5	89399.7	9245	9202	Climb
58	284	491.6	1239.3	89479.8	9249	9245	Climb
60	280	497.6	1252.9	89623.5	9253	9210	Climb
58	280	499.3	1256.7	89705.3	9257	9253	Climb
60	252	530.2	1346.2	90811.3	9261	9226	Climb
56	284	493.0	1242.6	89557.8	9266	9249	Climb
54	284	494.3	1251.9	89716.9	9271	8646	Cruise
68	296	487.8	1228.4	88957.8	9276	9144	Accel.
70	296	484.0	1220.9	88827.8	9281	9147	Cruise
68	300	485.1	1221.5	88861.8	9282	9156	Dive
70	292	493.5	1241.6	89085.5	9286	8995	Cruise
70	300	487.5	1228.2	88845.4	9294	9281	Decel.
66	288	499.6	1255.1	89376.4	9296	9158	Climb
68	284	506.2	1269.8	89534.2	9298	9162	Climb
66	284	509.7	1277.6	89648.6	9302	9298	Climb
66	276	498.0	1254.0	89565.4	9306	9169	Climb
70	276	493.6	1245.0	89394.7	9310	9025	Accel.
70	280	512.0	1282.7	89674.6	9312	9162	Accel.
66	268	506.1	1277.7	89874.5	9314	9174	Climb
68	264	510.6	1288.3	89988.7	9319	9174	Accel.
70	264	511.1	1289.6	89963.9	9322	9172	Accel.
66	264	514.9	1298.4	90122.6	9234	9319	Climb
62	300	487.8	1225.9	89015.6	9330	9037	Cruise
62	292	494.6	1245.0	89250.2	9334	9244	Cruise
64	300	489.6	1230.1	88947.5	9338	9047	Cruise
64	296	495.6	1243.6	89139.4	9343	9181	Cruise

Table 4.2. Portion of an Optimal Trajectory as Determined by Program in Table 1.1

ALT.	MACH x 10^{-2}	DIST.	TIME	WEIGHT	PPT	NPT	OPT. CONT.
56	272	508.6	1278.3	90089.2	9443	9410	Climb
58	272	507.1	1274.7	90008.9	9410	9405	Accel.
58	276	504.2	1268.1	89876.7	9405	9401	Climb
60	276	502.5	1264.1	89794.0	9401	9218	Climb
62	276	500.4	1259.4	89706.1	9218	9215	Climb
64	276	598.0	1253.8	89611.3	9215	9206	Accel.
64	280	593.1	1242.9	89441.9	9206	9164	Climb
66	280	490.3	1236.3	89339.2	9164	9022	Climb
68	280	486.7	1228.4	89224.0	9022	9012	Accel.
68	284	479.4	1212.6	89007.9	9012	8847	Accel.
68	288	472.1	1196.1	88793.9	8847	8825	Accel.
68	292	465.0	1180.7	88582.1	8825	8821	Accel.
68	296	458.0	1165.7	88372.8	8821	8682	Cruise
68	296	448.6	1145.8	88201.4	8682	8667	Cruise

take about 1 hour of computing time for each 1,000 nm in range on the IBM 7090.

Reduction of the Dimensionality of High-Speed Memory Requirement from n to m — If $m < n$, i.e., if there are fewer control variables than state variables, then a substantial saving in high-speed memory requirement can be achieved by transforming the state space so that the control vector changes only m of the state variables. As a result, $n - m$ of the state variables in the next state can be fixed and values of minimum cost at the next state need be stored only as a function of m of the state variables.

The conditions under which the transformation can be made are closely related to conditions for controllability. For the

the scalar control case (m=1), the transformation consists of rewriting (4.1) in the form

$$x_1(k+1) = x_2(k) ,$$

$$x_2(k+1) = x_3(k) ,$$

$$\vdots$$

$$x_{n-1}(k+1) = x_n(k) ,$$

$$x_n(k+1) = \Phi_n[x_1(k),\ldots,x_n(k),u(k)] .$$

For fixed values of $x_2(k),x_3(k),\ldots,x_n(k)$, the state variables $x_1(k+1),x_2(k+1),\ldots,x_{n-1}(k+1)$ are also fixed. Thus at these fixed values, optimal control at all values of $x_1(k)$ can be computed by storing the minimum costs as a function of $x_n(k+1)$ only. The point at which optimal control is computed at stage k and the corresponding points at stage k+1 for which the minimum cost must be stored in high-speed memory are shown, for an example where m = 1, n = 2 in Figure 4.9.

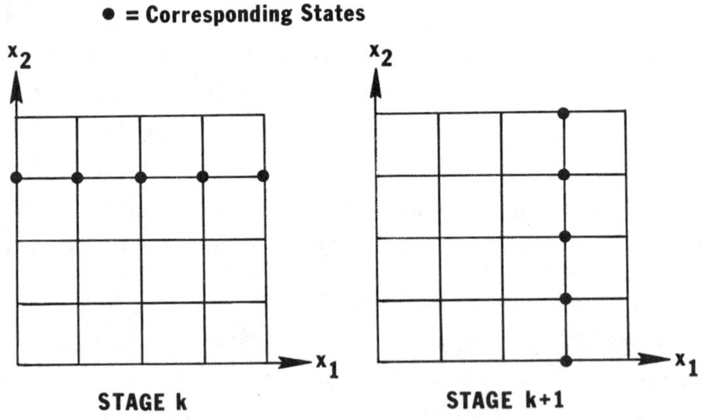

Figure 4.9. States at Stage k+1 for Which I(x,k+1) Must Be Available in Order to Compute Optimal Control for the Indicated States at Stage k

The case of multiple input systems $(m>1)$ is somewhat more complex, although the same basic principles can be applied.

To illustrate the general idea, consider the case $m=2$ and assume the state variables are separated into disjoint sets $\{x_1,\ldots,x_k\}$, $\{x_{k+1},\ldots,x_n\}$ such that the system dynamics decompose into the two subsystems

$$\dot{y}_1 = f_1(x,u) ,$$

$$\dot{y}_2 = f_2(x,y) ,$$

where

$$y_1 = (x_1,\ldots,x_k)' ,$$

$$y_2 = (x_{k+1},\ldots,x_n)' .$$

Employing a relabelling of variables, we obtain

$$\begin{aligned}
\dot{x}_1 &= x_2 , \\
\dot{x}_2 &= x_3 , \\
&\vdots \\
\dot{x}_k &= f_1(x_1,x_2,\ldots,x_n,u_1,u_2) , \\
\dot{x}_{k+1} &= x_{k+2} \\
\dot{x}_{k+2} &= x_{k+3} \\
&\vdots \\
\dot{x}_n &= f_2(x_1,x_2,\ldots,x_n,u_1,u_2) .
\end{aligned}$$

It is evident from the above structure that the same technique used for the scalar input case may now be employed. The secret of successful use of the above "shift-vector" procedure lies in finding a suitable transformation of the system state variables to obtain the above system structure. Fortunately, two important cases when this can be easily accomplished frequently arise in practice:

(1) when the system dynamics are given in "autoregressive" form

$$g(x,\dot{x},\ldots,x^{(n)}) = p(u,\dot{u},\ldots,u^{(r)}) ,$$

with g and p appropriate functions, and

(2) when the dynamics are linear

$$\dot{x} = Fx + Gu ,$$

and the pair (F,G) is completely controllable.

In the latter case, an even stronger type of decomposition takes place since it may be shown that a suitable linear transformation of state and control variables will yield new F and G matrices of the following structure

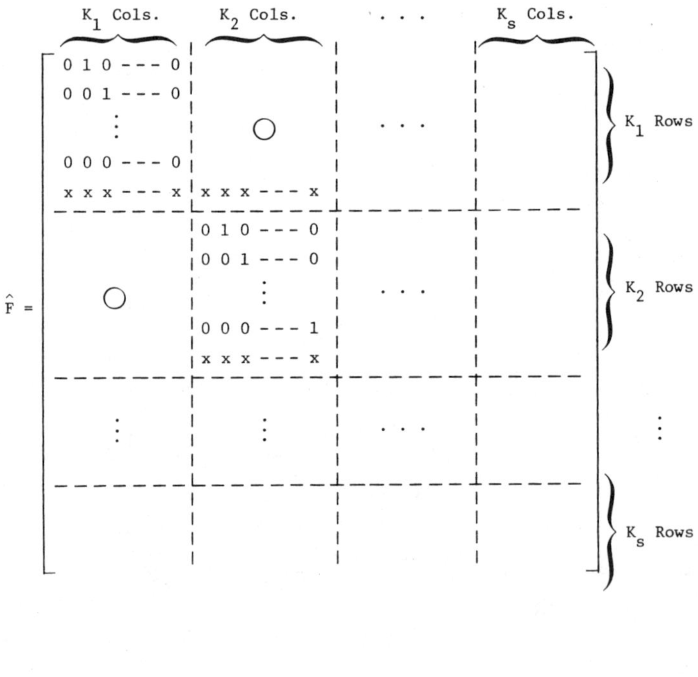

$$\hat{G} = \left[\begin{array}{c|c|c|c} \begin{matrix}0\\0\\\vdots\\0\\1\end{matrix} & \bigcirc & & \bigcirc \\ \hline \bigcirc & \begin{matrix}0\\0\\\vdots\\0\\1\end{matrix} & & \bigcirc \\ \hline \vdots & \vdots & \vdots & \vdots \\ \hline \bigcirc & \bigcirc & & \begin{matrix}0\\0\\\vdots\\0\\1\end{matrix} \end{array}\right] \begin{array}{l} \left.\vphantom{\begin{matrix}0\\0\\\vdots\\0\\1\end{matrix}}\right\} K_1 \text{ Rows} \\ \left.\vphantom{\begin{matrix}0\\0\\\vdots\\0\\1\end{matrix}}\right\} K_2 \text{ Rows} \\ \quad\vdots \\ \left.\vphantom{\begin{matrix}0\\0\\\vdots\\0\\1\end{matrix}}\right\} K_s \text{ Rows} \end{array}$$

where $K_i \geq 0$, $\sum_{i=1}^{s} K_i = n$, and "x" denotes a non-zero number, in general. The integers $\{K_i\}$ are called the Kronecker indices of the pair (F,G) and are determined entirely by F and G.

The importance of the above structure is clear since not only do we obtain a decomposition of the state vector into s blocks of size K_i, $i = 1,\ldots,s$, but the structure of G shows that only one component of the control vector influences each block. Thus, we have only a one-dimensional search which must be carried out in the control space, thereby enabling us to employ the Fibonacci techniques discussed earlier. In addition, such a decomposition of the state space will prove particularly useful if parallel computers are available as each minimization in the control space may be carried out concurrently.

We note, in passing, that if the original dynamics are nonlinear, the Kronecker technique may be used by first linearizing the dynamics about a nominal trajectory-control pair. We shall have more to say about linearization procedures later.

Forward Dynamic Programming — If the minimum cost function is redefined to be the minimum cost to reach a given state and stage from the initial state, an iterative functional equation analogous to (4. 9) can be derived. In this case the calculations proceed forward in k rather than backward.

For many applications this solution is more desirable than the backward dynamic programming solution. This is particularly true when the initial state is fixed and the terminal state and/or stage is free. The optimum final state can be selected by searching over all admissible final states and, if desired, adding a terminal cost function. The terminal cost function can be quite flexible, and the effect of using different functions can easily be seen without repeating the dynamic programming calculation. One class of problems where this procedure has considerable appeal is real-time dispatching in which the initial state is fixed and the human operator is allowed some freedom in determining the final state.

A computational procedure analogous to that given earlier can be used. However, this necessitates inversion of the system equations, i.e., finding $x(k)$ such that

$$\Phi[x(k),u(k),k] = x(k+1) \quad ,$$

for a given $x(k+1)$ and $u(k)$. An alternative procedure is to use the system equations forward in time, establish a tentative minimum cost for each new next state computed, compare this cost whenever another trajectory arrives at the same next state, and store the tentative minimum costs as actual minimum costs when all controls have been applied for all quantized present states.

PROCEDURES FOR OBTAINING THE OPTIMAL TRAJECTORY FROM A SINGLE INITIAL STATE — *Direct Iteration* —In order to begin this approach, all available information about the system is utilized to obtain a nominal trajectory. The set of admissible states X(k) is then adjusted to cover a region about this trajectory, and a normal dynamic programming computation is made. This calculation is usually done at a great saving in both storage and time requirements over the case where X(k) covers the entire admissible state space.

If it is found that the optimal trajectory from the initial state is unsatisfactory, e.g., if it leaves the set X(k) at some stage, then the region is readjusted and a new computation is made. Iterations are performed until a satisfactory trajectory is obtained.

The successful application of a direct iterative procedure depends to a large extent on how close the nominal trajectory is to the true optimal trajectory. In many applications, a good nominal trajectory can readily be found, while in other cases this is a very difficult task.

A direct iterative procedure of particular interest when a good nominal cannot be found is to choose a large initial region for X(k) and to use a coarse quantization in both state and control variables. Successive iterations then decrease the size of the region and the quantizations are made finer.

In a direct approach based on state increment dynamic programming, the region is specified by choosing only certain blocks to be processed. The variation of t according to (4.36) is useful for insuring that the constraints on X(k) are not violated when the region becomes small.

The big danger, of course, in using the coarse grid approach to determine the nominal trajectory is that a very sharp minimum may be missed and instead we obtain a local minimum. For example, in searching for the minimum of the function $f(x)$ depicted in Figure 4.10, over the grid $0, \Delta, 2\Delta, \ldots,$ the sharp global minimum would be missed and we would obtain the local minimum at 2Δ. Although most realistic problems will have a more satisfactory behavior, we must bear such phenomena in mind when contemplating choice of grid size.

QUASILINEARIZATION — Another computational procedure based on the existence of a nominal trajectory is quasilinearization. The basic idea behind this technique is to make a linear expansion of the performance criterion about the nominal trajectory and then to use the closed-form solution discussed in Part A. A number of variations of this method have been studied, and computer programs for implementing many of them are available.

SUCCESSIVE APPROXIMATIONS — In this approach a nominal trajectory is again assumed. For the case $m = n$, i.e., where there are as many control variables as state variables, all but one of the state variables is held fixed. This state variable is then used in a one-dimensional dynamic programming problem, where the control vector is n-dimensional, except that holding the sequences of the other $n-1$ state variables fixed imposes $n-1$ equality

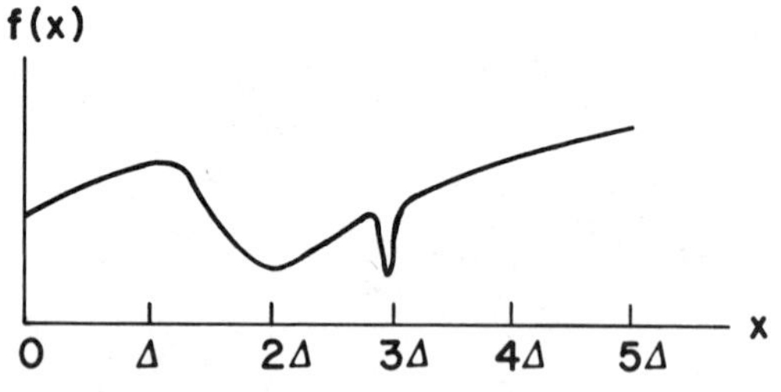

Figure 4.10
Search for a Minimum of f(x)

constraints on the control variables. The performance criterion and constraints remain the same. After this problem has been solved, the optimal sequence of states for this state variable is found, and the procedure is repeated with a different state variable. Iterations continue until convergence is obtained. In this manner the solution of an n-dimensional dynamic programming problem is reduced to solving a sequence of one-dimensional problems.

The philosophy behind the method is to break up a problem with several control variables into sub-problems in which there is only one control variable. Thus, for a given number of state variables, the size of the subproblem, and hence the computational requirements, becomes smaller as the number of control variables is increased. The method is most clearly explained when there are as many control variables as state variables, i.e., the case of complete state-transition control discussed above. For clarity, application of the method to this case will be discussed first, and the extension to other problems will be indicated later in this section.

For the discrete-time case the system equations are

$$x(k+1) = g[x(k),u(k),k] ,$$

the performance criterion is

$$J = \sum_{k=0}^{K} \ell[x(k),u(k),k] ,$$

the constraints are

$$x \in X(k) ,$$
$$u \in U(x,k) ,$$

and the initial state is given by

$$x(0) = c ,$$

with the dimension of both x and u taken to be n.

It is now assumed that a nominal trajectory from $x(0) = c$ is given. The sequence of states along this trajectory is denoted as $\{x^{(0)}(k)\}$, where $x^{(0)}(0) = c$.

Next, one of the n state variables, say x_{i1}, is selected. A dynamic programming problem is then solved in which the sequence of all other states, namely $\{x_i^{(0)}(k)\}$, $i \neq i_1$, is held fixed. Thus, the problem has only one state variable, x_{i1}. The control vector, on the other hand, still has n components. However, the constraint that the sequence of (n-1) of the states must remain fixed imposes (n-1) equality constraints on the control vector, so that effectively there is only one control variable as well. The performance criterion remains the same, and the constraints are as above. The solution to this optimization problem is a new control sequence, $\{u^{(1)}(k)\}$, and a new trajectory $\{x^{(1)}(k)\}$.

Next, a different state variable, say x_{i2}, is selected and the above optimization procedure repeated. This continues until every state variable has been selected for optimization at least once. Termination occurs when the same control sequence and trajectory are obtained for optimization with respect to any of the state variables.

The advantage of this procedure is that the solution of an n-dimensional dynamic programming problem is reduced to the solution of a sequence of one-dimensional problems. Thus, the computational requirements in both time and storage increase linearly with n rather than exponentially. The computational savings in high-dimensional problems can be truly astronomical.

The disadvantage of this technique is that convergence to the true optimum cannot be guaranteed in all cases. However, it can be shown that convergence is always monotonic. Also, if the nominal trajectory is sufficiently close to the true optimum trajectory, convergence to this optimum is obtained. Bellman has suggested a number of techniques for improving the likelihood of finding the true optimum, such as starting from a number of different nominal trajectories. In addition, for certain classes of problems convergence to the true optimum can be rigorously proved (see Supplementary Problem No. 4.2).

The same basic ideas are also useful in problems where there are fewer control variables than state variables. The approach is to break up the problem into several subproblems, where each subproblem contains fewer than n state variables and one or more control variables. For example, a problem with 12 state variables and 3 control variables might be broken up into 3 subproblems, each with 4 state variables and 1 control variable.

EXAMPLE

4.4 As an example of this approach, consider the problem with system equation

$$x_1(k+1) = x_1(k) + x_2(k) + u_1(k) ,$$

$$x_2(k+1) = x_2(k) + u_2(k) ,$$

performance criterion

$$J = \sum_{k=0}^{4} [x_1^2(k) + x_2^2(k) + u_1^2(k) + u_2^2(k)] + 2.5[x_1(5) - 2]^2 + 2.5[x_2(5)-1]^2 ,$$

constraints

$$0 \le x_1 \le 2 ,$$

$$-1 \le x_2 \le 1 ,$$

$$-1 \le u_1 \le 1 ,$$

$$-1 \le u_2 \le 1 ,$$

and initial condition

$$x_1(0) = 2 ,$$

$$x_2(0) = 1 .$$

The state variables are quantized in uniform increments of $\Delta x_1 = \frac{1}{2}$ and $\Delta x_2 = \frac{1}{4}$, respectively. The control variables are not quantized, but are allowed to take on any admissible value such that for any given quantized present state the next state is also a quantized state. This eliminates the need for interpolation.

The nominal trajectory is specified in Table 4.3.

Optimization is first performed with x_2 as the state variable. The sequence of states x_1 is held at the values in Table 4.3. The control variable u_1 is adjusted as a function of x_2 so as to maintain this sequence, i.e.,

$$u_1(k) = x_1^{(0)}(k+1) - x_1^{(0)}(k) - x_2(k) .$$

The optimization problem thus has system equation

$$x_2(k+1) = x_2(k) + u_2(k) ,$$

Table 4.3 Nominal Trajectory for Example

k	x_1	x_2	u_1	u_2
0	2	1	-1	-1
1	2	0	-1	0
2	1	0	-1	0
3	0	0	1	0
4	1	0	1	1
5	2	1		

Cost = 18.

performance criterion,

$$J = \sum_{k=0}^{4} \left\{ [x_1^{(0)}(k)]^2 + x_2^2(k) + [x_1^{(0)}(k+1) - x_1^{(0)}(k) - x_2(k)]^2 + u_2^2(k) \right\} + 2.5[x_1^{(0)}(5)-2]^2 + 2.5[x_2(k)-1]^2 ,$$

and constraints

$$-1 \le x_2(k) \le 1$$

$$-1 \le u_2(k) \le 1$$

$$-1 \le x_1^{(0)}(k+1) - x_1^{(0)}(k) - x_2(k) \le 1 .$$

The third constraint, which appears as a state variable constraint on $x_2(k)$, arises from the control variable constraint on $u_1(k)$ and the restriction that the sequence $\{x_1^{(0)}(k)\}$ is maintained.

The result of this optimization is shown in Figure 4.11. As before, the optimal control is shown below and to the right of the quantized state, while the minimum cost is above and to the right. In addition, as an aid to understanding these calculations, other information is included in this figure. Quantized states for which the third constraint is not met are denoted by x's. The sequence $\{x_1^{(0)}(k)\}$ is shown along the k axis. The optimal trajectory in the (x_2,k) plane starting from $x_2(0) = 1$ is traced out on the figure.

The complete new trajectory from $x_1(0) = 2$, $x_2(0) = 1$ is listed in Table 4.4. Note that the sequences $\{x_2(k)\}$ and $\{u_2(k)\}$ are significantly changed from the nominal trajectory. Note also that although the sequence $\{x_1(k)\}$ is still the same, the sequence $\{u_1(k)\}$ has been changed. This latter change results from the effects of the variations in $\{x_2(k)\}$ on the equation for u. Finally, note that the cost has been reduced in one iteration from 18 to 16.34.

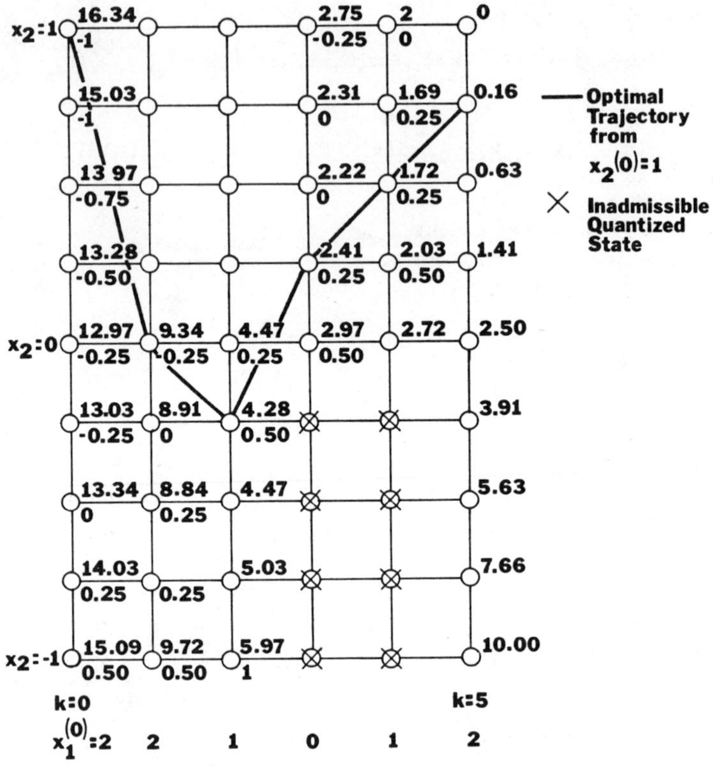

Figure 4.11. Solution to the Optimization Problem Solved on the First Iteration

Table 4.4. Trajectory After One Iteration in Example

k	x_1	x_2	u_1	u_2
0	2.00	1.00	-1.00	-1.00
1	2.00	0.00	-1.00	-0.25
2	1.00	-0.25	-0.75	0.50
3	0.00	0.25	0.75	0.25
4	1.00	0.50	0.50	0.25
5	2.00	0.75		

Cost = 16.34.

On the next iteration, the sequence $\{x_2^{(1)}(k)\}$ is held fixed and the optimization is performed with x_1 as the state variable. For this optimization the system equation is

$$x_1(k+1) = x_1(k) + x_2^{(1)}(k) + u_1(k) ,$$

the performance criterion is

$$J = \sum_{k=0}^{4} x_1^2(k) + [x_2^{(1)}(k)]^2 + u_1^2(k) + [u_2^{(1)}(k)]^2$$

$$+ 2.5[x_2(5)-2]^2 + 2.5[x_2^{(1)}(5)-1]^2 ,$$

and the constraints are

$$0 \leq x_1(k) \leq 2 ,$$

$$-1 \leq u_1(k) \leq 1$$

Because the system equation for $x_2(k)$ does not involve $x_1(k)$, it is true that

$$u_2(k) = x_2^{(1)}(k+1) - x_2^{(1)}(k) = u_2^{(1)}(k) .$$

Thus, in this case the optimization with respect to $x_1(k)$ does not affect the control $u_2(k)$, and there is no constraint on $x_1(k)$ due to constraints on $u_2(k)$.

The results of this optimization appear in Figure 4.12. The resulting trajectory from $x_1(0) = 2$, $x_2(0) = 1$ appears in Table 4.5. In this iteration the cost has been reduced from 16.34 to 15.47.

In the third iteration the optimization is again with x_2 as a state variable. The problem formulation is as before. The only difference between this problem and the optimization problem in iteration 1 is that the sequence $\{x_1(k)\}$ comes from Table 4.5

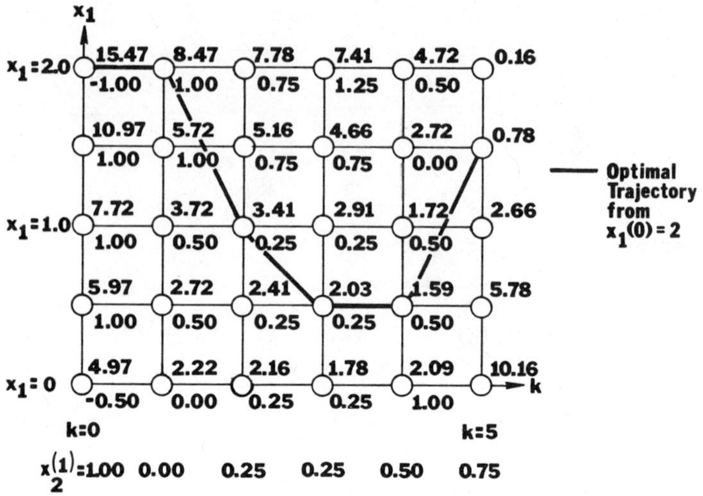

Figure 4.12. Solution to the Optimization Problem on the Second Iteration

Table 4.5. Trajectory After Two Iterations in Example

k	x_1	x_2	u_1	u_2
0	2.00	1.00	-1.00	-1.00
1	2.00	0.00	-1.00	-0.25
2	1.00	-0.25	-0.25	0.50
3	0.50	0.25	-0.25	0.25
4	0.50	0.50	0.50	0.25
5	1.50	0.75		

Cost = 15.47

rather than from the original nominal trajectory. The results of the optimization are shown in Figure 4.13. The new trajectory from $x_1(0) = 2$, $x_2(0) = 1$ appears in Table 4.6.

On the fourth iteration the optimization is with x_2 as the state variable. The results of this optimization appear in Figure 4.14. In this case the same optimial trajectory as on

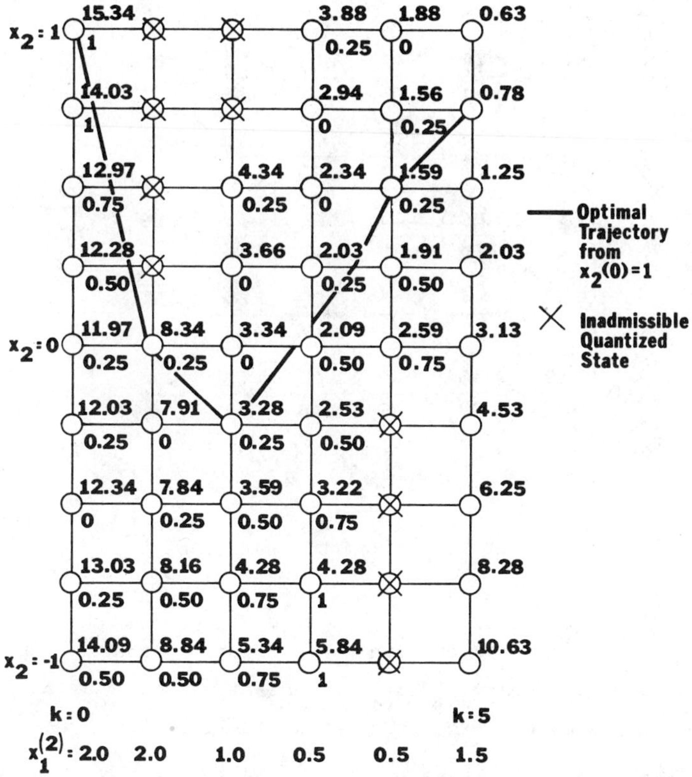

Figure 4.13. Solution to the Optimization Problem Solved on the Third Iteration

Table 4.6. Trajectory After Three Iterations in Example; Optimal Trajectory

k	x_1	x_2	u_1	u_2
0	2.00	1.00	-1.00	-1.00
1	2.00	0.00	01.00	-0.25
2	1.00	-0.25	-0.25	0.25
3	0.50	0.00	0.00	0.50
4	0.50	0.50	0.50	0.25
5	1.50	0.75		

Cost = 15.34

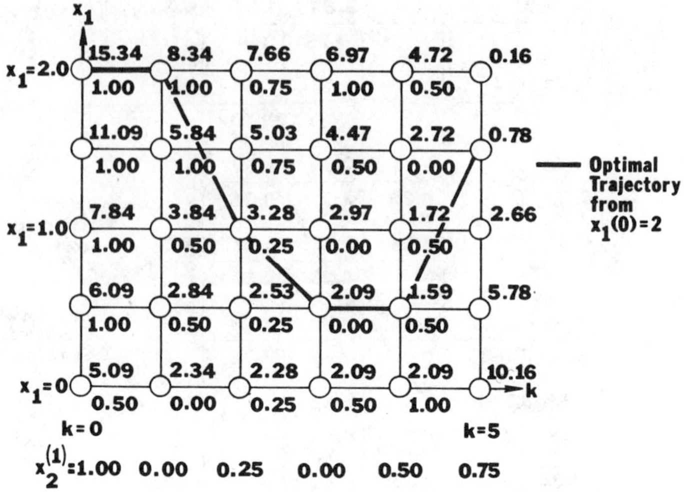

Figure 4.14. Solution to the Optimization Problem Solved on the Fourth Iteration

the previous iteration is obtained. Further iterations do not change this trajectory. Thus in three iterations the procedure has converged to the trajectory shown in Table 4.6.

In this case the problem is of sufficiently low dimensionality that an exact solution can be obtained by the conventional dynamic programming procedure. When the same increment sizes are used, the same optimal trajectory is obtained. Thus, in this problem convergence to the true optimum solution is obtained.

USE OF A LAGRANGE MULTIPLIER — In problems with certain types of constraints, it is possible to eliminate state variables by defining Lagrange multipliers. This reduction in dimensionality makes possible substantial savings in both storage and time, as we have already indicated in Part I.

Procedures for Obtaining a Steady-State Solution

APPROXIMATION IN FUNCTION SPACE — Under certain conditions, e.g., if Φ and L do not explicitly depend on k and if $K \to \infty$, then $I(x,k)$ does not depend on k. The functional equation then becomes

$$I(x) = \min_u \{L(x,y) + I[\Phi(x,u)]\} . \tag{4.62}$$

The problem is thus effectively single state; however, the solution is complicated by the fact that the minimum cost function appears both inside and outside the minimization.

One method of obtaining a solution to this problem is to make an appropriate guess of $I(x)$, say $I^{(0)}(x)$, and solve for a sequence of minimum cost functions according to the relation

$$I^{(j+1)}(x) = \min_u \{L(x,y) + I^{(j)}[\Phi(x,y)]\} . \tag{4.63}$$

If $I^{(0)}(x)$ is a close approximation to the true $I(x)$, this procedure may be expected to converge to the proper function. However, the conditions under which convergence can be guaranteed are quite restrictive.

APPROXIMATION IN POLICY SPACE — An alternative method, which has better convergence properties, is to guess an optimal policy $\hat{u}^{(0)}(x)$. The corresponding minimum cost function $I^{(0)}(x)$ is then computed by a direct iteration according to the relation

$$I^{(0,i+1)}(x) = L[x,\hat{u}^{(0)}(x)] + I^{(0,i)}\{\Phi[x,\hat{u}^{(0)}(x)]\} . \tag{4.64}$$

The initial guess $I^{(0,0)}(x)$ is usually

$$I^{(0,0)}(x) = 0 . \tag{4.65}$$

When $I^{(0)}(x)$ has been found from iteration of (4.33), a new policy $\hat{u}^{(1)}(x)$ is found by solving

$$I^*(x) = \min_u \{L(x,u) + I^{(0)}[\Phi(x,u)]\} . \qquad (4.66)$$

The policy $\hat{u}^{(1)}(x)$ for a given value of x is determined as the value of u for which the minimum is attained in (4.66). However, $I^*(x)$ is not $I^{(1)}(x)$, the minimum cost function corresponding to policy $\hat{u}^{(1)}(x)$, because $I^{(0)}[\Phi(x,u)]$ appears inside the brackets in (4.66). Another direct iteration as in (4.64) is thus necessary.

In general, the minimum cost function $I^{(j)}(x)$ corresponding to the policy $\hat{u}^{(j)}(x)$ is found by iteration of

$$I^{(j,i+1)}(x) = L[x,u^{(j)}(x)] + I^{(j,i)}[\Phi(x,\hat{u}^{(j)}(x))] , \qquad (4.67)$$

with initial guess

$$I^{(j,0)}(x) = 0 . \qquad (4.68)$$

A new policy $\hat{u}^{(j+1)}(x)$ is then formed from

$$I^*(x) = \min_u \{L(x,u) + I^{(j)}[I(x,u)]\} . \qquad (4.69)$$

Convergence to the true optimum can be proved for many important cases.

Distributed Parameter Problems

In this section we wish to discuss some applications of dynamic programming to distributed parameter control processes. We shall see that there is no formal difficulty in extending the basic ideas used in the finite-dimensional case to the infinite-dimensional setting as far as analytic results are concerned. The significant obstacle is dimensionality, which may be treated by a variety of techniques already discussed.

Consider the problem of minimizing

$$J(u,v) = \int_0^T \int_0^1 (u^2 + v^2)\, dx\, dt ,$$

where u and v are connected by the equation

$$u_t = u_{xx} + v ,$$

$$u(x,0) = g(x) , \qquad 0 < x < 1 ,$$

$$u(0,t) = u(1,t) = 0 , \qquad t > 0 ,$$

with an analogous problem for the case where u is ruled by a differential-difference equation.

Stimulated by particular questions of this genre, let us pose the more general problem of minimizing the functional

$$J(u,v) = \int_0^T \left(\int_R g(u,v)\, dA \right) dt ,$$

where $u = u(p)$, $v = v(p)$, $p \in R$, with R a region bounded by B and u and v are related by the equation

$$u_t = L(u) + H(u,v) , \qquad t > 0 ,$$

$$u_{t=0} = c(p) , \qquad p \in R , \tag{4.70}$$

$$u_{p \in B} = g(p) , \qquad t > 0 ,$$

and $L(u)$ is a linear operator with properties we shall bequeath as we need them.

As in the finite-dimensional case, we view this variational problem as a multistage decision process. The state vector may be taken to be the function $c(p)$ with the decision variable the function v.

We introduce the return functional

$$I[c(p),T] = \min_{v} J(u,v) ,$$

and wish to obtain a functional equation for f which will yield the optimal policy.

The functional equation for $I[c(p),T]$ will involve a conventional partial derivative with respect to T as well as a new concept, a functional derivative with respect to $c(p)$. Let us briefly discuss this important tool of functional analysis. Consider the difference quotient

$$D(f,w,\Delta) = \frac{I[c(p) + w(p),T] - I[c(p),T]}{\Delta} ,$$

and suppose that the limit of this quantity exists as $\Delta \to 0$. Let us further assume that the limit is a linear functional of w having the form

$$\int_R \left(\frac{\delta I}{\delta c}\right) w \, dA ,$$

defined in some interval $0 \leq T \leq T_0$, in the region R, and for some class of functions embracing c and w. The coefficient of w is called the functional partial derivative.

We suppose that L is a linear operator such that (4.70) can be converted into the relation

$$u(p,t) = F_1[c(p),t] + F_2[h(u,v),t] ,$$

with F_1 and F_2 linear operators determined by L.

For small t, say $t = \Delta$, we further suppose that

$$F_1[c(p),\Delta] = c(p) + F_3[c(p)]\Delta + \cdots ,$$

$$F_2[h(u,v),\Delta] = h(u,v)\Delta + \cdots ,$$

$$= h[c(p),v]\Delta + \cdots ,$$

where the three dots indicate terms of order Δ^2. Writing

$$\int_0^T = \int_0^\Delta + \int_\Delta^T ,$$

we have, proceeding completely formally,

$$I[c(p),T] = \min_v \left[\Delta \int_R g[c(p),v]\, dA + I[c(p) + F_3[c(p)]\Delta + h[c(p),v]\Delta, T - \Delta] + 0(\Delta) \right] . \tag{4.71}$$

Using the functional partial derivative described above, we may write

$$I[c(p) + F_3[c(p)]\Delta + h[c(p),v]\Delta, T - \Delta]$$

$$= \Delta \int_R \left[F_3[c(p)] + h[c(p),v], \frac{\delta I}{\delta c} \right] dA - \frac{\partial I}{\partial T} + 0(\Delta) .$$

Hence, (4.71) becomes in the limit as $\Delta \to 0$

$$\frac{\partial I}{\partial T} = \min_v \left[\int_R g[c(p),v]dA + \int_R \left(F_3[c(p)] + h[c(p),v], \frac{\partial I}{\partial c} \right) dA \right] .$$

The initial condition is

$$I[c(p),0] = 0 .$$

The minimization with respect to v in some cases may be carried out by calculus. In general, it will require the calculus of variations. When the minimization operation is performed, we obtain an equation of the form

$$\frac{\partial I}{\partial T} = \varphi\left(\frac{\delta I}{\delta c}\right), \quad I[c(p),0] = 0 ,$$

a partial differential-functional equation.

The procedure is readily carried through in detail for the case where the describing equation is

$$u_t = L(u) + v ,$$

and the criterion functional is quadratic in u and v. References to a good deal of work of this nature will be found at the end of this chapter.

The most often used method for dealing with distributed parameter systems is to convert the partial differential equations to a set of ordinary differential equations and then to apply the methods developed for these equations. In theory, one partial differential equation would require an infinite number of ordinary differential equations to represent it exactly. However, in any practical application it is necessary to use an approximation based on a finite number of equations.

A general procedure for making this approximation is to use a lumped model for the distributed parameter system. In this approach all the independent variables except time are quantized, and the space containing these variables is subdivided into small cells. The dependent variables are assumed to be fixed within a cell except for a variation with time. It is then possible to write a set of coupled ordinary differential equations, where the values of the dependent variables in each cell are the state variables for the system. Unfortunately, computational difficulties often arise when a large number of cells are required to represent the distributed parameter system with sufficient accuracy. Nonetheless, this procedure can conceptually be applied to almost any distributed parameter system.

In order to illustrate the general method, attention will be focused on problems with a single dependent variable y, a single control variable u, and two independent variables z and t. It will be assumed for the moment that the highest partial derivative with respect to t is $\partial y(z,t)/\partial t$. This partial differential equation can then be written as

$$\frac{\partial y}{\partial t} = f\left[u,\ z,\ t,\ y,\ \frac{\partial y}{\partial z},\ \frac{\partial^2 y}{\partial z^2},\ \ldots\right], \tag{4.72}$$

where f is a nonlinear scalar functional. The first step in the procedure is to quantize z into n increments of size Δz. These quantized values of z can then be indexed as

$$\begin{aligned} z_i &= z^- + (i-1)\Delta z , \qquad (i = 1,2,\ldots,n) , \\ (n-1)\Delta z &= z^+ - z^- , \end{aligned} \tag{4.73}$$

where

z^- = minimum value of z ,

z^+ = maximum value of z .

The values of y and u at $z = z_i$ and t can then be denoted as $y_i(t)$ and $u_i(t)$, respectively, i.e.,

$$y_i(t) = y(z_i,t) , \tag{4.74}$$

$$u_i(t) = u(z_i,t) .$$

The partial derivatives on the right side of Eq. (4.72) can then be approximated by finite difference equations. If central differences are used then the approximations become

$$\frac{\partial y(z_i,t)}{\partial z} \cong \frac{y(z_{i+1},t) - y(z_{i-1},t)}{2\Delta z} ,$$

$$\frac{\partial^2 y(z_i,t)}{\partial z^2} \cong \frac{y(z_{i+1},t) - 2y(z_i,t) + y(z_{i-1},t)}{(\Delta z)^2} .$$

$$\vdots$$

Using these expressions and the definitions in Eqs. (4.73)-(4.75), the single partial differential equation, Eq. (4.72), is transformed to the n ordinary differential equations

$$\dot{y}_i = f_i(y_1,y_2,\ldots,y_n,u_1,u_2,\ldots,u_n,t) , \quad (i = 1,2,\ldots,n) . \tag{4.76}$$

If an n-dimensional state vector x is defined as

$$x_i = y_i , \quad (i = 1,2,\ldots,n) , \tag{4.77}$$

and if the n-dimensional vector u is defined to have components u_i, $i = 1,2,\ldots,n$, the set of equations in Eq. (4.76) can be written more compactly as

$$\dot{x} = f(x,u,t) .$$

These equations are now in the standard form to apply dynamic programming and other optimization techniques. This procedure can easily be extended to the case where higher partial derivatives are present.

The same approach can be applied to the performance criterion. If the original performance criterion is

$$J = \int_{t_0}^{t_f} \ell\left(u, z, t, y, \frac{\partial y}{\partial z}, \frac{\partial^2 y}{\partial z^2}, \ldots\right) dt ,$$

where ℓ is a scalar functional, then use of Eqs. (4.73)-(4.75) and (4.77) converts the function to

$$J = \int_{t_0}^{t_f} \ell(x,u,t)dt . \qquad (4.78)$$

Similarly, constraints can be put in the form

$$x \in X(t) ,$$

$$u \in U(x,t) .$$

REDUCTION OF THE HIGH-SPEED MEMORY REQUIREMENTS IN THE OPTIMIZATION OF A LUMPED APPROXIMATION TO A DISTRIBUTED-PARAMETER SYSTEM — In many applications involving lumped approximation to partial differential equations, the change in the variables in a given cell depends only on the values of variables in neighboring cells; in fact, the partial differential equation is often derived from such relations in the first place. If such a relation cannot be easily deduced on physical grounds, then it can be obtained by using central differencing techniques involving only values in neighboring cells.

Another condition that is frequently met in practice is that the control, u, is applied not everywhere in the system, but only on a boundary.

If both these conditions are satisfied for the problem described by the partial differential equation in Eq. (4.72), then the corresponding system of differential equations, Eq. (4.78), can be written in the special form

$$\begin{aligned} \dot{x}_1 &= f_1(x_1,x_2,u,t) , \\ \dot{x}_i &= f_i(x_{i-1},x_i,x_{i+1},t) , \qquad (i=2,3,\ldots,n-1) , \\ \dot{x}_n &= f_n(x_{n-1},x_n,t) , \end{aligned} \tag{4.80}$$

where the control u is applied at $z = z^-$.

This structure, which occurs in a great many problems, can be exploited to obtain significant reductions in the high-speed memory requirement. First, the differential equations in Eq. (4.80) are approximated by the equations

$$\begin{aligned} x_1(t+\delta t) &= x_1(t) + f_1[x_1(t),x_2(t),u(t),t]\delta t , \\ x_i(t+\delta t) &= x_i(t) + f_i[x_{i-1}(t),x_i(t),x_{i+1}(t),t]\delta t , \\ &\qquad (i=2,3,\ldots,n-1) , \\ x_n(t+\delta t) &= x_n(t) + f_n[x_{n-1}(t),x_n(t)]\delta t . \end{aligned} \tag{4.81}$$

If δt is a fixed value Δt, as in the conventional procedure, then the following difference equations can be written

$$\begin{aligned} x_1(t+\Delta t) &= g_1[x_1(t),x_2(t),u(t),t] , \\ x_i(t+\Delta t) &= g_i[x_{i-1}(t),x_i(t),x_{i+1}(t),t] , \\ &\qquad (i=2,3,\ldots,n-1) , \\ x_n(t+\Delta t) &= g_n[x_{n-1}(t),x_n(t),t] , \end{aligned} \tag{4.82}$$

where the functions g_i are easily determined from Eq. (4.81) with $\delta t = \Delta t$. This approximation is referred to in the literature as an explicit integration formula.

An alternative state description can be given for this problem: the variables $x_1, x_2, \ldots, x_n$ are replaced by the values of x_n at t, $t+\Delta t$, $t+2\Delta t$, ..., $t+(n-1)\Delta t$. The new state vector $\tilde{x}$ can then be defined to have components

$$\tilde{x}_i(t) = x_n[t+(i-1)\Delta t] , \qquad (i = 1,2,\ldots,n) .$$

The system equations are then

$$\begin{aligned} \tilde{x}_i(t+\Delta t) &= \tilde{x}_{i+1}(t) , \qquad (i = 1,2,\ldots,n-1) , \\ \tilde{x}_n(t+\Delta t) &= \tilde{g}[\tilde{x}(t),u(t),t] . \end{aligned} \tag{4.83}$$

The function $\tilde{g}$ is computed in the following steps. First, it is noted that for a given initial condition, $\tilde{x}(0)$, it is possible to compute $x(0)$ directly. This is due to the nature of the explicit integration formula. In Figure 4.15 the quantized values of z and t for which values of x are used in computing $x_i(t+\Delta t)$ are indicated. These values are $x_{i-1}(t)$, $x_i(t)$, and $x_{i+1}(t)$. The set of all four points forms a triangle, as shown in the figure. If the values $x_i(0)$, $i = 1,2,\ldots,n$, are known,

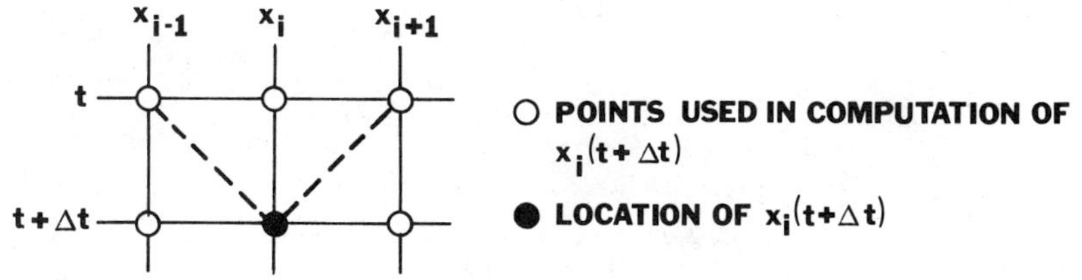

Figure 4.15. Representation of Explicit Integration Formula

then values of $x_i(\Delta t)$ for $i = 2,3,\ldots,n$ can immediately be obtained by use of Eq. (4.82).

Next, $x_i(2\Delta t)$, $i = 3,4,\ldots,n$, can be obtained from $x_i(\Delta t)$, $i = 2,3,\ldots,n$. In general, $x_i(k\Delta t)$, $i = k+1,\ldots,n$, can be determined from the original set $x_i(0)$, $i = 1,2,\ldots,n$. The values of z and t where $x_i(t)$ is known thus form the large triangle in the (z,t) plane shown in Figure 4.16. The desired values of $x_n(t)$, $t = 0, t, \ldots, (n-1)\Delta t$ are on the right-hand side of this initial triangle.

The second step in the computation of $\tilde{g}$ is related to computing additional triangles as t increases. In general, if

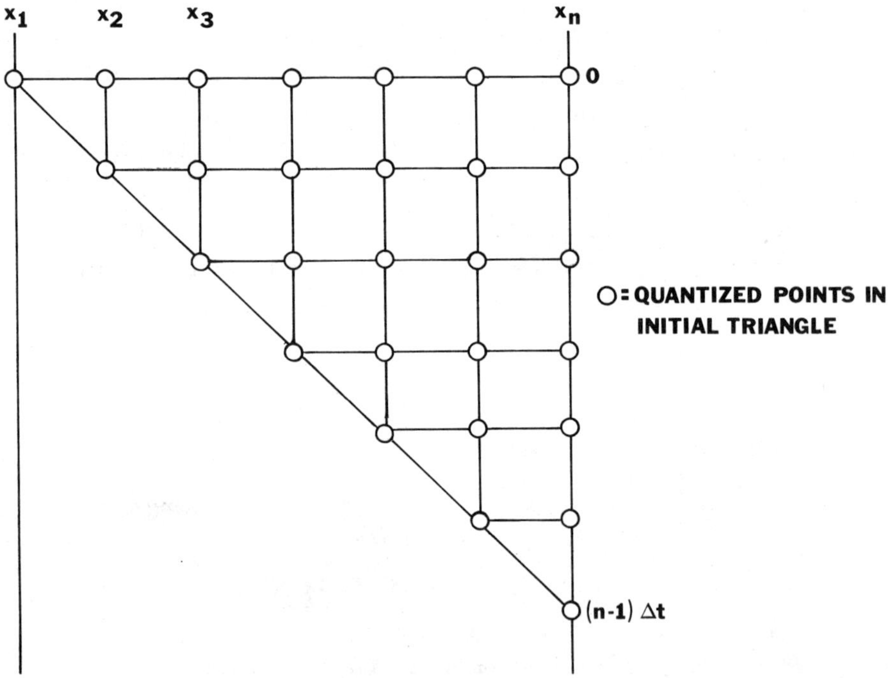

Figure 4.16. Initial Triangle $x_i(t)$ is Determined from x(0)

$x(t)$ is known, a triangle of values, $x_i[t+(i-1)\Delta t]$, $i = 2,3,\ldots,n$, is known. The next triangle is computed for a given value of u by applying the equations

$$x_1(t+\Delta t) = g_1[x_1(t),x_2(t),u(t),t]$$

$$x_i(t+i\Delta t) = g_1[x_{i-1}[t+(i-1)\Delta t],x_i[t+(i-1)\Delta t],$$
$$x_{i+1}[t+(i-1)\Delta t],t+(i-1)\Delta t] \ , \quad i = 2,3,\ldots,n-1 \ ,$$

$$x_n(t+n\Delta t) = g_n[x_{n-1}[t+(n-1)\Delta t],x_n[t+(n-1)\Delta t],t+(n-1)\Delta t] \ . \tag{4.84}$$

The quantities on the right-hand side of Eq. (4.84) are obtained either from points within the present triangle or else from previously computed points along the new triangle. The computations are pictured in Figure 4.17, where the quantized points in the triangle at t are connected by solid lines, and the quantized points in the new triangle are connected by a dotted line.

Finally, when the new triangle has been computed, the desired value $x_n(t+\Delta t) = x_n(t+n\Delta t)$, has been computed as the last of Eqs. (4.84).

The entire set of operations in Eq. (4.83), once the initial triangle has been computed, involves about the same amount of computations as Eq. (4.82). However, the form of Eq. (4.83) has some special advantages. In particular, for a given set of values $\tilde{x}_2(t),\tilde{x}_3(t),\ldots,\tilde{x}_n(t)$, the values of $\tilde{x}_1(t+\Delta t),\tilde{x}_2(t+\Delta t),\ldots,\tilde{x}_{n-1}(t+\Delta t)$ are immediately determined. The single control variable thus affects only the variable $\tilde{x}_n(t+\Delta t)$. As a result, in searching for the minimum cost at the next state, $I[\tilde{x}(t+\Delta t),t+ t]$, it is necessary only to bring in a one-dimensional array of values, namely the set with values $\tilde{x}_i(t+\Delta t)$ fixed at $\tilde{x}_{i+1}(t)$, $i = 1,2,\ldots,(n-1)$, but with $\tilde{x}_n(t+\Delta t)$ allowed to vary over its entire range. It is then possible to compute a one-dimensional array of values of $I(\tilde{x},t)$ based on this one-dimensional array of values of

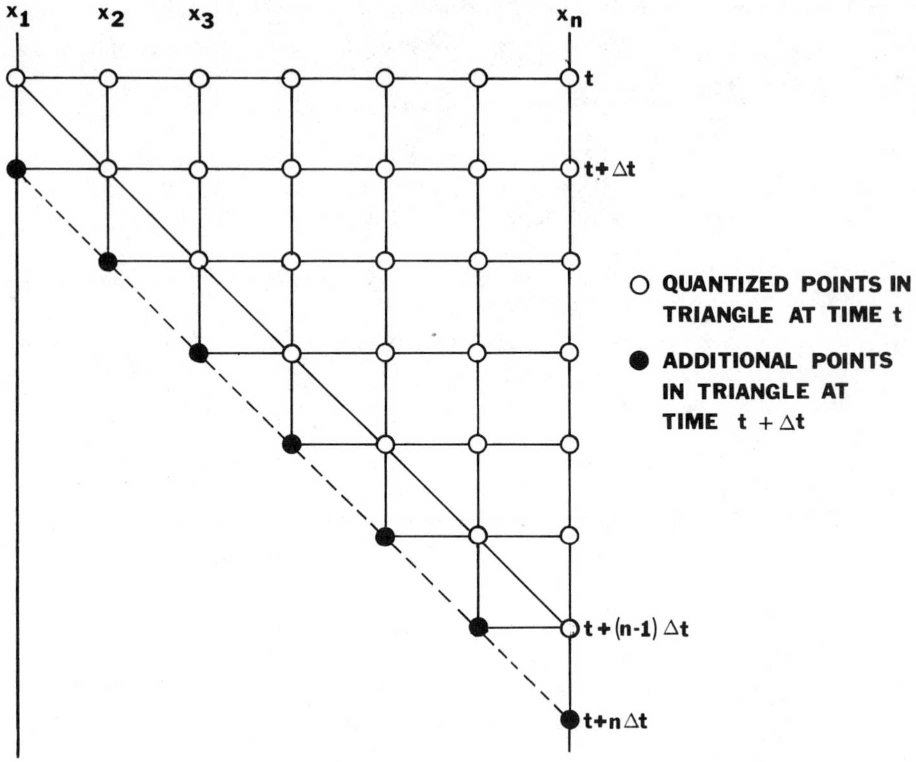

Figure 4.17. Computation of a New Triangle at $t+\Delta t$ Based on the Triangle at t

$I(\tilde{x},t+\Delta t)$. Consequently, the high-speed memory requirement is N_n, the number of quantization levels in the nth coordinate, rather than

$$\prod_{i=1}^{n} N_i = (N_n)^n .$$

As a useful by-product, interpolations of $I(\tilde{x},t+\Delta t)$ are in one dimension rather than in n dimensions.

This result is illustrated in Figure 4.18 for a two-dimensional example. For any state $\tilde{x}(t+\Delta t)$ in the shaded seventh column on

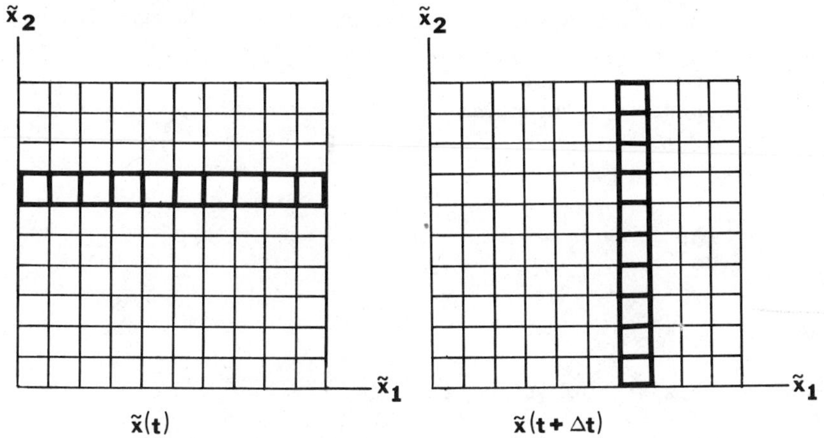

Figure 4.18. Reduction of the High-Speed Memory Requirement in a Two-Dimensional Problem to the Value for a One-Dimensional Problem

the right, the previous state $\tilde{x}(t)$ must lie in the corresponding shaded seventh row on the left; this is true because the fixed value $\tilde{x}_1(t+\Delta t)$ must be the same as $\tilde{x}_2(t)$. This figure also illustrates that the next state lies in the same one-dimensional set for an entire one-dimensional set of present states; in two dimensions, for every next state $\tilde{x}(t+1)$ in a given column, the present state must lie in a corresponding row. Thus, by putting values of the minimum cost at $(t+\Delta t)$ for a one-dimensional set into the high-speed memory, it is possible to determine optimal control and minimum cost at t for a corresponding one-dimensional set.

EXAMPLE

4.5 Short-Term Optimization of Natural Gas Pipeline Networks

In this problem it is desired to find the operating policy for the compressors in a pipeline such that over a 24-hour period, the cost of compressor operation is minimized while the required flow of gas is delivered at a pressure higher than contract pressure.

The partial differential equations for the flow of gas in the pipeline are

$$\frac{\partial P}{\partial t} = -c_1 \frac{\partial Q}{\partial z} ,$$

$$\frac{\partial P^2}{\partial z^2} = -c_2 Q^2 ,$$

where $P(z,t)$ = pressure inside the pipeline,

$Q(z,t)$ = instantaneous mass flow rate in the pipeline,

z = spatial coordinate along the pipeline,

t = time,

c_1, c_2 = constants that depend on the pipeline characteristics.

These equations can be written as the single equation

$$\frac{\partial x}{\partial t} = \varphi(x) \frac{\partial^2 x}{\partial z^2} ,$$

where

$$x(z,t) = P^2(z,t) ,$$

$$\varphi(x) = \frac{c_1}{\sqrt{c_2}} \left(\frac{x}{\left[\frac{\partial x}{\partial z}\right]} \right)^{1/2} .$$

For a single pipeline of length L, with a compressor located at $z = 0$ and a demand point at $z = L$, the boundary conditions are

$$x(0,t) = u^2(t) ,$$

$$\frac{\partial x(L,t)}{\partial z} = -c_2 d^2(t) ,$$

where $u(t)$ = control variable

= pressure generated by the compressor at the input to the pipeline,

$d(t)$ = flow required at the end of the pipeline.

If the pipeline is divided into n sections of length z, as in Figure 4.19, the central difference approximation introduced above can be applied. After suitable manipulation, the system equations for the explicit integration formula become

$$x_1(t+\Delta t) = u^2(t+\Delta t) \; ,$$

$$x_i(t+\Delta t) = x_i(t) + \frac{\Delta t}{(\Delta z)^2}\,\frac{c_1}{\sqrt{c_2}}\left(\frac{(2\Delta z)x_i(t)}{x_{i+1}(t) - x_{i-1}(t)}\right)^{1/2}$$

$$\cdot\; [x_{i+1}(t) - 2x_i(t) + x_{i-1}(t)] \; , \quad i = 2,3,\ldots,n-1 \; ,$$

$$x_n(t+\Delta t) = x_{n-1}(t+\Delta t) - c_2(\Delta z)\; d^2(t+\Delta t) \; .$$

The performance criterion is generally taken to be compressor fuel cost, which is proportional to compressor energy. Under reasonable assumptions this criterion can be written as

$$J = \sum_{t=0}^{K\Delta t} \left(\frac{x_1(t) - x_2(t)}{c_2(\Delta z)}\right)^{1/2} \left[c_3\left(\frac{x_1(t)}{P_0^2}\right)^{c_4} - 1\right] \; ,$$

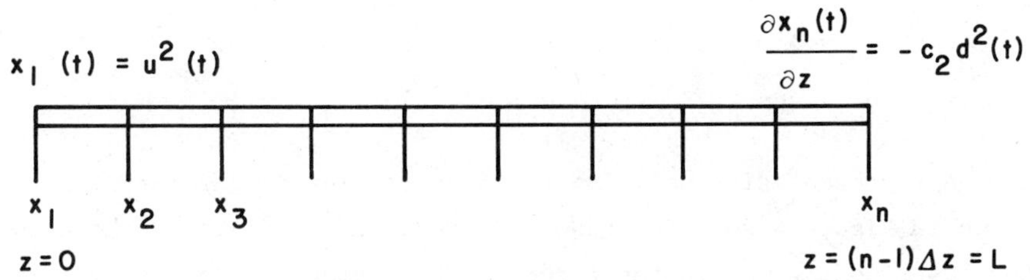

Figure 4.19. A Single Pipeline with Compressor at $z = 0$ and Demand Point at $z = L$

where $K\Delta t$ = time interval over which optimization is performed,

P_0 = suction pressure at input to compressor,

c_3, c_4, c_5 = constants for the specific pipeline.

The constraints are that the flow required is to be supplied at a pressure greater than or equal to contract pressure, P_{con}, and that pressure inside the pipeline can never exceed a value P_M, determined from safety considerations. The constraint on flow required is already incorporated into Eq. (4.85). The constraint on contract pressure can be written as

$$x_n(t) \geq P_{con}^2 .$$

The constraint on maximum pressure in the pipeline is

$$x_i(t) \leq P_M^2 , \quad i = 1,2,\ldots,n .$$

A computer program based on the theory of the preceding section was run for a particular pipeline. The number of sections was determined by setting $n = 4$.* The constants for the program were

$$t = 10 \text{ minutes} , \qquad c_1 = 113.414 ,$$

$$K = 18 , \qquad c_2 = 22.631\ 10^{-9} ,$$

$$L = 15.95 \text{ miles} , \qquad c_3 = 1.0088 ,$$

$$P_{con} = 450 \text{ psi} , \qquad c_4 = 0.1085 .$$

$$P_M = 500 \text{ psi} ,$$

A representative solution for a particular demand curve is shown in Figure 4.20. The flow rate required, $d(t)$, is shown as the lower solid line in the figure. The corresponding input flow

*Because of the way the boundary conditions were handled, the number of state variables in the program was not $n = 4$, but 2.

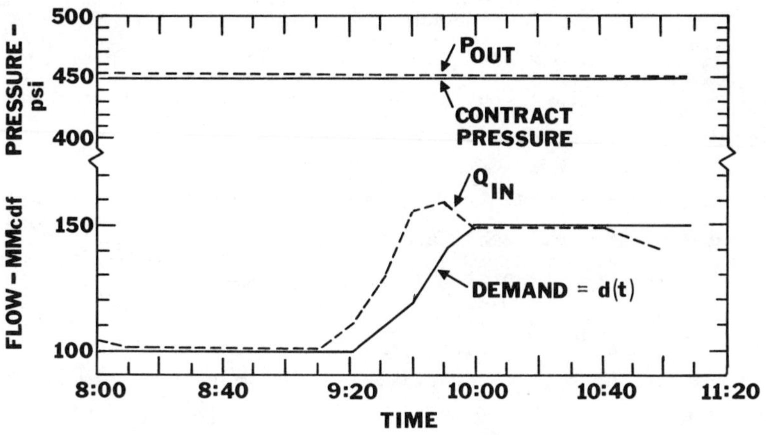

Figure 4.20. Solution to the Pipeline Optimization Problem

rate $Q_{in}(t)$ is shown as the lower dotted line. This latter quantity is plotted rather than input pressure, $\sqrt{x_1(t)}$, because in this program it was preferable for numerical reasons to use input flow rate as the control variable, instead of input pressure.

A very interesting result of this program is that output pressure, $P_{out}(t) = \sqrt{x_n(t)}$, was very close to contract pressure, P_{con}, over the entire run. This same phenomenon was observed in many other cases. Thus, a near optimal policy is clearly to adjust the compressor operation so as to maintain P_{out} as near P_{con} as possible.

Other Computational Considerations

ADAPTIVE DATA PROCESSING. By varying increment sizes, nominal trajectories, and other program inputs, it is possible to cut down substantially the number of computations for a given problem. These variations can either be preprogrammed or made in real time by the programmer/operator.

NEW CONCEPTS IN COMPUTER ARCHITECTURE. Improvements in computer hardware and software are constantly increasing the scope of dynamic programming. Some recent ideas for computer reorganization, especially in the areas of parallel processing and cellular logic, could lead to very large reductions in the computer time and storage required for solving large problems.

CLEVER PROBLEM FORMULATION. Let us explore the use of parallelism in much greater detail and develop some algorithms specifically suited to this type of computing architecture.

Characteristics of a Parallel Computer — Before turning to the algorithms, let us briefly describe what we mean by a parallel computer. For our purposes, a machine with the following characteristics is needed.

It is capable of processing many data streams simultaneously using the same instruction stream. We assume that it is made up of a set of processing elements (PE's), each having the capability of executing logical and arithmetic instructions. The PE's are controlled by the central processing unit (CPU) which broadcasts the same sequence of instructions to all PE's.

Each PE is connected to its own rapid access memory and also to at least two of its nearest neighbors with which it is able to communicate. Figure 4.21 illustrates this overall desired architecture.

The ILLIAC IV machine at the NASA Ames Research Center has the above characteristics, with 64 PE's. However, since it is being designed as a general purpose machine, it is a great deal more complex than is necesary for execution of the algorithms to be described here.

Parallel States, Decisions, and Stages Algorithms — Evaluation of the optimal value function at all states and stages generally involves three nested interactive loops, as follows: (a) iterate over all stages; (b) iterate over all stages at a stage; (c) iterate over all decisions; evaluate right side of iterative Eq. (4.8)

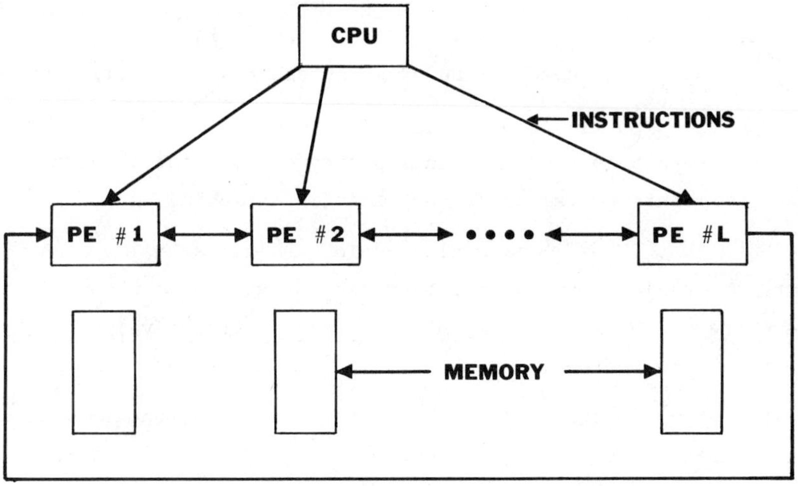

Figure 4.21. Parallel Computer Architecture

using one decision, compare answer with previous value, and save minimum.

The following algorithms each eliminate one of the above loops.*

Parallel States — If the number of processing elements in a parallel machine is equal to or greater than $(QS_i)^{n_i}$ for each i, then each PE can do the computations for one quantized state value. Hence, all the computations at one stage are done in parallel by the following procedure:

(1) Compute right side of Eq. (4.8) using V_{i+1} stored in PE, and save.

---

*In what follows, we let QS_i = the number of quantized states at time i, n_i = the number of state vector components at time i QD_i = the number of quantized controls at time i, m_i = the number of control vector components at time i, $V_i(\bullet)$ = values of the optimal function $I(\bullet,i)$.

(2) Route V_{i+1} to neighboring PE as shown in Figure 4.22.

(3) Compute second value of right side of Eq. (4.8) using new V_{i+1} and compare answer with result of step (1), saving the minimum.

(4) Return to step (2) and repeat procedure for all admissible quantized values of the decision variable.

Note that $V_i(x_i)$ is evaluated in parallel at all states x_i during L evaluations, comparisons, and storage of minimum right side of recursive formula, with L routes of stored values $V_{i+1}(x_{i+1})$ in between.

Depending on the form of the state transition function, if a given quantized state x_i and decision u_i does not yield a

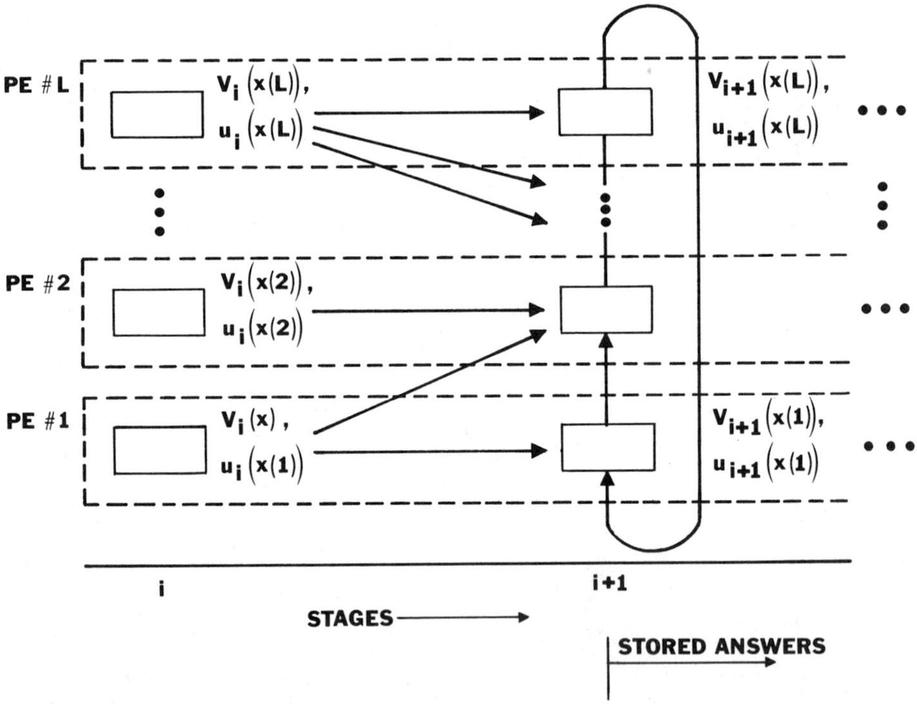

Figure 4.22. Schematic Diagram of Parallel States Algorithm

quantized next state x_{i+1}, then one of the following schemes can be used, depending on the problem being solved: (a) interpolation between two surrounding values of $V_{i+1}(x_{i+1})$ to obtain the proper $V_{i+1}(x_{i+1})$; (b) interpolation between successive decision values to obtain proper quantized next state x_{i+1}; or (c) round off to the next higher or lower quantized state value.

Parallel Decisions — If the number of processing elements in a parallel machine is equal to or greater than $(QD_i)^{m_i}$ for each i, then the optimum decision at each quantized state can be found efficiently by the following parallel scheme:

(1) Each PE computes right side of Eq. (4.8) for one value of the decision variable, all at the same state value, as shown in Figure 4.23.

(2) The optimum decision for that state is then found by the binary route and comparison routine depicted in Figure 4.24.

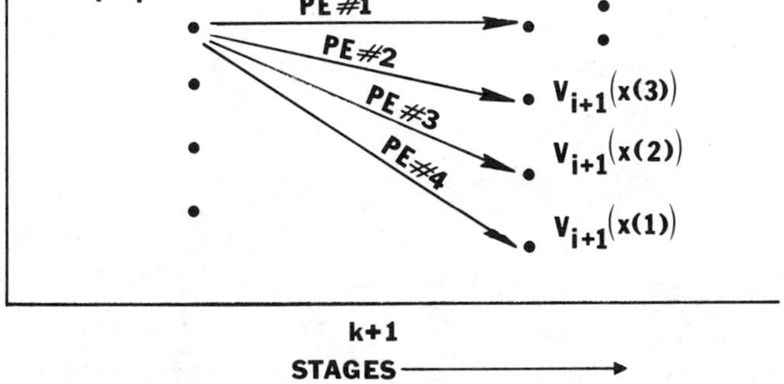

Figure 4.23. Parallel Decisions Algorithm

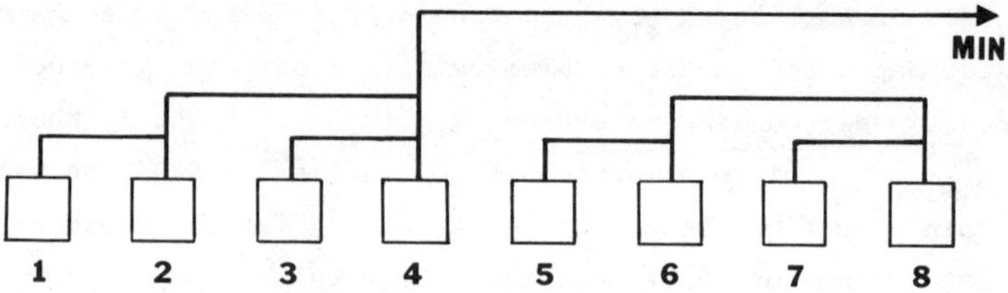

Figure 4.24. Binary Minimization Procedure

Each PE performs routings of +1,+2,+4,... PE's with a comparison and storage of the minimum performance after each routing. When M routes and comparisons have been made, each PE has the optimum performance $V_i(x_i)$ and corresponding decision stored in it.

Comparisons with the Serial Algorithm — It is very easy to make comparisons between the previous two algorithms and straightforward serial computation, since in one case the state loop was eliminated, and in the other the decision loop was eliminated. Figures 4.25 and 4.26 illustrate the time savings of the parallel schemes over serial computation. In both cases, the actual position of the curves is governed by the amount of time necessary to route information between PE's. In the case of parallel states, this amount of overhead simply grows linearly with the number of states. In the parallel decisions scheme, the overhead computation is a function of $\log_2$ (number of decisions) and, hence, diminishes the efficiency of the parallel algorithm more for a small number of decisions than for a large number.

Parallel Stages and States — Replacement of the stage loop in the evaluation of the recursive formula is not as straightforward as the state and decisions loops. However, in certain special cases the following parallel stages and states algorithm may quickly converge to an optimal solution. The required number of PE's is equal to the product of the number of states at a stage times the

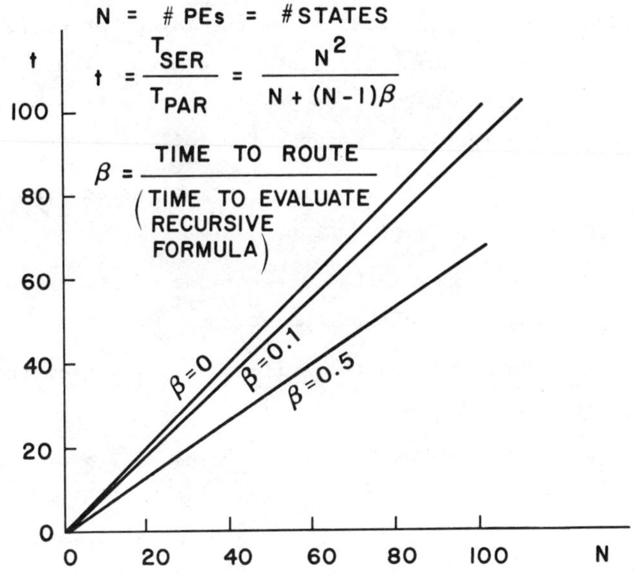

Figure 4.25. Parallel States Time Savings

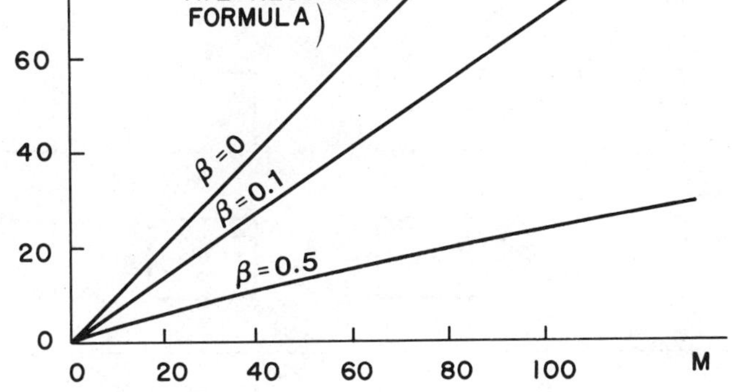

Figure 4.26. Parallel Decisions Time Savings

number of stages. Figure 4.27 shows the layout of PE's, each PE corresponding to one quantized state at a stage. Each PE is connected to its neighbors to which possible state transitions might take place. The algorithm executes the following steps:

(1) Each PE is given a nominal optimal value function $V_i(x_i)$ to start the computation.

(2) Each PE computes the index of one of its neighbors using a quantized decision choice.

(3) Each PE retrieves the nominal $V_i(x_i)$ from the proper neighbor chosen in step (2).

(4) Each PE computes an updated value of its own nominal $V_i(x_i)$ using the right side of Eq. (4.8), compares the answer with the previously stored value, and retains the minimum value.

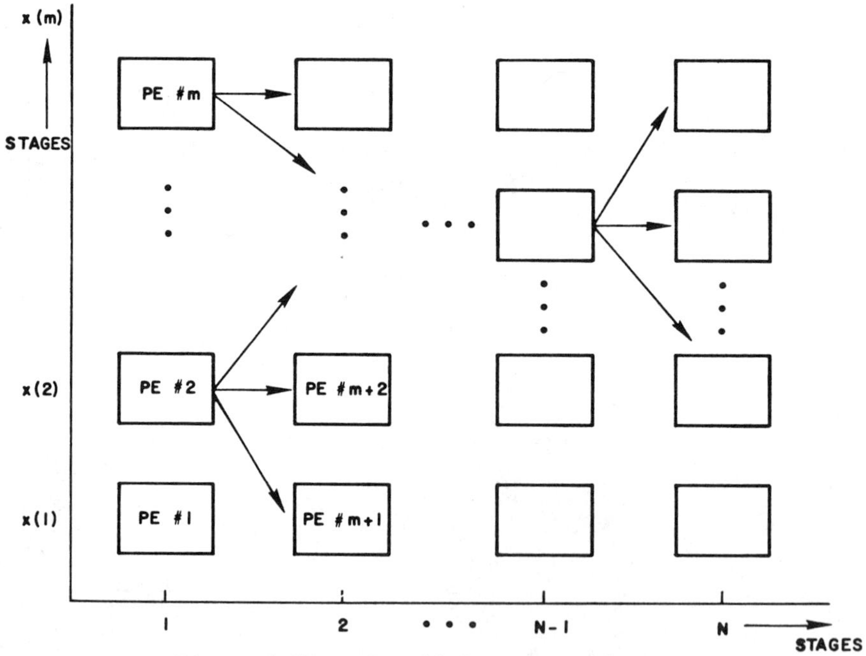

Figure 4.27. Parallel Stages and States

(5) Return to step (2), at least until all decisions are used once, and until the difference between successive stored values of $V_i(x_i)$ is small.

Convergence of the above algorithm to an optimum at all states and stages is guaranteed in at most n cycles through the decision choices. In that time, the optimum answers will have passed from stage n down to stage 1. With well-chosen starting values, convergence may be expected to occur much sooner.

Parallelism with Shift Vectors — Any process with scalar input that can be represented by the following nth-order difference equation (discretized nth-order differential equation)

$$x_{i+n} = f_i(x_{i+n-1},\ldots,x_{i+1},x_i,u_i) ,$$

may also be written in the equivalent shift vector form as we have already seen,

$$x^1_{i+1} = x^2_i ,$$

$$x^2_{i+1} = x^3_i ,$$

$$\vdots$$

$$x^n_{i+1} = f_i(x^1_i,\ldots,x^n_i,u_i) ,$$

where the entire shift vector is $x_i = (x^1_i,\ldots,x^n_i)^T$.

The advantage of using shift vectors with dynamic programming is that the usual n-dimensional problem is reduced to a number of parallel one-dimensional problems. Figure 4.9 illustrates this for a two-dimensional state. The shift vector equations are

$$x^1_{i+1} = x^2_i ,$$

$$x^2_{i+1} = f_i(x^1_i,x^2_i,u_i) , \qquad i = 1,2,\ldots,n .$$

Thus, for any fixed value of x_i^2, the value of x_{i+1}^1 is fixed and consequently all transitions from the line $x_i^2 = C$ lie along the line $x_{i+1}^1 = C$, C a constant. Therefore, to solve Eq. (4.8) along any line at stage i, only the stored data along its corresponding line at stage i+1 is needed in high-speed memory. Furthermore, all parallel lines at stage i, corresponding to all quantized values of x_i^2, can be processed simultaneously. Figure 4.28 depicts this scheme. The same is true for the n-dimensional case except that a PE is needed for each value of the quantized vector $(x_i^2,\ldots,x_i^n)$, as shown in Figure 4.29.

Hence, given a parallel processor with $(QS)^{n-1}$ PE's, the computation time necessary to solve an n-dimensional problem using shift vectors is the same as the time for solving a scalar state problem with the same quantization, plus the slight additional time needed to transpose the stored answers between stages.

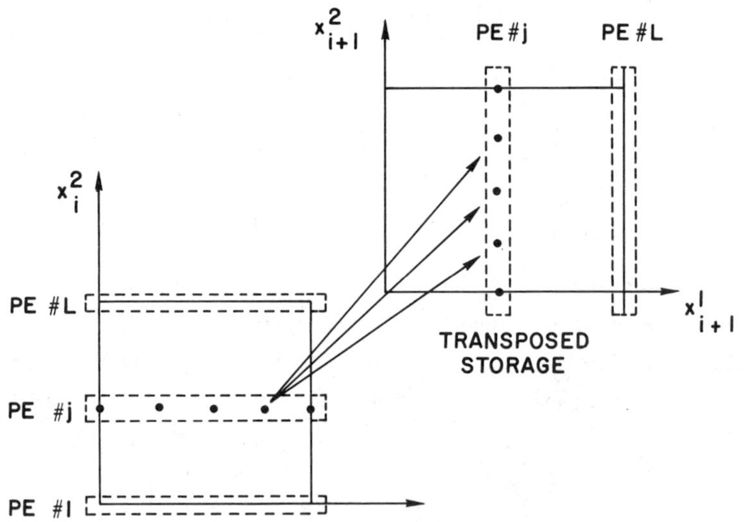

Figure 4.28. Parallel Two-Dimensional Case

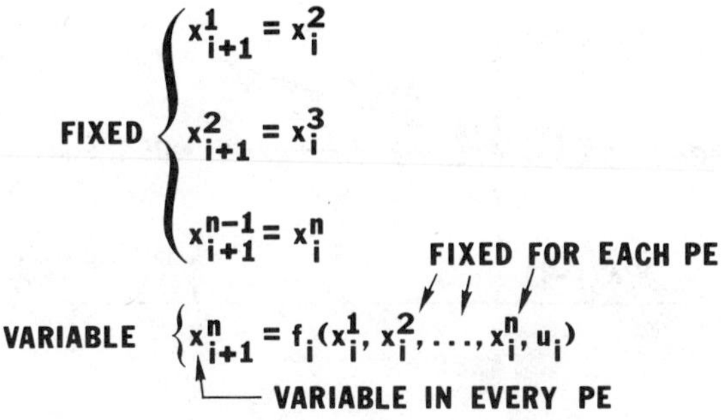

Figure 4.29. Required PE's = $(QS)^{n-1}$ for Parallel Solution in n-Dimensional Case

State Increment Dynamic Programming in Parallel — In order to solve Eq. (4.3) by state increment DP, the time increment δt is allowed to vary until either one increment of change takes place in one of the components of the state, using a particular quantized decision value, or until $\delta t = \Delta t$. This is signified by the formula

$$\delta t(j) = \min_{i=1,\dots,n} [|x^i|/|f^i(x,u(j),t)|,\Delta t] .$$

Hence, all candidate right sides of Eq. (4.8) can be evaluated by interpolating between values of $I(x,t)$ stored at the nearest neighbors.

The state increment method can be used within the parallel stages and states algorithm, which was previously described. The advantage of using the state increment method is that far fewer connections and data transfers between PE's are necessary. The necessary communication is illustrated in Figure 4.30.

Successive Approximation — The successive approximation algorithm successively solves lower-dimensional problems and, thus, improves the value of the performance by optimizing with respect to one or a small number of decision variables at a time. This procedure can be carried out in parallel by allowing each PE to optimize

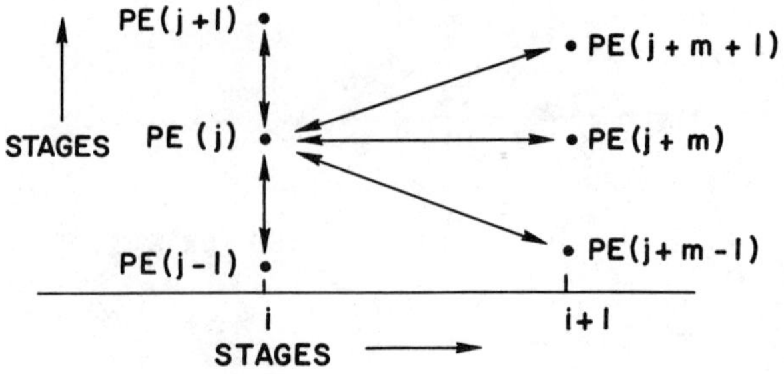

Figure 4.30. Communication Links for State Increment Dynamic Programming in Parallel

with respect to one decision variable or group of decision variables $(U^1,\ldots,U^m)$. Figure 4.31 shows the overall computational process.

The lower-dimensional subproblem which is solved by PE_j is defined as

$$\min_{\{u_i^j\}} \sum_{i=1}^{N} h_i(x_i^j, u_i^j) ,$$

where

$$x_{i+1}^j = f_i^j(x_i^1, x_i^2, \ldots, x_i^j, \ldots, x_i^n, u_i^j) ,$$

$$x_1 = C , \quad x_i^j \in X^j, \quad u_i^j \in U^j .$$

Here, the sequences $\left\{x_i^k,\ i=1,2,\ldots,N, k\neq j\right\}$ are fixed, which imposes n-1 constraints on the sequence $\left\{u_i^k, k\neq j\right\}$, that is,

$$u_i^k = \alpha_i(\{x_i^k\}) ,$$

$$x_{i+1}^k = f_i^k(x_i^1,\ldots,x_i^n,u_i^k), \quad k\neq j .$$

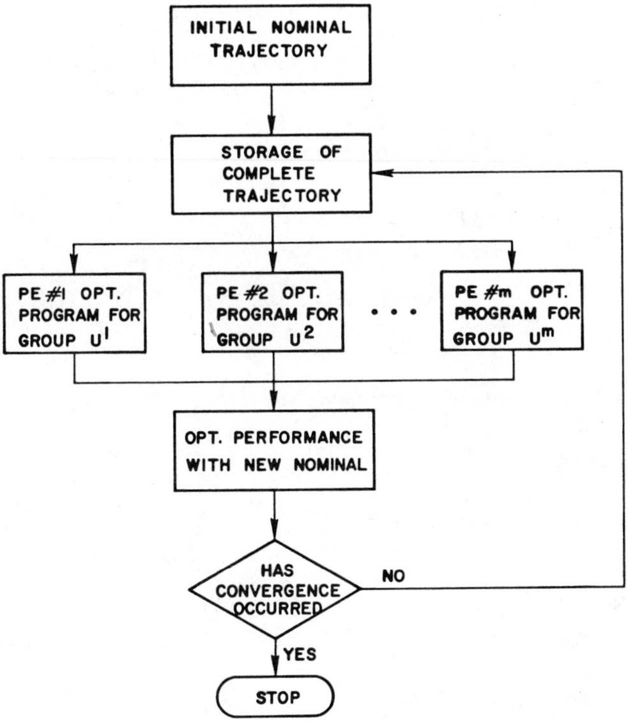

Figure 4.31. Flowchart for Successive Approximations in Parallel

Also, the functions $\{h_i\}$ are determined from the original functions $\{g_i\}$ by fixing the sequences $\{x_i^k, k \neq j\}$.

In the final analysis, careful choice of state, control, and stage variables, selection of an appropriate performance criterion, and use of judicious approximations are the most important factors in the successful application of dynamic programming. Unless the mathematical problem formulation is both a concise and a meaningful representation of the actual situation, the application of sophisticated computational procedures will yield little benefit.

SOLVED PROBLEMS

4.1 Consider the problem of minimizing

$$J = \left(x(T),\ Bx(T)\right) + \int_0^T (u,u)\, dt ,$$

where

$$\dot{x} = Fx + u , \quad x(0) = c .$$

Assume the matrix B has the special form

$$B = \left[\begin{array}{c|c} b_1 & 0 \\ \hline 0 & 0 \end{array}\right] , \quad b_1 > 0 .$$

Derive a <u>scalar</u> Riccati equation characterizing the optimal control law for the problem.

<u>Solution</u>: Following the procedure sketched in the text, we let

$$I(c,T) = \min J ,$$

and, utilizing the linear-quadratic problem structure, we have

$$I(c,T) = \left(c,\ R(T)c\right) .$$

However, in light of the special form of B, only the first component of x enters into the cost. Thus, we write

$$x_1(T) = (d,c) + \int_0^T \left(\alpha(s),\ u(s)\right) ds ,$$

where d = the transpose of the first row of e^{AT},

$\alpha(s)$ = the transpose of the first row of $e^{A(T-S)}$.

Hence,

$$J = b_1 \left[(d,c) + \int_0^T \big(\alpha(s),u(s)\big)\, ds \right]^2 + \int_0^T (u,u)\, dt .$$

To deal with the above problem, we consider the more general problem of minimizing the quadratic form

$$J(r,a) = b_1 \left[r + \int_a^T \big(\alpha(s),u(s)\big) ds \right]^2 + \int_a^T (u,u)\, dt ,$$

where r is a scalar parameter, $|r| < \infty$, and $0 \leq a \leq T$. Let

$$\varphi(r,a) = \min_u J(r,a) .$$

Arguing as before, we easily obtain the equation for φ

$$\frac{\partial \varphi}{\partial a} = \min_v \left[(v,v) + (\alpha(a),v) \frac{\partial \varphi}{\partial r} \right] ,$$

$$\varphi(r,T) = b_1 r^2 .$$

The form of the problem immediately suggests that φ has the structure

$$\varphi(r,a) = r^2 \rho(a) ,$$

where $\rho(a)$ is a function to be determined. Substituting this structure into the equation for φ, we obtain

$$-r^2 \dot{\rho}(a) = \min_v \left[(v,v) + (\alpha(a),v)\, 2r\rho(a) \right] .$$

Carrying out the minimization over v, we have

$$v_{min} = -\alpha(a)\, \rho(a) r .$$

Thus,

$$-r^2 \dot{\rho}(a) = \big(\alpha(a),\alpha(a)\big) \rho^2(a) r^2 - 2r^2 \rho^2(a) \big(\alpha(a),\alpha(a)\big) ,$$

or

$$-\dot{\rho}(a) = -\rho^2(a)\Big(\alpha(a),\alpha(a)\Big) ,$$

$$\rho(T) = b_1 .$$

4.2 It is possible to modify the basic dynamic programming computational procedure so that n "best" solutions, best in the sense of the performance criterion, are obtained, rather than just the single best solution. The concept is to calculate the n lowest minimum costs at each state and stage, rather than just the single minimum cost, and to consider the n minimum costs at each next state when performing the basic minimization. In other words, the basic equation becomes

$$I(x,k;n) = \underset{u,n'}{\mathrm{Min}(n)} \{L(x,u,k) + I[g(x,y,k),k+1]\}$$

where $\underset{u,n'}{\mathrm{Min}(n)}$ denotes the operator which takes the n lowest values of the quantity in brackets, allowing all n' values of the minimum cost at each next state, and where $I(x,k;n)$ denotes the n lowest values of minimum cost at x and k. We find the three best solutions (n=3) for the problem with performance criterion

$$J = \sum_{k=0}^{4} [x^2(k) + u^2(k)] + 2.5[x(5)-2)^2 ,$$

system equation

$$x(k+1) = x(k) + u(k) ,$$

and constraints

$$0 \leq x \leq 2 ,$$

$$-1 \leq u \leq 1 .$$

We quantize in unit increments, so that

$$X = \{0,\ 1,\ 2\}\ ,$$

$$U = \{-1,\ 0,\ 1\}\ .$$

Solution: We have

$$I^{(n)}(\underline{x},k) = \underset{\underline{u},n'}{\mathrm{Min}^{(n)}} \{L(\underline{x},\underline{u},k) + I^{(n')}[\underline{\Phi}(\underline{x},\underline{u},k),k+1]\}\ ,$$

$$n \text{ and } n' = 1,2,\ldots,\bar{n}(=3)\ ,$$

with

$$L(x,u,k) = x^2(k) + u^2(k)\ ,$$

$$\Phi(x,y,k) = x(k) + u(k)\ .$$

Utilizing the $\underset{u,n'}{\mathrm{Min}^{(n)}}(\ldots)$ operation described above, we readily compute the three best trajectories by the usual procedures. The optimal policies and costs are as shown in Figure 4.32.

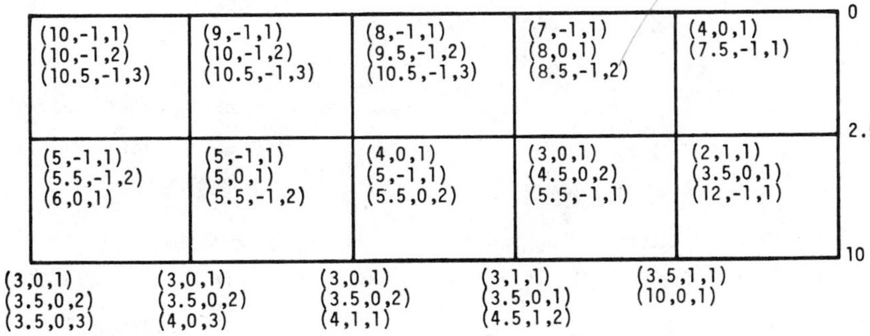

Figure 4.32. Solution to Sample Problem for $\bar{n} = 3$

4.3 The Airline Scheduling Problem. The airline scheduling problem has already been formulated in Volume I. In this problem a computational algorithm for this problem based on the successive approximation procedure is described.

To review, the airline scheduling problem can be formulated as follows.

Given:

(1) A specification of the airline system that includes the flight time for each origin-destination pair.

(2) A set of scheduled flights, for each of which the origin, destination, departure time and net benefit are specified.

(3) A set of allowable ferries (unscheduled empty flights) and their net cost.

(4) A fleet of available aircraft.

(5) A set of constraints on how the aircraft can be operated.

Find:

The assignment of aircraft to scheduled flights and ferries such that the total benefit of scheduled flights minus the cost of ferries is maximized, subject to the constraints.

In Volume I the net benefit of a scheduled flight was computed on the basis of passenger revenue and direct operating cost. Many other types of benefit and cost functions can be used without changing the method.

A representative schedule planning problem is shown in Tables 4.7 to 4.10 and Figure 4.33. The objective in this study was to determine an operating policy such that a minimum number of aircraft, all of the same type was required to meet a given schedule, and that for this number of aircraft the net benefit was maximized.

Table 4.7. Scheduled Flights and Corresponding Benefits

(All flights for a given origin-destination pair are listed in the same column. Origin-destination appears at the head of the column. For each scheduled flight the benefit is listed in the row corresponding to the departure time.

O/D	SAN FRANCISCO						CHICAGO									NEW YORK				
TIME OF DEPARTURE	1/2	1/3	1/4	1/6	1/8	1/9	2/1	2/3	2/4	2/5	2/6	2/7	2/8	2/9	2/10	3/1	3/2	3/4	3/5	3/6
0	100		30	150					100	80		30		55	30		55	150	80	30
1		150									55					150				
2							100						100							
3								55												
4																	55			
5			30												30					
6	100		30		55	55			100	80	55		100	55		150	55	150		30
7			30	150			100	55											80	30
8	100	150	30						100			30					55			
9			30								55			55		150	55	150		30
10			30		55		100	55		80		30								30
11	100		30						100											
12		150	30	150	55	55		55						55		150	55			30
13			30								55				30			150	80	
14			30				100	55					100							30
15			30														55		80	
16		150	30		55			55		80	55	30		55						
17	100		30					55									55			30
18		150	30	150	55	55	100	55					100			150		150		30
19			30						100		55			55			55		80	
20	100		30				100													30
21			30					55			55						55			
22			30									30								

Table 4.7 Scheduled Flights and Corresponding Benefits (Continued)

TIME OF DEPARTURE	NEW YORK 3/7	3/8	LOS ANGELES 4/1	4/2	4/3	4/5	4/6	4/8	4/9	4/10	MIAMI 5/2	5/3	5/4	5/6	5/7	WASHINGTON 6/1	6/2	6/3	6/4	6/5
0			100	150				80				80		80						
1		30								80	80					150		30		80
2	30																			
3																				
4									55										150	
5			30												80		55			
6	30		30		150	125	150					80	125	80				30		
7	30	150	30	100						80							55			
8			30					80			80							30	150	
9			30											80		150				
10			30	100			150	80	55								55	30		80
11			30												80					
12			30	100	150							80				150	55	30	150	
13			30				150				80		125							
14	30		30						55						80					80
15			30		150												55	30		
16			30	100														30		
17			30			125						80					55	30	150	
18	30	150	30		150		150			80	80			80		150		30		
19			30	100								80								
20	30		30								80									
21			30														55	30		80
22			30																	
23			30																	

Table 4.7 Scheduled Flights and Corresponding Benefits (Continued)

TIME OF DEPARTURE	WASHINGTON		DETROIT				SEATTLE					DENVER					KANSAS CITY			
	ORIGIN/DESTINATION																			
	6/7	6/10	7/2	7/3	7/5	7/6	8/1	8/2	8/3	8/4	8/9	9/1	9/2	9/4	9/8	9/10	10/2	10/4	10/6	10/9
0								100		80										
1		80	30														30	80		
2	30												55							
3																				
4					80															30
5						30				80	80			55						
6			30	30				100							80	30		80		30
7	30								150			55					30		80	
8			30	30	80		55						55	55						
9						30					80						30			
10							55												80	30
11				30									55							
12	30					30		100												
13		80		30						80		55			80	30	30			
14			30				55						55							
15														55				80		
16					80															
17	30			30							80		55			30				
18		80						100	150	80	80	55							80	30
19			30												80					
20	30						55						55	55						
21																30				
22																				
23																				

Table 4.8 Flight Times and Costs of Ferries (The flight time in hours is the first number, and the cost of the ferry in arbitrary units is the second number.)

DESTINATION	ORIGIN 1	2	3	4	5	6	7	8	9	10
1		4, -40	6, -60	1, -10	6, -60	6, -60	5, -50	2,-20	2, -20	3, -30
2	4, -40		2, -20	4, -40	3, -30	2, -20	1, -10	4, -40	2, -20	1, -10
3	6, -60	2, -20		6, -60	3, -30	1, -10	1, -10	6, -60	4, -40	3, -30
4	1, -10	4, -40	6, -60		5, -50	6, -60	5, -50	3, -30	2, -20	3, -30
5	6, -60	3, -30	3, -30	5, -50		3, -30	3, -30	7, -70	4, -40	3, -30
6	6, -60	2, -20	1, -10	6, -60	3, -30		1, -10	6, -60	4, -40	3, -30
7	5, -50	1, -10	1, -10	5, -50	3, -30	1, -10		4, -40	3, -30	2, -20
8	2, -20	4, -40	6, -60	3, -30	7, -70	6, -60	4, -40		3, -30	4, -40
9	2, -20	2, -20	4, -40	2, -20	4, -40	4, -40	3, -30	3, -30		1, -10
10	3, -30	1, -10	3, -30	3, -30	3, -30	3, -30	2, -20	4, -40	1, -10	

Table 4.9. Solution to the One-Dimensional Optimization Problem for the First Aircraft

The entry at a given airport and time denotes the maximum profit, the airport from which the optimal flight just came, and the accumulated number of flight houts. Underscored entries specify the optimal flight sequence.

	AIRPORT NUMBER									
TIME	1	2	3	4	5	6	7	8	9	10
0	0. 0. 0	0. 0. 0	0. 0. 0	0. 0. 0	0. 0. 0	0. 0. 0	0. 0. 0	0. 0. 0	0. 0. 0	0. 0. 0
1	0. 1. 0	0. 2. 0	0. 3. 0	30. 1. 1	0. 5. 0	30. 3. 1	30. 2. 1	0. 8. 0	0. 9. 0	30. 2. 1
2	60. 4. 2	60.10. 2	60. 6. 2	30. 4. 1	0. 5. 0	30. 6. 1	30. 7. 1	0. 8. 0	55. 2. 2	30.10. 1
3	60. 1. 2	60. 2. 2	80. 5. 3	80. 8. 3	80. 2. 3	80. 5. 3	90. 3. 3	80. 4. 3	55. 9. 2	50. 2. 3
4	70. 4. 4	110. 9. 4	80. 7. 4	110.10. 4	110. 6. 4	80. 7. 4	90. 7. 3	80. 8. 3	55. 9. 2	110. 6. 4
5	100. 4. 5	110. 2. 4	115. 2. 4	110. 5. 4	110. 5. 4	80. 7. 4	100. 2. 5	80. 8. 3	140.10. 5	110.10. 4
6	160. 2. 6	135. 3. 6	150. 4. 6	150. 3. 6	110. 5. 4	150. 1. 6	105. 3. 6	160. 2. 6	165. 4. 6	140. 2. 5
7	180. 4. 7	135. 7. 6	180. 6. 7	195. 9. 7	170. 7. 6	180. 3. 7	180. 3. 7	160. 8. 6	170.10. 6	195. 9. 7
8	225. 4. 8	225.10. 8	180. 3. 7	210. 1. 8	170. 5. 6	210. 3. 8	210. 6. 8	215. 1. 8	215. 1. 8	195.10. 7
9	240. 4. 9	240. 7. 9	240. 7. 9	255. 1. 9	215. 2. 9	210. 6. 8	255. 2. 9	245. 9. 9	215. 9. 8	215. 2. 9
10	285. 4.10	270. 9.10	280. 2.10	270. 1.10	260. 3.10	285. 7.10	255. 7. 9	245. 8. 9	215. 9. 8	275. 4.10
11	300. 4.11	295. 3.11	315. 6.11	315. 1.11	290. 7.11	310. 3.11	300. 2.11	290. 4.11	305.10.11	275.10.10
12	345. 4.12	340. 6.12	330. 7.12	330. 1.12	290. 5.11	310. 6.11	305. 3.22	340. 1.12	335. 8.12	295. 9.12
13	360. 4.13	360. 9.13	340. 6.12	375. 1.13	365. 6.13	360. 3.13	340. 6.12	350. 4.13	335. 9.12	365. 9.13
14	405. 4.14	385. 3.14	395. 2.14	390. 1.14	365. 5.13	360. 6.13	370. 5.14	350. 8.13	400. 1.14	390. 2.14
15	420. 4.15	400. 7.15	395. 3.14	435. 1.15	365. 5.13	425. 3.15	425. 3.15	415. 9.15	400. 9.14	390. 9.15
16	465. 4.16	455. 9.16	455. 6.16	450. 1.16	420. 3.15	425. 6.15	425. 7.15	415. 8.15	445. 4.16	440. 6.16
17	480. 4.17	480. 6.17	455. 6.16	495. 1.17	440. 6.16	445. 3.17	485. 2.17	415. 8.15	445. 9.16	445. 2.17
18	525. 4.18	480. 2.17	515. 7.18	510. 1.18	475. 3.17	510. 2.18	485. 7.17	520. 1.18	510. 2.18	475. 9.17
19	540. 4.19	510. 3.18	540. 6.19	555. 1.19	535. 2.10	545. 3.19	545. 3.19	520. 8.18	510. 9.18	500. 9.19
20	585. 4.20	575. 7.20	540. 3.19	570. 1.20	535. 5.19	545. 6.19	545. 7.10	580. 1.20	580. 1.20	500. 9.19
21	585. 1.20	575. 2.20	540. 3.19	570. 4.20	535. 5.19	570. 3.20	575. 6.20	580. 8.20	580. 9.20	500.10.19
22	585. 1.20	575. 2.20	540. 3.19	570. 4.20	535. 5.19	570. 6.20	575. 7.20	580. 8.20	580. 9.20	500.10.19
23	585. 1.20	575. 2.20	540. 3.19	570. 4.20	535. 5.19	570. 6.20	575. 7.20	580. 8.20	580. 9.20	500.10.19
24	585. 1.20	575. 2.20	540. 3.19	570. 4.20	535. 5.19	570. 6.20	575. 7.20	580. 8.20	580. 9.20	500.10.19

Table 4.10.* Optimum Schedules for the 47 Aircraft

Aircraft 1

HOUR	ORG	DEST	PROFIT	HRS
0	3	6	30	1
1	6	10	80	4
4	10	9	30	5
5	9	4	55	7
7	4	1	30	8
8	1	4	30	9
9	4	1	30	10
10	1	4	30	11
11	4	1	30	12
12	1	4	30	13
13	4	1	30	14
14	1	4	30	15
15	4	1	30	16
16	1	4	30	17
17	4	1	30	18
18	1	4	30	19
19	4	1	30	20

Total profit = 585

Aircraft 2

HOUR	ORG	DEST	PROFIT	HRS
0	2	10	30	1
1	10	4	80	4
4	4	9	55	6
6	9	10	30	7
7	10	2	30	8
8	2	7	30	9
9	7	6	30	10
10	6	2	55	12
12	2	9	55	14
14	9	2	55	16
16	2	7	30	17
17	7	3	30	18
18	3	7	30	19
19	7	2	30	20

Total profit = 570

Aircraft 3

HOUR	ORG	DEST	PROFIT	HRS
0	3	4	150	6
6	4	1	30	7
7	1	4	30	8
8	4	1	30	9
9	1	4	30	10
10	4	1	30	11
11	1	4	30	12
12	4	1	30	13
13	1	4	30	14
14	4	1	30	15
15	1	4	30	16
16	4	1	30	17
17	1	4	30	18
18	4	1	30	19
19	1	4	30	20

Total profit = 570

Aircraft 4

HOUR	ORG	DEST	PROFIT	HRS
0	2	7	30	1
1	7	2	30	2
2	2	1	100	6
6	1	4	30	7
7	4	10	80	10
10	10	9	30	11
11	9	2	55	13
13	2	6	55	15
15	6	2	55	17
18	2	3	55	19
20	3	6	30	20

Total profit = 550

Legend*
HOUR = Time of Departure
ORG = Origin
DEST = Destination
PROFIT = Net Benefit of a Scheduled Flight or Cost of a Ferry
HRS = Accumulated No. of Hours Flown

Table 4.10. Optimum Schedules for the 47 Aircraft (Continued)

Aircraft 5

HOUR	ORG	DEST	PROFIT	HRS
0	4	3	150	6
6	3	6	30	7
7	6	7	30	8
8	7	3	30	9
9	3	6	30	10
10	7	5	80	13
13	5	2	80	16
16	2	3	55	18
18	3	6	30	19
20	6	7	30	20

Total profit = 545

Aircraft 6

HOUR	ORG	DEST	PROFIT	HRS
0	5	3	80	3
4	3	2	55	5
6	2	9	55	7
8	9	4	55	9
10	4	6	150	15
16	6	3	30	16
17	3	6	30	17
18	6	3	30	18
19	3	2	55	20

Total profit = 540

Aircraft 7

HOUR	ORG	DEST	PROFIT	HRS
0	1	6	150	6
6	6	3	30	7
7	3	7	30	8
8	7	5	80	11
11	5	7	80	14
14	7	2	30	15
16	2	9	55	17
18	9	1	55	19
20	1	4	30	20

Total profit = 540

Aircraft 8

HOUR	ORG	DEST	PROFIT	HRS
0	3	2	55	2
2	2	8	100	6
6	8	2	100	10
10	2	7	30	11
11	7	3	30	12
12	3	2	55	14
14	2	3	55	16
17	3	2	55	18
19	2	6	55	20

Total profit = 535

Aircraft 9

HOUR	ORG	DEST	PROFIT	HRS
0	2	4	100	4
5	4	1	30	5
6	1	9	55	7
8	9	2	55	9
10	2	3	55	11
12	3	6	30	12
14	6	5	80	15
17	5	3	80	18
20	3	7	30	19
22	7	3	30	20

Total profit = 545

Aircraft 10

HOUR	ORG	DEST	PROFIT	HRS
0	2	9	55	2
2	9	2	55	4
6	2	5	80	7
9	5	6	80	10
12	6	7	30	11
13	7	3	30	12
15	3	5	80	15
18	5	2	80	18
21	2	3	55	20

Total profit = 545

Table 4.10. Optimum Schedules for the 47 Aircraft (Continued)

Aircraft 11

HOUR	ORG	DEST	PROFIT	HRS
1	1	3	150	6
7	3	6	30	7
8	6	4	150	13
14	4	9	55	15
17	9	10	30	16
18	10	6	80	19
21	6	3	30	20

Aircraft 12

HOUR	ORG	DEST	PROFIT	HRS
0	5	6	80	3
4	6	4	150	9
10	4	8	80	12
13	8	4	80	15
16	4	2	100	19
22	2	7	30	20

Total profit = 520

Aircraft 13

HOUR	ORG	DEST	PROFIT	HRS
0	1	4	30	1
1	4	10	80	4
4	10	2	-10	5
5	2	10	30	6
6	10	9	30	7
7	9	1	55	9
10	1	8	55	11
12	8	2	100	15
16	2	6	55	17
18	6	10	80	20

Total profit = 505

Aircraft 14

HOUR	ORG	DEST	PROFIT	HRS
1	10	2	30	1
3	2	3	55	3
6	3	2	55	5
8	2	4	100	9
12	4	2	100	13
16	2	4	100	17
20	4	1	30	18
21	1	4	30	19
23	4	1	30	20

Total profit = 530

Aircraft 15

HOUR	ORG	DEST	PROFIT	HRS
0	8	4	80	3
4	4	1	-10	4
5	1	4	30	5
6	4	6	150	11
12	6	3	30	12
13	3	5	80	15
18	5	6	80	18
21	6	2	55	20

Total profit = 495

Aircraft 16

HOUR	ORG	DEST	PROFIT	HRS
1	6	5	80	3
5	5	7	80	6
8	7	2	30	7
9	2	6	55	9
12	6	2	55	11
14	2	1	100	15
18	1	9	55	17
20	9	4	55	19
22	4	1	30	20

Total profit = 540

Table 4.10. Optimum Schedules for the 47 Aircraft (Continued)

Aircraft 17

HOUR	ORG	DEST	PROFIT	HRS
0	1	2	100	4
6	2	6	55	6
8	6	3	30	7
10	3	6	30	8
12	6	1	150	14
18	1	3	150	20

Total profit = 515

Aircraft 18

HOUR	ORG	DEST	PROFIT	HRS
0	4	2	100	4
6	2	4	100	8
10	4	2	100	12
14	2	8	100	16
18	8	4	80	19
21	4	1	30	20

Total profit = 510

Aircraft 19

HOUR	ORG	DEST	PROFIT	HRS
1	6	1	150	6
8	1	2	100	10
12	2	3	55	12
14	3	7	30	13
16	7	5	80	16
20	5	2	80	19

Total profit = 495

Aircraft 20

HOUR	ORG	DEST	PROFIT	HRS
0	8	2	100	4
6	2	8	100	8
10	8	1	55	10
12	1	9	55	12
14	9	10	-10	13
15	10	4	80	16
18	4	10	80	19
22	10	2	30	20

Total profit = 490

Aircraft 21

HOUR	ORG	DEST	PROFIT	HRS
1	2	6	55	2
5	6	2	55	4
7	2	3	55	6
9	3	2	55	8
11	2	4	100	12
15	4	3	150	18
21	3	2	55	20

Total profit = 525

Aircraft 22

HOUR	ORG	DEST	PROFIT	HRS
0	2	5	80	3
6	5	6	80	6
9	6	1	150	12
16	1	8	55	14
18	8	3	150	20

Total profit = 515

Table 4.10. Optimum Schedules for the 47 Aircraft (Continued)

Aircraft 23

HOUR	ORG	DEST	PROFIT	HRS
1	3	1	150	6
7	1	6	150	12
13	6	10	80	15
18	10	9	30	16
19	9	8	80	19

Total profit = 490

Aircraft 24

HOUR	ORG	DEST	PROFIT	HRS
0	4	8	80	3
7	8	3	150	9
13	3	4	150	15
19	4	2	100	19

Total profit = 480

Aircraft 25

HOUR	ORG	DEST	PROFIT	HRS
0	3	5	80	3
6	5	3	80	6
9	3	6	-10	7
10	6	3	30	8
12	3	1	150	14
18	1	6	150	20

Total profit = 480

Aircraft 26

HOUR	ORG	DEST	PROFIT	HRS
1	5	2	80	3
5	2	7	-10	4
6	7	2	30	5
7	2	1	100	9
11	1	2	100	13
16	2	5	80	16
19	5	3	80	19

Total profit = 460

Aircraft 27

HOUR	ORG	DEST	PROFIT	HRS
1	6	3	30	1
2	3	7	30	2
6	7	3	30	3
8	3	2	55	5
10	2	5	80	8
13	5	4	125	13
18	4	3	150	19

Total profit = 500

Aircraft 28

HOUR	ORG	DEST	PROFIT	HRS
2	6	7	30	1
5	7	3	-10	2
6	3	1	150	8
12	1	6	150	14
18	6	1	150	20

Total profit = 470

Aircraft 29

HOUR	ORG	DEST	PROFIT	HRS
6	3	4	150	6
12	4	3	150	12
18	3	1	150	18

Total profit = 450

Aircraft 30

HOUR	ORG	DEST	PROFIT	HRS
1	4	1	30	1
5	1	4	-10	2
6	4	5	125	7
12	5	3	80	10
15	3	2	55	12
17	2	3	55	14
19	3	5	80	17

Total profit = 415

Table 4.10. Optimum Schedules for the 47 Aircraft (Continued)

Aircraft 31

HOUR	ORG	DEST	PROFIT	HRS
5	8	9	80	3
9	9	10	30	4
10	10	6	80	7
13	6	3	-10	8
14	3	6	30	9
15	6	3	30	10
16	3	6	-10	11
17	6	3	30	12
18	3	4	150	18

Total profit = 410

Aircraft 32

HOUR	ORG	DEST	PROFIT	HRS
4	7	5	80	3
8	5	2	80	6
11	2	10	-10	7
12	10	2	30	8
13	2	10	30	9
14	10	9	-10	10
15	9	4	55	12
18	4	6	150	18

Total profit = 405

Aircraft 33

HOUR	ORG	DEST	PROFIT	HRS
6	9	8	80	3
9	8	9	80	6
12	9	8	80	9
17	8	9	80	12
20	9	2	55	14

Total profit = 375

Aircraft 34

HOUR	ORG	DEST	PROFIT	HRS
5	7	6	30	1
6	6	3	-10	2
7	3	8	150	8
14	8	1	55	10
17	1	2	100	14
21	2	6	55	16

Total profit = 380

Aircraft 35

HOUR	ORG	DEST	PROFIT	HRS
6	5	4	125	5
11	4	1	-10	6
12	1	3	100	12
18	3	8	150	18

Total profit = 365

Aircraft 36

HOUR	ORG	DEST	PROFIT	HRS
6	1	2	100	4
10	2	1	100	8
16	1	3	150	14

Total profit = 350

Aircraft 37

HOUR	ORG	DEST	PROFIT	HRS
6	10	4	80	3
10	4	9	55	5
13	9	1	55	7
18	1	8	55	9
20	8	1	55	11
23	1	4	30	12

Total profit = 330

Table 4.10. Optimum Schedules for the 47 Aircraft (Continued)

Aircraft 38

HOUR	ORG	DEST	PROFIT	HRS
6	1	8	55	2
8	8	1	55	4
12	1	4	-10	5
13	4	6	150	11
21	6	5	80	14

Total profit = 330

Aircraft 39

HOUR	ORG	DEST	PROFIT	HRS
7	10	6	80	3
12	6	4	150	9
19	4	1	-10	10
20	1	2	100	14

Total profit = 320

Aircraft 40

HOUR	ORG	DEST	PROFIT	HRS
6	3	7	30	1
8	7	3	-10	2
9	3	4	150	8
15	4	9	-20	10
17	9	2	55	12
20	2	1	100	16

Total profit = 305

Aircraft 41

HOUR	ORG	DEST	PROFIT	HRS
8	1	3	150	6
16	3	6	-10	7
17	6	2	55	9
19	2	4	100	13

Total profit = 295

Aircraft 42

HOUR	ORG	DEST	PROFIT	HRS
6	4	3	150	6
16	3	6	-10	7
17	6	4	150	13

Total profit = 290

Aircraft 43

HOUR	ORG	DEST	PROFIT	HRS
7	3	5	80	3
14	5	7	80	6
17	7	2	-10	7
18	2	1	100	11
22	1	4	30	12

Total profit = 280

Aircraft 44

HOUR	ORG	DEST	PROFIT	HRS
9	3	1	150	6
16	1	4	-10	7
17	4	5	125	12

Total profit = 265

Aircraft 45

HOUR	ORG	DEST	PROFIT	HRS
5	8	4	80	3
8	4	8	80	6
18	8	2	100	10

Total profit = 260

Table 4.10. Optimum Schedules for the 47 Aircraft (Concluded)

Aircraft 46

HOUR	ORG	DEST	PROFIT	HRS
7	6	2	55	2
9	2	9	55	4
12	9	10	30	5
17	10	2	-10	6
18	2	8	100	10

Total profit = 230

Aircraft 47

HOUR	ORG	DEST	PROFIT	HRS
7	4	2	100	4
11	2	7	-10	5
12	7	6	30	6
17	6	7	30	7
18	7	2	-10	8
19	2	9	55	10
21	9	10	30	11

Total profit = 225

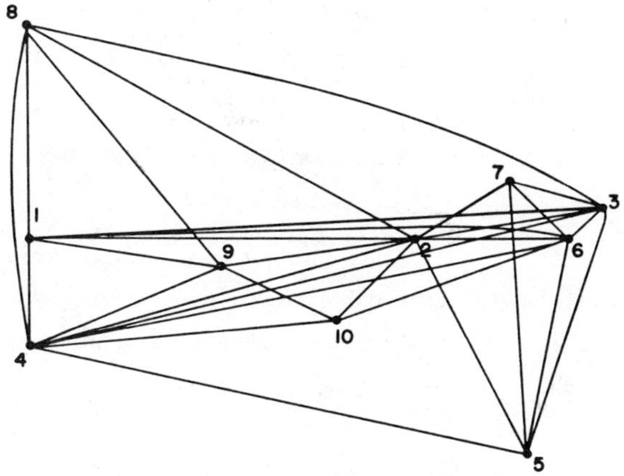

AIRPORTS

1 SAN FRANCISCO	6 WASHINGTON
2 CHICAGO	7 DETROIT
3 NEW YORK	8 SEATTLE
4 LOS ANGELES	9 DENVER
5 MIAMI	10 KANSAS CITY

Figure 4.33. System Map (Allowed origin-destination pairs are indicated by a line between the two airports)

Solution: The airline system in the study contains 10 airports. These airports are shown in Figure 4.33; airport pairs between which there are scheduled flights are connected by lines. The schedule was designed so that 60 aircraft could serve all scheduled flights without any ferries. The complete schedule of 305 flights is shown in Table 4.7. Each column in the table contains entries expressing net benefit, in arbitrary units, of the scheduled flights between a given origin-destination pair. The entry for a particular flight appears in the row corresponding to the time of departure. The flight times, in hours, are shown in Table 4.8. Because ferries are allowed between all airport pairs, this table contains an entry for every possible combination. The cost of a ferry, in the same arbitrary units as the benefit, is taken to be 10 units for each hour of flying time. There are no constraints on the initial or final locations of the aircraft. However, each aircraft is constrained to fly at most 20 hours during the optimization period (24 hours).

The forward dynamic programming algorithm described in this chapter was applied to this problem.

The results of the first one-dimensional optimization problem appear in Table 4.9. Forward dynamic programming was used to generate the table. Entries at each airport and hour specify the maximum benefit, the optimum airport from which the airplane just came (the present airport if the optimum policy was for the aircraft to be on the ground for the previous one-hour time period), and the accumulated flying hours. The latter entry is necessary for imposing the 20-hour limit on flying hours for each aircraft.

Since there is no constraint on the final location of the aircraft, the terminal state is selected on the basis of maximum net benefit. Examination of the bottom row of the table shows that the maximum value appears in the column corresponding to airport 1 (San Francisco). The scheduling can then be recovered by tracking backwards in time on the basis of the previous airport entries in this table. The resulting sequence for this problem is specified by the circled entries in Table 4.9. Note that when optimum flights are longer than one hour, there are no circled entries in the table for this time period; for example, because the circled entry at airport 10 (Kansas City), time 4, specifies the previous airport as airport 6 (Washington), which is a three-hour flight from airport 10, the next entry in the sequence appears at airport 6, time 1, and there are no circled entries at times 2 and 3. Since there is no constraint on the initial location of this aircraft, the first airport in the sequence can be any of the 10; in this case, it happens to be airport 3 (New York).

In this problem it is found that 47 aircraft suffice to meet all scheduled flights. The schedules for all 47 aircraft are shown in Table 4.4. Thus, as a result of this procedure 13 less aircraft are required than originally provided. The number of ferries required is 24, and the cost of the ferries is 250 units (corresponding to 25 hours of flying time). The total net benefit, including the cost of ferries, is 20,850 units; the reduction from 60 to 47 aircraft thus entails a cost in ferries of a little more than 1% of the total net benefit.

There is no guarantee that this solution is the absolute optimum. However, there is no obvious way to modify this solution so as to either reduce the number of aircraft required or to increase the net benefit with the 47 aircraft. The aircraft are clearly being used efficiently; 23 of the aircraft are flown for a maximum of 20 hours, only 15 are flown for less than 18 hours, and the minimum number of hours flown is 10. Improvements over this schedule might be obtained by iterations of a successive approximations procedure, but it is doubtful that the increase in net benefit would be significant.

If the departure times of scheduled flights are allowed to change, it is possible that a reduction in the number of ferries and/or required aircraft might be obtained. Techniques for doing this have been developed and applied to other examples, in some cases with significant improvements. However, these methods have not been tried here.

It is instructive to note the computational requirements of solving this problem according to the standard procedure of Part I. If intermediate states are defined as shown in Part I, a total of 409 discrete states are obtained. If a simultaneous optimization involving 47 aircraft is attempted, the high-speed memory requirement is $(409)^{47} \approx 10^{80}$, a truly astronomical figure. Computing time and the low-speed memory requirement are also prohibitively large. Thus, the computational difficulties of this problem are reduced from totally impossible levels to those of a routine calculation on a medium-size computer.

SUPPLEMENTARY PROBLEMS

4.1 Consider the problem of minimizing

$$J = a_1 x(T)^2 + a_2 \dot{x}(T)^2 + \ldots + a_k x^{(k)}(T)^2 + \int_0^T u^2 \, dt ,$$

where

$$x^{(n)} + b_1 x^{(n-1)} + \ldots + b_n x = u , \qquad x^{(i)}(0) = c_i ,$$

$$i = 0, 1, \ldots, n-1 .$$

Show that this problem can be solved utilizing a matrix Riccati equation of size $(k-1) \times (k-1)$.

4.2 Consider the set of discrete-time optimal control problems for which the number of control variables equals the number of state variables. Show that the dynamic programming successive approximation technique converges to the optimal solution in the following cases:

(a) states are bounded, controls are unbounded, and the elimination of controls by inverting the dynamics yields an objective function which is convex in the states;

(b) the controls are bounded, states are unbounded, and the elimination of states results in an objective function which is convex in the controls and the initial state.

(c) states and controls are bounded and the objective function is quadratic.

HINT: Relate the control problem to a convex or quadratic programming problem.

REFERENCES

Dynamic Programming Fundamentals: [B-15], [B-41], [H-3], [H-23], [J-1], [K-13], [N-4].

New Computational Procedures: [A-22], [B-34], [B-88], [K-6], [K-27], [K-35], [L-5], [L-6], [L-8], [L-14], [L-21], [M-31], [W-22], [W-23], [W-24].

CHAPTER 5

DYNAMIC PROGRAMMING AND OPERATIONS RESEARCH PROBLEMS

INTRODUCTION

Part I of this work has been heavily flavored, in both theoretical and practical aspects, with problems from the realm of what is now termed optimal control theory. No doubt this bias reflects the personal inclinations of the authors; however, the theory of dynamic programming historically arose in an attempt to come to grips with quite a different class of problems — a type of problem now categorized as being within the province of the field of operations research. As a result, in an attempt to present a balanced picture of the subject and its applications, we now turn to this important subset of applied mathematics as further illustration of the scope and content of the theory of dynamic programming.

The types of problems we treat in the present chapter fall into three not particularly distinct groups: allocation problems, scheduling problems, and inventory problems. The names of the classes suggest the flavor of the type of problem included, although the lines of demarcation between categories are not sharply defined and small modifications of the physical situation and mathematical assumptions can easily take a problem from one group to another.

In the multistage allocation processes, we are concerned with the situation in which a certain finite amount of resource x is to be distributed among N activities in order to achieve an optimal return. Some of the examples discussed in Part I were of this nature, without being explicitly labelled as such. From this general description it is clear that many significant problems in long-range planning, financial investment programs, advertising campaigns, and so on may be subsumed under the heading of allocation problems.

Optimal scheduling of men and materials through a sequence of processing points is one of the primary goals of any industrial concern. To insure optimal operation and utilization of facilities, we must expend our resources of men, machinery, money, and time in the most efficient manner. We shall see how a dynamic programming approach leads to one possible mathematical specification of this principle.

The final class of problems which we address are questions of optimal inventory control. We have a warehouse which stocks certain items for delivery to customers. A cost is incurred if items are not available when requested. For example, the cost may be the ill will of the customer, a fixed breach of contract cost, or both. On the other hand, if the warehouse overstocks certain items other types of costs accrue, namely, the costs of storage and accounting for the item and costs for the item becoming technologically obsolescent. The central question is how, in the face of uncertain demands, the warehouse should be stocked in order to achieve an optimal balance between the above cited costs. The functional equation technique of dynamic programming allows us to make some inroads into the solution of this difficult question.

MULTISTAGE ALLOCATION PROCESSES

Of primary concern in this section is the following problem: maximize the function of n variables

$$J(x_1,x_2,\ldots,x_n) = \sum_{i=1}^{n} \ell_i(x_i) , \tag{5.1}$$

over the region

$$\sum_{i=1}^{n} x_i \leq X , \tag{5.2}$$

$$x_i \geq 0 . \tag{5.3}$$

The motivation for consideration of this particular class of problems is to regard X as an amount of resource available for distribution among N activities. An allocation x_i to activity i brings a return of $\ell_i(x_i)$. Thus, we wish to partition X among the N activities to maximize the total return. We assume that the structure of the ℓ_i may be such that the optimal allocation policy may consist in holding some of the resource in reserve, i.e., strict inequality in (5.2) may hold. Also, it is assumed that the x_i are given in some commensurable unit.

To obtain the relevant functional equation describing the optimal allocation rule, as well as the maximum value of J, we observe that the optimal value of J is dependent upon two parameters N, the number of activities and X, the total amount of resources available for allocation. We write

$$\max_{\sum_{i=1}^{n} x_i \leq X} J(x_1,\ldots,x_n) = I_n(X) .$$

Assume x_n has been allocated to activity n. Then an amount $X - x_n$ is left for possible allocation among the remaining $n-1$ activities. According to the principle of optimality, this resource must be optimally allocated if the whole process is to be optimal. Thus,

$$I_n(X) \geq \ell_n(x_n) + I_{n-1}(X - x_n) ,$$

for any choice of x_n. If we assume the g_i are all continuous functions, then an optimal policy certainly exists. Consequently, equality holds for some x_n such that $0 \le x_n \le X$. Hence, we have the functional equation

$$I_n(X) = \max_{0 \le x_n \le X} [\ell_n(x_n) + I_{n-1}(X - x_n)] , \quad n \ge 2 . \tag{5.4}$$

Clearly, if we have only one activity,

$$I_1(X) = \max_{0 \le x_1 \le X} [\ell_1(x_1)] . \tag{5.5}$$

An extremely important special case of one-dimensional allocation processes of the foregoing type is that in which the individual return functions are concave, i.e.,

$$\ell_i[\alpha x_i + (1-\alpha)y_i] \le \alpha \ell_i(x_i) + (1-\alpha)\ell_i(y_i) , \quad 0 \le \alpha \le 1 .$$

The reason for the importance of this class of problems is twofold: (1) they occur very frequently in practice, and (2) there exists an extremely efficient algorithm for obtaining the optimal solution, almost by inspection. The algorithm we refer to is the so-called "maximal marginal return" procedure which basically states that at any stage of the process we make our allocation so as to maximize the immediate marginal return.

The algorithm can best be understood by working a simple example. Consider a three-activity problem, in which the performance criterion to be maximized is

$$J = \sum_{i=1}^{3} \ell_i(x_i) ,$$

subject to the constraint on total resources

$$\sum_{i=1}^{3} x_i \le 8 ,$$

and to the constraints on resource allocated to each activity

$$0 \le x_i \le 5 , \quad i = 1,2,3 .$$

The resources x_i are to be added in unit increments, so that the allowed values of x_i are

$$x_i = 0,1,2,3,4,5 , \quad i = 1,2,3 .$$

Having computed the incremental performance gains (the marginal returns), the solution procedure is trivial. We list the marginal returns in decreasing order and assign resources to the corresponding activities until the total resource available is exhausted. For example, see Table 5.1. Thus, we allocate one unit to activity 3 since it has the marginal return 0.45. Eliminating 0.45 from our list, for the second allocation we have the example shown in Table 5.2. Hence, the maximal marginal return is 0.35 and we allocate one unit to activity 2. Crossing 0.35 off our list, we proceed as before, finally obtaining the policy shown in Table 5.3.

In summary, we can state the maximal marginal return algorithm as:

(1) Initialize the <u>solution list</u> by applying the lower bound resource to each activity.

(2) Initialize the <u>candidate list</u> by placing <u>in order</u> the incremental performance on each activity for adding Δx to the lower bound resource.

(3) Update the solution list by taking highest incremental performance from the top of candidate list. Note the activity corresponding to this incremental performance.

(4) Update the candidate list by placing in the proper order on this list the incremental performance for adding Δx more resource to the activity noted in step (3).

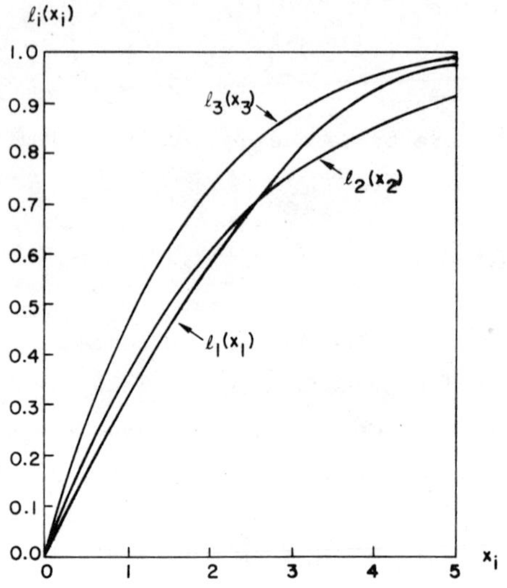

RETURN FUNCTIONS FOR THREE-ACTIVITY EXAMPLE

x_1	$\ell_1(x_1)$
0	0.00
1	0.30
2	0.55
3	0.75
4	0.92
5	0.98

x_2	$\ell_2(x_2)$
0	0.00
1	0.35
2	0.58
3	0.73
4	0.85
5	0.92

x_3	$\ell_3(x_3)$
0	0.00
1	0.45
2	0.71
3	0.85
4	0.95
5	0.99

INCREMENTAL PERFORMANCE GAINS FOR THREE-ACTIVITY EXAMPLE

x_1	$\Delta\ell_1(x_1)$
1	0.30
2	0.25
3	0.20
4	0.17
5	0.06

x_2	$\Delta\ell_2(x_2)$
1	0.35
2	0.23
3	0.15
4	0.12
5	0.07

x_3	$\Delta\ell_3(x_3)$
1	0.45
2	0.26
3	0.14
4	0.10
5	0.04

Table 5.1.

Candidate List

INCREMENTAL PERFORMANCE	ACTIVITY
0.45	
0.35	
0.30	

Solution List

AMOUNT OF RESOURCE	CUMULATIVE PERFORMANCE	INCREMENTAL PERFORMANCE	ACTIVITY FOR PRESENT ALLOCATION
0	0	—	—
1			
2			
3			
4			
5			
6			
7			
→ 8			
9			
10			
11			
12			
13			
14			

Table 5.2.

Candidate List

INCREMENTAL PERFORMANCE	ACTIVITY
0.35	2
0.30	1
0.26	3

Solution List

AMOUNT OF RESOURCE	CUMULATIVE PERFORMANCE	INCREMENTAL PERFORMANCE	ACTIVITY FOR PRESENT ALLOCATION
0	0	—	—
1	0.45	0.45	3
2			
3			
4			
5			
6			
7			
→ 8			
9			
10			
11			
12			
13			
14			
15			

Table 5.3.

Candidate List

INCREMENTAL PERFORMANCE	ACTIVITY
—	—
—	—
—	—

Solution List

AMOUNT OF RESOURCE	CUMULATIVE PERFORMANCE	INCREMENTAL PERFORMANCE	ACTIVITY FOR PRESENT ALLOCATION
0	0	—	—
1	0.45	0.45	3
2	0.80	0.35	2
3	1.10	0.30	1
4	1.36	0.26	3
5	1.61	0.25	1
6	1.84	0.23	2
7	2.04	0.20	1
→ 8	2.21	0.17	1
9	2.36	0.15	2
10	2.50	0.14	3
11	2.62	0.12	2
12	2.72	0.10	3
13	2.79	0.07	2
14	2.85	0.06	1
15	2.89	0.04	3

(5) Continue steps (3) and (4) until the total resource is used.

A not particularly challenging analytic exercise is the extension of (5.4)-(5.5) to the situation where there are several different types of noncommensurable resources to allocate such as men, money, machines, etc. For example, suppose we have two types of resources X and Y which are to be allocated among n activities to achieve an optimal return. Suppose that $\ell_i(x_i,y_i)$ represents the return from activity i if x_i amoung of resource 1 and y_i amount of resource 2 are assigned. Then the mathematical formulation of the problem is

$$\max J = \sum_{i=1}^{n} \ell_i(x_i,y_i) ,$$

over the region

$$\sum_{i=1}^{n} x_i \leq X , \quad \sum_{i=1}^{n} y_i \leq Y ,$$

$$x_i,y_i \geq 0 , \quad i = 1,2,\ldots,n .$$

Denoting the maximum value of J by $I_n(X,Y)$, we have no difficulty in obtaining the functional equation

$$I_n(X,Y) = \max_{\substack{0 \leq x_n \leq X \\ 0 \leq y_n \leq Y}} [\ell_n(x_n,y_n) + I_{n-1}(X-x_n,Y-y_n)] , \quad n \geq 2 , \tag{5.6}$$

$$I_1(X,Y) = \max_{\substack{0 \leq x_1 \leq X \\ 0 \leq y_1 \leq Y}} \ell_i(x_1,y_1) . \tag{5.7}$$

Equations (5.4)-(5.5) (or their multidimensional counterparts (5.6)-(5.7)) are our basic equations for the treatment of most allocation problems. However, it should be noted that, under appropriate circumstances, other criteria of the type introduced in Part I may take the place of the additive criteria utilized above. Also, a number of variants of the basic problem may appear such as stochastic effects, infinite stage processes, additional side constraints, and so forth. Some of these topics will be illustrated in the worked examples and exercises.

SCHEDULING PROBLEMS

A commonly encountered problem in many situations is to determine optimal (or even feasible) schedules of units through a sequence of processing stations. A somewhat related problem is a so-called "smoothing" problem in which it is desired to determine a purchasing schedule such that supply levels fluctuate as little as possible while still meeting various demands.

Perhaps the simplest version of the first type of problem is that in which we have n items which must be processed on two machines, 1, 2, in this order. If we are given the processing times for each item on each machine, the problem is to arrange the items in some sequence such that the overall processing time is minimized. A dynamic programming formulation of this problem is the following:

Let

p_i = the time necessary to process item i on machine 1,

q_i = the time necessary to process item i on machine 2, $i = 1,2,\ldots,n$.

Introduce the function

$I(p_1,q_1,p_2,q_2,\ldots,p_n,q_n;t)$ = the amount of time used in processing the n items, when machine 2 is committed t time units ahead, the processing times are p_i,q_i, respectively, and an optimal scheduling policy is used.

If item i is processed first, we apply the principle of optimality to obtain

$$I(p_1,q_1,\ldots,p_n,q_n;t) = \min_i \,[p_i + I(p_1,q_1,\ldots,0,0,\ldots,p_n,q_n;$$
$$\max(t-p_i,0))] ,$$

where the (0,0) combination occurs where (p_i,q_i) was before.

The optimal scheduling policy is obtained from this relation by considering the effect of interchanging the order of two succeeding items. Processing first item i then item j, we have

$$I(p_1,q_1,p_2,q_2,\ldots,p_n,q_n;t) = p_i+p_j+I(q_1,q_1,\ldots,0,0,\ldots,$$
$$0,0,\ldots,p_n q_n;t_{ij}) ,$$

where

$$t_{ij} = q_j + \max[q_i + \max(t-p_i,0) - p_j,0]$$
$$= q_j + q_i - p_j - p_i + \max[t,\max(p_i+p_j-q_i,p_i)] .$$

Thus, we see that if

$$\max[p_i+p_j-q_i,p_i] > \max[p_i+p_j-q_j,p_j] ,$$

we should interchange the order from (i,j) to (j,i). Later we shall make use of the above result to determine an effective algorithm for the optimal ordering of the n items.

Another type of scheduling problem that frequently appears in industrial contexts is the question of replacement of worn-out and/or technologically obsolescent equipment and tools by modern devices. To simplify our discussion, we shall assume that we have a single piece of equipment which produces a known amount of revenue per year $r(t)$, requires a known amount of maintenance $u(t)$, and may be exchanged for a new machine with the trade-in value $c(t)$ being known.

Suppose that at the beginning of each year we are allowed only the choice of keeping the current machine for another year, or trading it in on a new machine. Then we introduce the function

$I(t)$ = the total return obtained starting with a machine of age t and employing an optimal policy.

Then $I(t)$ satisfies the equation

$$I(t) = \max \begin{cases} K: & r(t) - u(t) + aI(t+1) \\ B: & r(0) - u(0) - c(t) + aI(1) \end{cases},$$

where the constant a, $0 < a < 1$, has been introduced as a discount factor to keep the total return finite, with K representing the decision to keep the current machine, while B means to buy a new machine.

The above equation for I has a closed form solution of the following type: it is evident that the optimal policy is to keep a new machine for z years, and then replace it by a new machine. Thus, using the preceding equation, we obtain

$$\begin{aligned} I(0) &= r(0) - u(0) + aI(1) , \\ I(1) &= r(1) - u(1) + aI(2) , \\ &\vdots \\ I(z-1) &= r(z-1) - u(z-1) + aI(z) , \\ I(z) &= -c(z) + r(0) - u(0) + aI(1) . \end{aligned}$$

The solution of this set of linear algebraic equations for $I(1)$ gives

$$I(1) = \frac{\begin{bmatrix} r(1) - u(1) + a[r(2)-u(2)]+\ldots+a^{z-2} \\ [r(z-1)-u(z-1)]+[r(0)-u(0)]a^{z-1} \end{bmatrix} -a^{z-1}c(z)}{1--a^{z}} .$$

The unknown time z is now chosen to maximize $I(1)$, since maximizing $I(1)$ clearly maximizes $I(0)$.

More complex models involving upkeep and return costs which take account of technological improvements, as well as overhaul costs, will be examined in the solved problems.

INVENTORY PROBLEMS

The basic situation for the determination of optimal ordering policies in an inventory control context is as follows: at certain times we may order supplies of a set of items, where the cost of ordering depends upon the number of each item ordered, as well as upon some fixed costs, which are independent of the number ordered. At other times, demands are made upon the stocks of items. We will be interested in the case where only probability distribution functions for these demands are known. Further, to insure nontrivial policies we assume that a penalty is levied if the demand for an item exceeds the supply.

The simplest process of the above sort is where we are stocking only one item. We assume that the orders are made at a finite number of equally spaced times and are immediately filled. After an order has been made and filled, a demand is made. The demand is satisfied as well as possible, with excess demand leading to a penalty cost.

Assume that the following functions are completely known:

(1) $\varphi(s)ds$ = the probability that the demand will lie between s and $s+ds$,

(2) $k(z)$ = the cost of ordering z items to increase the stock level,

(3) $p(z)$ = the cost of ordering z items to meet an excess, z, of demand over supply.

Let x be the stock level at the beginning of the process, and assume there are n stages. At any stage, the stock level, x, and the number of stages remaining, n, characterize the problem.

Define

$I_n(x)$ = the expected total cost for an n-stage process starting with an initial stock level x and using an optimal ordering policy.

For a single-stage process, we have

$$I_1(x) = \min_{y \geq x} \left[k(y-x) + \int_y^{\infty} p(s-y)\varphi(s)\,ds \right].$$

In general, for $n \geq 2$, we have

$$I_n(x) = \min_{y \geq x} \left[k(y-x) + \int_y^{\infty} p(s-y)\varphi(s)\,ds + I_{n-1}(0) \int_y^{\infty} \varphi(s)\,ds + \int_0^{y} I_{n-1}(y-s)\varphi(s)\,ds \right]. \tag{5.8}$$

The analogous problem for an infinite time period with discounted future costs leads to the functional equation

$$I(x) = \min_{y \geq x} \left[k(y-x) + \int_y^{\infty} p(s-y)\varphi(s)\,ds + aI(0) \int_y^{\infty} \varphi(s)\,ds + \int_0^{y} I(y-s)\varphi(s)\,ds \right], \tag{5.9}$$

where $0 < a < 1$, is the discount factor. We shall examine various properties of Eqs. (5.8)-(5.9) in the solved problems.

A typical smoothing problem is the following: we have a warehouse which is required to meet a set of known demands for supplies. A cost is incurred for not meeting the demands. But another cost is incurred if the warehouse is overstocked. Additionally, there is a third type of cost associated with changing supply levels. The problem is to determine the stock levels so as to minimize the total cost of the process.

To mathematically formulate the foregoing problem, we assume for simplicity, that the warehouse stocks only one type of item. It will be evident from what follows how to handle the multi-item situation, analytically, at least.

Let

$$r_k = \text{demand at time } k,$$

$$x_k = \text{supply level at time } k, \quad k = 1,2,\ldots,n,$$

with $x_0 = c$, a fixed initial level. Introduce the cost functions

$$\alpha_i(x_k - r_k) = \text{the cost incurred at time } k \text{ if } x_k > r_k,$$

$$\beta_k(x_k - x_{k-1}) = \text{the cost incurred at time } k \text{ if } x_k \neq x_{k-1}.$$

We assume that it is required to always meet the demand. Thus,

$$x_k \geq r_k, \quad k = 1,2,\ldots,n .$$

Our objective is to choose the $x_k \geq r_k$, $k = 1,2,\ldots,n$ to minimize the function

$$J = \sum_{k=1}^{n} [\alpha_k(x_k - r_k) + \beta_k(x_k - x_{k-1})] ,$$

which represents the total cost of the process.

Since the individual cost functions α_k, β_k may be dependent on the time, we define the modified criterion function

$$J_M = \sum_{k=M}^{n} [\alpha_k(x_k - r_k) + \beta_k(x_k - x_{k-1})] ,$$

and let

$$I_M(c) = \min_{x_k} J_M , \quad M = 1,2,\ldots,n .$$

Then,

$$I_n(c) = \min_{x_n \geq r_n} [\alpha_n(x_n - r_n) + \beta_n(x_n - c)] ,$$

and

$$I_M(c) = \min_{x_M \geq r_M} [\alpha_M(x_M - r_M) + \beta_M(x_M - c) + I_{M+1}(x_M)] ,$$

$$M = 1,2,\ldots,n-1 .$$

ROUTING PROBLEMS

A convenient bridge between many problems of operations research and optimal control theory is the class of so-called "generalized" routing problems.

In abstract terms, we are interested in a discrete control process of the following type: let p be an element of a finite set S and let $T(p,q)$ be a family of transformations with the property that $T(p,q) \varepsilon S$ whenever $p \varepsilon S$, for all $q \varepsilon D$, a discrete decision space. The problem is to determine a sequence of decisions $q_1, q_2, \ldots,$ which transforms p, the initial state, into p_N, a specified state, in an optimal manner.

Actually, in many cases, we are really interested in feasibility or, in the language of system theory, controllability (i.e., does there exist any sequence from D transferring p to p_N); however, in this chapter we narrow our sights and consider optimality.

A particular case of the foregoing type of problem is the following: consider a set of N points numbered $1,2,\ldots,N$, with N being the desired terminal point. We suppose that there exists a direct link between any two points i and j which requires a time $t_{ij} \geq 0$ to traverse, with $t_{ii} = 0$, $i = 1,2,\ldots,N$. The question is what path should we follow, starting at 1, passing through some subset of $\{2,3,\ldots,N-1\}$, to arrive at N in minimum time?

To treat this problem by dynamic programming, we imbed the original problem within the family of problems consisting of determining the optimal path from k to N for $k = 1,2,\ldots,N-1$. Let

$$I(k) = \text{minimum time to go from } k \text{ to } N, \quad k = 1,2,\ldots,N-1.$$

and set

$$I(N) = 0.$$

Then, we immediately obtain

$$I(k) = \min_{j \neq k} [t_{kj} + I(j)] .$$

The above functional equation is implicit, rather than explicit and requires the type of analysis given in Part I for its solution. One way to circumvent these difficulties is to use the method of successive approximations.

If we write

$$I^{(n)}(k) = \min_{j \neq k} [t_{kj} + I^{(n-1)}(j)] ,$$

$$I^{(o)}(k) = t_{1k} ,$$

then we immediately have a feasible computational approach to the problem. Furthermore, it is clear that the approximate function $I^{(n)}(k)$ has intuitive significance: it is the minimal cost for a path starting at k, ending at N, and containing at most n intermediate stops. This interpretation, together with the finiteness of N, immediately shows that the above "approximation in function space" will converge to the optimal solution in at most N-2 iterations.

Now let us briefly consider the computational feasibility of the successive approximation method sketched above. To calculate the set of values $\{I^{(n)}(k)\}$, $k=1,2,\ldots,N$, we require the previous set $\{I^{(n-1)}(k)\}$ and the matrix $T=[t_{ij}]$. If N is large, e.g., $N=10^4$, storage of T involves 10^8 values, an unbearable amount. However, the situation is improved somewhat if we note that, for fixed k, only one row of T needs to be available in high-speed storage. Thus, we actually need storage locations for only 2N values. Of course, we pay a time-penalty for shifting numbers to and from low-speed storage but this cost may be acceptable compared with not being able to treat the problem at all.

SOLVED PROBLEMS

5.1 Consider the problem of maximizing $J=\sum_{i=1}^{2n} \ell(x_i)$, over the region

$$R_1:\ \sum_{i=1}^{n} x_i = y_1\ , \qquad R_2:\ \sum_{i=N+1}^{2n} x_i = y_2\ .$$

Show that the optimal value function for a process of length $n=2^k$ can be calculated in (k+1) stages.

Solution: Introduce the sequence of functions

$$f_n(y_1) = \max_{R_1}\ [\ell(x_1)+\ell(x_2)+\ldots+\ell(x_n)]\ ,$$

$$h_n(y_2) = \max_{R_2}\ [\ell(x_{n+1})+\ldots+\ell(x_{2n})]\ .$$

After calculating the functions $f_n(y_1)$, $h_n(y_2)$, we may determine the solution to the original problem by maximizing

$$f_n(y_1) + h_n(y_2) \quad ,$$

where $y_1 + y_2 = x$, $y_1, y_2 \geq 0$.

Utilizing this result plus the fact that the stage returns ℓ are all identical, it is easy to see that

$$I_{2^k}(x) = \max_{0 \leq y \leq x} \left[I_{2^{k-1}}(y) + I_{2^{k-1}}(x-y) \right] \quad , \quad k = 1,2,\ldots \; .$$

Thus, a process of length 2^k can be solved in $(k+1)$ stages.

(The importance of the foregoing result is that with parallel computers at our disposal, the functions $f_n(y_1)$, $h_n(y_2)$ may be calculated simultaneously.)

5.2 Assume that at time 0 we have X pounds of a certain chemical. At each of n subsequent periods, if we sell d_k pounds of the chemical at time k we obtain a return $r(d_k)$, and if we use z_k pounds of the chemical for production in period k, the yield is az_k pounds of the same chemical at the end of the period, $a > 1$. Determine the amount of the chemical to be sold and used for production to maximize the total profit assuming $r(d_k) = bd_k$.

Solution: From the linearity of the criterion and the system dynamics, as well as the physical situation, it is intuitively clear that the optimal policy will consist of retaining all chemical for production of additional chemicals, and then selling the total amount at the last stage.

To see this result mathematically, let

$$J = \sum_{i=1}^{n} bd_i \ .$$

Introduce the functions

$$\begin{aligned} I_k(X) &= \max J \\ &= \text{total return from a k-stage process, when we begin with } X \text{ pounds of chemical and follow an optimal allocation policy, } k=1,2,\ldots,n. \end{aligned}$$

It is clear that the sequence $\{I_k(X)\}$ satisfies the relation

$$I_k(X) = \max_{0 \le d_k + z_k \le X} [bd_k + I_{k-1}(X-d_k+az_k)] \ , \quad k \ge 2 \ ,$$

$$I_1(X) = \max_{0 \le d_1 + z_1 \le X} [bd_1] \ .$$

After working out the first few members of the sequence, we hypothesize that the structure of the optimal return is

$$I_k(X) = Xb(1+a)^{k-1} \ ,$$

and the optimal policy is

$$z_k = \begin{cases} X, & k=2,3,\ldots,n \ , \\ 0, & k=1 \ , \end{cases}$$

$$d_k = \begin{cases} 0, & k=2,3,\ldots,n \ , \\ X, & k=1 \ . \end{cases}$$

To establish these results, we proceed by induction. The results are clearly true for $k=1$, since in a single-stage process there is nothing to be gained by expending any resources for the purpose of producing additional chemical. Assume that

$$I_{k-1}(X) = Xb(1+a)^{k-2},$$

and that the optimal policy is $z_{k-1}=X$, $d_{k-1}=0$. Then

$$I_k(X) = \max_{0 \leq d_k + z_k \leq X} [bd_k + b(1+a)^{k-2}(X - bd_k + az_k)] .$$

By the linearity of $I_k(X)$, it is clear that the maximum occurs at either $d_k = X$, $z_k = 0$ or at $d_k = 0$, $z_k = X$. To decide which, we must compare the quantities

$$b - b^2(1+a)^{k-2} \quad \text{and} \quad ab(1+a)^{k-2},$$

or, upon cancelling a factor b in each, the quantities

$$1 - b(1+a)^{k-2} \quad \text{and} \quad a(1+a)^{k-2} .$$

But,

$$1 - b(1+a)^{k-2} < 1 - 2^{k-2}b < 1 < a(1+a)^{k-2},$$

since $a > 1$. Thus, the optimal choice is $d_k = 0$, $z_k = X$, which when substituted into the expression for $I_k(X)$ yields

$$\begin{aligned} I_k(X) &= b(1+a)^{k-2}X + ab(1+a)^{k-2}X \\ &= Xb(1+a)^{k-1}, \end{aligned}$$

which establishes our claim regarding the structure of the optimal value function and the optimal policy.

5.3 Formulate as a dynamic programming problem

$$\min[\max \ell_k(v_k)] \;, \quad k = 1,2,\ldots,n \;,$$

subject to

$$\sum_{k=1}^{n} v_k = M \;, \quad v_k \geq 0 \;.$$

Solution: Let $J = \max_{1 \leq k \leq N} [\ell_k(v_k)]$. Introduce the functions

$I_n(M)$ = minimum value of J for an n-stage process which starts with an amount M to allocate and an optimal policy is used.

Using the obvious fact that for any allocation policy $(v_1, v_2, \ldots, v_n)$, J takes its maximum value either for $k = 1$ or for $2 \leq k \leq n$, we may employ the principle of optimality to assert

$$I_n(M) = \max\left\{ \min_{0 \leq v_n \leq M} [\ell_n(v_n)] \;, \; \min_{0 \leq v_n \leq M} I_{n-1}[(M - v_n)] \right\} \;,$$

$$I_1(M) = \min_{0 \leq v_1 \leq M} \ell_1(v_1) \;.$$

5.4 A company has six salesmen and three market areas:

(a) Determine a dynamic programming functional equation for determining the number of salesmen to allocate to each area in order to maximize profit.

(b) Assume the following table gives the profit from each market area as a function of the number of salesmen allocated. Numerically determine the optimal profit and allocation policy.

MARKET AREA	NUMBER OF SALESMEN						
	0	1	2	3	4	5	6
1	30	40	50	60	70	80	90
2	52	58	61	66	74	82	88
3	30	42	51	60	65	77	89

<u>Solution</u>: (a) Let $\ell_k(v_k)$ represent the profit realized from allocating v_k salesmen to area k. The only resource we have available, X, is salesmen. Thus, we introduce

$I_k(X)$ = the total profit from the k areas when we have X salesmen to allocate and an optimal policy is used, $k = 1,2,3$.

Our earlier remarks show that

$$I_k(X) = \max_{v_k=0,1,\ldots,X} [\ell_k(v_k) + I_{k-1}(X-v_k)] \; , \quad k = 2,3 \; ,$$

with

$$I_1(X) = \max_{v_1=0,1,\ldots,X} [\ell_1(v_1)] \; .$$

(b) Computational exercise.

5.5 Assume we have a resource x which may be utilized in a number of ways. If y is a parameter specifying a particular use, let $L(x,y)$ be the immediate return, and $D(x,y)$ the cost in resources. If $I(x)$ is the total return from repeated use of an initial resource x, obtained using an optimal allocation policy, we have

$$I(x) = \max_y [L(x,y) + I(x-D(x,y))] .$$

Assuming $D(x,y)$ is small compared to x, for all y, determine an expression for $I'(x)$ and give the interpretation of this result.

Solution: Since D is small compared with x, for all y, we have

$$I[x-D(x,y)] = I(x) - D(x,y)I'(x) + o(x) ,$$

where $o(x)$ denotes terms of higher order in x. Thus,

$$I(x) = \max_y [L(x,y) + I(x) - D(x,y)I'(x)] + o(x) ,$$

$$I'(x) = \max_y \frac{L(x,y)}{D(x,y)} .$$

We see that $I'(x)$ represents the optimal marginal return function, i.e., the optimal ratio of return for resources invested.

5.6 We are given initially x dollars and a quantity y of serum, together with the prerogative of purchasing additional amounts of the serum at specified times $t_1 < t_2 < \ldots$. At the kth purchasing opportunity, t_k, a quantity $c_k z$ of serum may be purchased for z dollars, where c_k is a monotone increasing function of k. Given the probability that an epidemic occurs between t_k and t_{k+1} and the condition that if an epidemic occurs we may only use the amount of serum on hand, we wish to determine the purchasing policy that maximizes the overall probability of successfully combating an epidemic, given the probability of success with a quantity ω of serum available. Determine a dynamic programming formulation of this problem.

Solution: Let

p_k = the probability that an epidemic occurs between t_k and t_{k+1}, assuming that no epidemic has previously occurred,

$\varphi(\omega)$ = the probability of successfully combating the epidemic with a quantity ω of serum,

$I_k(x,y)$ = the overall probability of success using an optimal purchasing policy from t_k on, given x dollars and an initial quantity y of serum on hand.

Applying by now familiar ideas, we see

$$I_k(x,y) = \max_{0 \leq z \leq x} [p_k \varphi(y+c_k z) + (1-p_k) I_{k+1}(x-z, y+c_k z)] .$$

5.7 The problem of designing an efficient water distillation plant for heavy water production involves minimizing

$$J = \ell(a_1) + \frac{\ell(a_2)}{a_1} + \frac{\ell(a_3)}{a_1 a_2} + \ldots + \frac{\ell(a_m)}{a_1 a_2 \ldots a_{m-1}} ,$$

where the a_i are subject to

(a) $a_i \geq 1$,

(b) $a_1 a_2 \ldots a_m = x$.

Derive a functional equation suitable for solution of this problem.

Solution: Let

$I_k(x)$ = minimum value of J over k stages, where an optimal allocation policy is used and an initial amount x is available.

As a result of choosing a_k, the amount available is reduced to $\frac{x}{a_k}$ and a cost $\ell(a_k)$ is incurred. In addition, the optimal cost for the remaining stages is reduced by a factor $\frac{1}{a_k}$, in accordance with the definition of J. Putting these remarks together with the principle of optimality, we have

$$I_k(x) = \min_{a_k \geq 1} \left[\ell(a_k) + \frac{1}{a_k} I_{k-1}\left(\frac{x}{a_k}\right) \right] .$$

5.8 An Aircraft Routing Problem. In this problem a computational procedure for the optimal short-term allocation of aircraft to meet a fixed schedule is discussed. This problem is commonly referred to as the aircraft routing problem. The same basic procedure can be applied to similar problems, such as crew scheduling, optimal planning of a time table, and optimal scheduling of maintenance. It also has applications to a wide range of other scheduling problems.

In aircraft routing, a limited resource, namely the fleet of airplanes, is to be utilized in such a way that the best system performance is obtained while meeting certain constraints. System performance can be measured by the net profit of the airline, the percentage of scheduled flights that are completed on time, or some other relevant measure. The constraints can include crew availability, regular maintenance of the aircraft at specified locations, the unavailability of a given airport due to weather conditions, and any other desired conditions.

The formulation of this problem in terms suitable for the application of dynamic programming proceeds as follows. First, the time interval over which the optimization is to be performed is quantized into increments; typically, this period is 24 hours and the time increment is one hour. The state variables for the system are then taken to be the positions of each of the airplanes. A complete description of position specifies the origin and destination of the flight as well as how far the airplane is along the flight path. This information is conveniently summarized in a state diagram such as the one shown in Figure 5.1. Here it is assumed that there are three airports, labeled as nodes 1, 2, and 3. The flight time from 1 to 2 is taken to be 4 hours; from 2 to 3 it is 2 hours; and from 1 to 3 it is 6 hours. If the flight time between two airports is k hours, then $(k-1)$ intermediate states are defined between them; in this manner, if the airplane takes off at a quantized time, then at each of the next k hours it passes from one state to the next until it arrives after k hours at the destination. Two sets of intermediate states are defined between each pair of airports, corresponding to flights in both directions.

For the moment, it is assumed that once an airplane takes off for a given destination, it flies on to that particular airport.

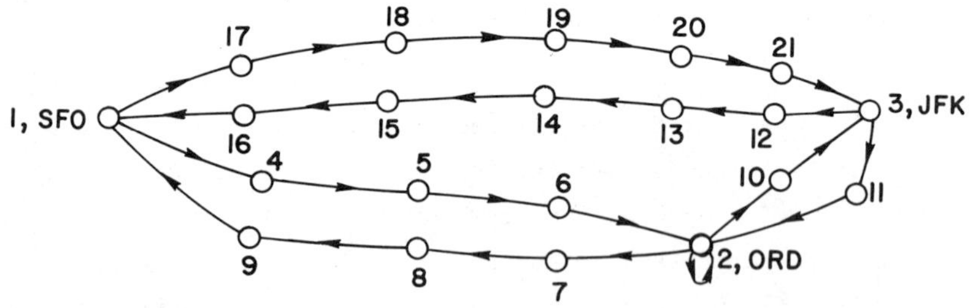

Figure 5.1. State Diagram for Aircraft Routing Problem

Consequently, the only time that a decision can be made concerning a specific airplane is when it is on the ground at an airport. Then the airplane has three choices: it can stay where it is or it can take off for either of the other two airports. Except for these decisions, the next state for each airplane is completely determined by knowing the present state.

The n-dimensional state vector x consists of the number of the state for each of the n airplanes. The n-dimensional control vector u specifies which airport each airplane heads for if it is presently on the ground. The system equations then take the form

$$x(k+1) = g[x(k),u(k),k] \ .$$

If the ith airplane obeys the state diagram in Figure 5.1, then the ith component of g has the form shown in Table 5.1.

Table 5.1. State Transition Function $x_i(k+1) = g_i[x_i(k),u_i(k),k]$ for System Shown in Figure 5.1

$x_i(k)$ PRESENT STATE	$u_i(k)^*$ CONTROL	$x_i(k+1) = g_i[x_i(k),u_i(k),k]$ NEXT STATE
1	1	1
	2	4
	3	17
2	1	7
	2	2
	3	10
3	1	12
	2	11
	3	3
6	—	2
9	—	1
10	—	3
11	—	2
16	—	1
21	—	3
All Other States, j	—	(j+1)

*A dash in the control column indicates that the next state is independent of the control; in this problem this means that the airplane is in the air and has no choice of control.

A general cost function of the following form can be used:

$$J = \sum_{k=0}^{K} \ell[x(k),u(k),k] .$$

As already mentioned, this function can be maximum profit, minimum disruption of service, etc. Constraints, such as those mentioned previously, can easily be implemented. These constraints simply cut down the number of alternatives that an airplane has; for example, if it is due for maintenance at airport 1 after 7 more flying hours and it is presently at airport 2, it has no choice but to fly to 1.

With minor modifications, the program can handle nonintegral times of flight between airports, nonintegral departure times, the possibility of diversion to other airports, varying flight times due to wind, and other constraints and conditions that arise in the actual operation of an airline.

A computer program using forward dynamic programming to optimize the system of Figure 5.1 has been written. One input to the program is the schedule of flights over the time period, including estimates of the number of passengers on each flight. The present state of each airplane in the fleet is also given. Finally, the constraints are specified. The program then uses forward dynamic programming to determine the order of flights that each aircraft should make in order to optimize the performance criterion within the given constraints.

In Tables 5.2 through 5.6, the results for an actual problem are summarized. The performance criterion is taken to be total net profit, defined as total revenue minus total operating cost. The fleet for this problem consists of two airplanes; one of which has a capacity of 100 passengers and an operating cost of \$1,250/hour, and the other a capacity of 60 passengers and an operating cost of \$750/hour. Both airplanes are assumed to fly at the speed on which the state transition diagram in Figure 5.1 is based. The fare

Table 5.2. Fares and Flight Times Between Airports

ORIGIN/DESTINATION PAIR	FARE	FLIGHT TIME
1(SFO) - 2(ORD)	$r_{12} = r_{21} = \$100$	$t_{12} = t_{21} = 4$ hours
1(SFO) - 3(JFK)	$r_{13} = r_{31} = \$150$	$t_{13} = t_{31} = 6$ hours
2(ORD) - 3(JFK)	$r_{23} = r_{32} = \$\ 50$	$t_{23} = t_{32} = 2$ hours

Table 5.3. Schedule for the Day, Including Expected Passenger Loads

DESTINATION \ ORIGIN	1(SFO)	2(ORD)	3(JFK)
1(SFO)	X	8:00-90 20:00-85	6:00-80
2(ORD)	0:00-40 16:00-90	X	6:00-65 22:00-60
3(JFK)	12:00-110	2:00-45 20:00-90	X

Table 5.4. Maximum Profits at $k = 24$

$x_2(24)$	$x_1(24)$ 1(SFO)	2(ORD)	3(JFK)
1(SFO)	$17,000	$22,500	$22,000
2(ORD)	23,000	19,500	22,000
3(JFK)	21,500	21,000	20,500

Table 5.5. Optimum Schedule for Airplanes with $x_1(24) = 1$, $x_2(24) = 2$ (Maximum Profit With No Final Constraints)

AIRPLANE 1

DEPARTS	TIME	ARRIVES	TIME	PASSENGERS	PROFIT
2	2:00	3	4:00	45	-$ 250
3	6:00	2	8:00	65	750
2	8:00	1	12:00	90	4,000
1	12:00	3	18:00	100	7,500
3	18:00	2	20:00	0	- 2,500
2	20:00	1	24:00	85	3,500
				Total Profit, Airplane 1:	$13,000

AIRPLANE 2

DEPARTS	TIME	ARRIVES	TIME	PASSENGERS	PROFIT
1	0:00	2	4:00	40	$ 1,000
2	4:00	3	6:00	0	- 1,500
3	6:00	1	12:00	60	4,500
1	16:00	2	20:00	60	3,000
2	20:00	3	22:00	90	1,500
3	22:00	2	24:00	60	1,500
				Total Profit, Airplane 2:	$10,000
				TOTAL PROFIT:	$23,000

schedule is shown in Table 5.2 along with the time of flights between airports. The net profit obtained in a flight from Airport j to Airport m using airplane i at time k can be expressed as

$$\ell(j,m,i,k) = p_{jm}(k)\, r_{jm} - c_i\, t_{jm} ,$$

where
$i = 1,2$,
$j = 1,2,3$,
$m = 1,2,3$,
$k = 0,1,2,\ldots,24$,
$p_{jm}(k)$ = number of passengers going from j to m at time k: constrained to be less than p_i,

Table 5.6. Optimum Schedule for Airplanes with $x_1(24) = 2$, $x_2(24) = 1$ (Maximum Profit When Airplane 2 Must Go To Airport 1 for Maintenance at $k = 24$)

AIRPLANE 1					
DEPARTS	TIME	ARRIVES	TIME	PASSENGERS	PROFIT
2	2:00	3	4:00	45	-$ 250
3	6:00	2	8:00	65	750
2	8:00	1	12:00	90	4,000
1	12:00	3	18:00	100	7,500
3	22:00	2	24:00	60	500
			Total Profit, Airplane 1:		$12,000
AIRPLANE 2					
DEPARTS	TIME	ARRIVES	TIME	PASSENGERS	PROFIT
1	0:00	2	4:00	40	$ 1,000
2	4:00	3	6:00	0	- 1,500
3	6:00	1	12:00	60	4,500
1	16:00	2	20:00	60	3,000
2	20:00	1	24:00	60	3,000
			Total Profit, Airplane 2:		$10,000
			TOTAL PROFIT:		$22,500

the capacity of airplane i ($p_1 = 100$, $p_2 = 60$) ,

r_{jm} = fare from j to m ,

c_i = operating cost per hour of airplane i ($c_1 = \$1{,}250/\text{hr}$, $c_2 = \$750/\text{hr}$) ,

t_{jm} = flight time from j to m ,

$\ell(j,m,i,k)$ = net profit obtained if airplane i is flown from j to m at time k.

The profit is evaluated once when the flight takes off, but not again during the flight; also, the profit is 0 if the airplane stays at the same airport. These conditions are expressed by

$$\ell(j,m,i,k) = 0 \quad (j \neq 1,2,3 \text{ or } m = 1,2,3, \text{ all } i,k) ,$$

$$\ell(j,j,i,k) = 0 \quad (\text{all } i,j,k) .$$

The performance criterion can then be expressed in the usual form

$$J = \sum_{k=0}^{K} \ell(j,m,i,k) .$$

A typical schedule for this particular problem is shown in Table 5.3. The initial state is specified as $x_1(0) = 2$, $x_2(0) = 1$, i.e., airplane 1, the large airplane, is at airport 2 and airplane 2 is at airport 1. The program then determines the maximum profit for every possible state at the end of 24 hours. In this program the final state was constrained to have both airplanes on the ground at $k = 24$. The maximum profit for each of these nine states is shown in Table 5.4. If there is no additional constraint on the final state, then the final state having the largest maximum profit among these nine values is chosen. This maximum profit is \$23,000, corresponding to the state $x_1(24) = 1$, $x_2(24) = 2$; the schedule for each airplane which achieves this profit is shown in Table 5.5. Note that in this schedule the airplane flies empty or with so few passengers that it loses money in some flights, but that it then arrives at a state where it is able to make a net profit later on. Note also that in this case an airplane is provided for all ten flights.

If instead airplane 2 is constrained to be in state 1 at $k = 24$ in order to have some maintenance performed, then the maximum profit that can be obtained is \$22,500, corresponding to $x_1(24) = 2$, $x_2(24) = 1$. The schedule that results in this profit is shown in Table 5.6.

It is possible to print out the best N trajectories in order to compare their performance on the basis of quantities not directly measured by the profit, such as percentage of flights actually made. Also, if sometime during the day different constraints arise, such as the temporary mechanical failure of one of the airplanes, these can be entered into the program and a new optimization performed over the rest of the day, starting at the present positions of the airplanes. As already mentioned, the program can be expanded to include other constraints, other performance criteria, diversions to other airports, more airplanes, more airports, nonintegral flight times and departure times, etc. This program can be used for planning the aircraft utilization for a given day, for real-time dispatching when a new situation arises, or for off-line determination of an optimum time table (flight schedule) based on the distribution of passenger demand.

5.9 In the optimal equipment replacement problem, reformulate the previous recurrence relations to account for:

(a) purchase of a used machine,

(b) overhaul of the current machine.

Solution:

(a) Define the function

$\ell_n(t,x)$ = the cost of replacing a machine of age t by one of age x in year n.

Let

$I_n(t)$ = the value at year n of the overall return from a machine which is t years old, where an optimal replacement policy is employed for the remainder of the process.

(Note that we are now considering a finite stage process. The future will still be discounted as before.)

The function $I_n(t)$ satisfies

$$I_n(t) = \max \begin{cases} P: & r_n(0)-u_n(0)-c_n(t)+aI_{n+1}(1) , \\ U: & \max_x[r_n(x)-u_n(x)-\ell_n(t,x)+aI_{n+1}(x+1)] , \\ K: & r_n(t)-u_n(t)+aI_{n+1}(t+1) , \end{cases}$$

where P, U, K denote purchase of a new machine, purchase of a used machine, and keeping the incumbent machine, respectively.

(b) To include overhaul, we introduce the function

$I_n(t_1,t_2)$ = the value at year n of the overall return from a machine of age t_1, and last overhauled at age t_2, where an optimal replacement policy is employed for the remainder of the process.

All costs must now be given in terms of both the age t_1, and the age of the machine at its last overhaul t_2. Taking these remarks into account, we have

$$I_n(t_1,t_2) = \max \begin{cases} P: & r_n(0,0)-u_n(0,0)-c_n(t_1,t_2)+aI_{n+1}(1,0) , \\ K: & r_n(t_1,t_2)-u_n(t_1,t_2)+aI_{n+1}(t_1+1,t_2) , \\ \theta: & r_n(t_1,t_1)-u_n(t_1,t_1)-\theta_n(t_1,t_2) \\ & \quad + aI_{n+1}(t_1+1,t_1) , \end{cases}$$

where $\theta_n(t_1,t_2)$ is the cost at time n of overhauling a machine of age t_1, which was last overhauled at age t_2.

5.10 Suppose we have a quantity of capital x and a choice of production of n different products in varying quantities. Assume that there is an unlimited supply of labor and machines for the production of any items we choose, in any quantities we wish.

If we decide to produce an amount x_i of item i, we incur the following costs:

a_i = unit cost of raw materials required for item i,

b_i = unit cost of machine production of item i,

c_i = unit cost of labor for item i,

C_i = a fixed cost, independent of the amount of item i produced, if $x_i > 0$.

Thus, the cost of producing an amount x_i of item i is

$$g_i(x_i) = \begin{cases} (a_i + b_i + c_i)x_i + C_i \ , & x_i > 0 \ , \\ 0 \ , & x_i = 0 \ . \end{cases}$$

Let p_i = selling price of item i. We wish to maximize the total profit, given by

$$J = \sum_{i=1}^{n} p_i x_i \ ,$$

subject to the restrictions

$$\sum_{i=1}^{n} g_i(x_i) \le x \ , \qquad x_i \ge 0 \ .$$

Formulate as a dynamic program.

Solution: Let

$$I_n(x) = \max J .$$

Then

$$I_1(x) = \begin{cases} \dfrac{p_1(x - C_1)}{a_1 + b_1 + c_1} , & x \geq C_1 , \\ 0 , & 0 \leq x \leq C_1 , \end{cases}$$

with

$$I_n(x) = \max_{\substack{x_n \geq 0 \\ g_n(x_n) \leq x}} [p_n x_n + I_{n-1}(x - g_n(x_n))] , \quad n \geq 2 .$$

5.11 In the optimal inventory problem discussed in the text, modify the formulation given to account for:

(a) unbounded time period with discounted costs,

(b) partially expendable items — unbounded time,

(c) a one-period lag in supply — unbounded time,

(d) a two-period lag in supply — unbounded time.

Solution:

(a) Letting a be the factor which discounts future costs, $0 < a < 1$, we set

$I(x)$ = the expected total discounted cost starting with an initial supply x and using an optimal policy.

By the consideration of all possible eventualities, we have

$$I(x) = \min_{y \geq x} \left[k(y-x) + a\int_y^\infty p(x-y)\varphi(s)\,ds + aI(0)\int_y^\infty \varphi(s)\,ds + a\int_0^y I(y-s)\varphi(s)\,ds \right] .$$

(b) If we assume that some of the items supplied on demand may be partially recovered, so that a demand of s items results in a return of bs items, $0 \leq b \leq 1$, which may be used again, we have

$$I(x) = \min_{y \geq x} \left[k(y-x) + a\int_y^\infty p(s-y)\varphi(s)\,ds + a\int_y^\infty I(bs)\varphi(s)\,ds + a\int_0^y I(y-s+bs)\varphi(s)\,ds \right].$$

(c) We now assume that if we order a quantity z, it does not become available until one period later. If the current supply is x, and y was on order from the previous period, then $x+y$ will be available to meet the next demand. The relevant functional equation now takes the form

$$I(x) = \min_{z \geq 0} \left[kz + a\int_x^\infty p(s-x)\varphi(s)\,ds + aI(z)\int_x^\infty \varphi(s)\,ds + a\int_0^x I(x-s+z)\varphi(s)\,ds \right].$$

(d) If we have a two-period lag, then we have two state variables that describe the state of the process,

x = quantity of stock available to meet the next demand,
y = quantity to be delivered one period after that.

Thus, we define

$I(x,y)$ = expected total cost with x and y as above, using an optimal policy.

Then $I(x,y)$ satisfies the equation

$$I(x,y) = \min_{z \geq 0} \left[kz + a\int_x^\infty p(s-x)\varphi(s)\,ds + aI(y,z)\int_x^\infty \varphi(s)\,ds + a\int_0^x I(x-s+y,z)\varphi(s)\,ds \right].$$

5.12 Multipurpose Reservoir System. In this problem, the optimum operation over 24 hours of a multipurpose four-reservoir system is determined. The reservoir network, which contains both series and parallel connections, is shown in Figure 5.2. In the optimization, use of water for power generation, irrigation, flood control and recreation is considered. Interaction of the short-term optimization with longer-term operating policies is also taken into account.

The state variables for the problem are taken to be the amounts of water contained in each reservoir. The amount of water in the ith reservoir is denoted as x_i, $i = 1,2,3,4$, where each x_i is expressed in normalized units.

On the basis of potential use of the reservoir for recreation purposes, a minimum water level for each reservoir is specified; the amount of water needed to achieve this level is arbitrarily set as $x_i = 0$, and a constraint is imposed that the amount of water in each reservoir cannot drop below this value.

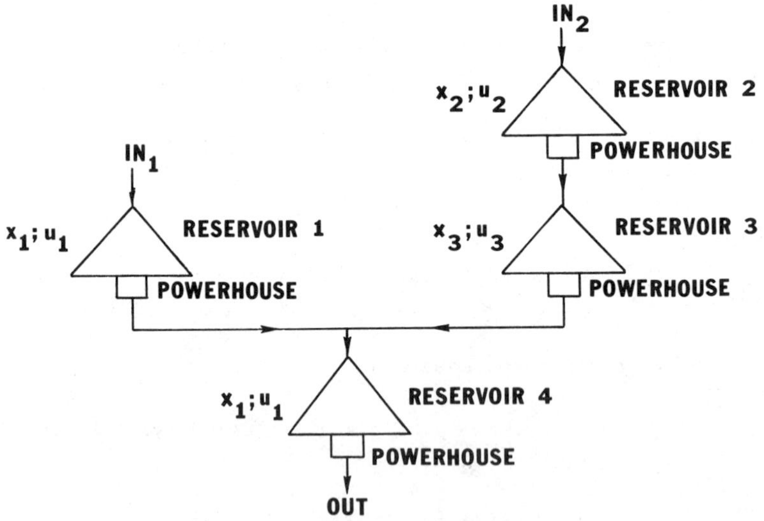

Figure 5.2. Network Configuration of Four-Reservoir Problem

On the basis of flood control considerations, a maximum water level for each reservoir is established. The amount of water needed to raise the level from minimum to the maximum value is then expressed in terms of the normalized units, and a constraint is imposed that each x_i cannot exceed this level. The particular constraints considered in this example are expressed as

$$0 \le x_i \le 10 ,$$
$$0 \le x_2 \le 10 ,$$
$$0 \le x_3 \le 10 ,$$
$$0 \le x_4 \le 15 .$$

The control variables are taken to be the flows out of the reservoirs. These quantities are also expressed in the same normalized units; the variables $u_i(k)$, $i = 1,2,\ldots,4$, specify the amount of water released from the <u>ith</u> reservoir over the <u>kth</u> time interval. In this example each time interval is two hours.

For each reservoir a maximum flow is determined by the capacity of the power generators, and a minimum flow is determined by considering the use of the down-stream flow for navigation, conservation, and municipal and industrial water supplies. For this example the constraints were

$$0 \le u_1 \le 3 ,$$
$$0 \le u_2 \le 4 ,$$
$$0 \le u_3 \le 4 ,$$
$$0 \le u_4 \le 7 .$$

The system equations express how the water flows between the reservoirs. They are

$$x_1(k+1) = x_1(k) - u_1(k) + IN_1 ,$$

$$x_2(k+1) = x_2(k) - u_2(k) + IN_2 ,$$

$$x_3(k+1) = x_3(k) - u_3(k) + u_2(k) ,$$

$$x_4(k+1) = x_4(k) - u_4(k) + u_3(k) + u_1(k) ,$$

$$(k = 0,1,\ldots,11).$$

The inflows IN_1 and IN_2 are assumed constant over the day as

$$IN_1 = 2 ,$$
$$IN_2 = 3 .$$

The performance criterion considers the use of water for both power generation and irrigation. It is assumed that there is a power generation station at each reservoir outflow. The benefit from the flow over a given two-hour period is assumed to be a linear function of the flow, i.e., the benefit from a flow out of reservoir i at time k is $c_i(k)u_i(k)$. The values of $c_i(k)$ are shifted in k with respect to each other to account for the transport delay of water between reservoirs. This delay is four hours from Reservoir 1 to Reservoir 4, four hours from Reservoir 2 to Reservoir 3, and two hours from Reservoir 3 to Reservoir 4. The values of $c_i(k)$, $i=1,2,3,4$ are shown in Table 5.7.

Irrigation benefits are considered only for the outflow from Reservoir 4. The benefit is again linear with flow; i.e., the

Table 5.7. Time Functions Used in Calculating Benefit

k	$c_1(k)$	$c_2(k)$	$c_3(k)$	$c_4(k)$	$c_5(k)$
0	1.1	1.4	1.0	1.0	1.6
1	1.0	1.1	1.0	1.2	1.7
2	1.0	1.0	1.2	1.8	1.8
3	1.2	1.0	1.8	2.5	1.9
4	1.8	1.2	2.5	2.2	2.0
5	2.5	1.8	2.2	2.0	2.0
6	2.2	2.5	2.0	1.8	2.0
7	2.0	2.2	1.8	2.2	1.9
8	1.8	2.0	2.2	1.8	1.8
9	2.2	1.8	1.8	1.4	1.7
10	1.8	2.2	1.4	1.1	1.6
11	1.4	1.8	1.1	1.0	1.5

benefit from flow $u_4(k)$ is $c_5(k)u_4(k)$. The function $c_5(k)$ is shown in Table 5.7.

The benefit function also includes a terminal cost for failing to reach a specified level for each reservoir at the end of the day. This function accounts for the long-term policy of filling or emptying the reservoir during a particular season. This function assesses a heavy penalty for having less than the specified amount of water at the end of the day, but gives no credit for having more than this amount. The particular function used is

$$\psi_i[x_i(12),m_i] = \begin{cases} -40[x_i(12)-m_i]^2 , & x_i(12) \le m_i , \\ 0 , & \text{otherwise} , \end{cases}$$

where m_i = desired level of reservoir i at end of the day (k=12).

The initial state is taken to be

$$x_1(0) = 5 ,$$
$$x_2(0) = 5 ,$$
$$x_3(0) = 5 ,$$
$$x_4(0) = 5 .$$

The desired final state is

$$m_1 = 5 ,$$
$$m_2 = 5 ,$$
$$m_3 = 5 ,$$
$$m_4 = 7 .$$

The dynamics and constraints are as above. The performance criterion is

$$J = \sum_{k=0}^{11} \sum_{i=1}^{4} c_i(k)u_i(k) + \sum_{k=0}^{11} c_5(k)u_4(k) + \sum_{i=1}^{4} \psi_i[x_i(12),m_i] ,$$

where $c_i(k)$, $i = 1,2,\ldots,5$ is specified in Table 5.7, $\psi_i[x_i(12),m_i]$ is shown above and m_i, $i = 1,2,3,4$ are as given.

The initial policy chosen is shown in Table 5.8. Basically, this policy consists of setting outflow equal to inflow at every time period so that the water level in each reservoir remains constant. The only exception to this policy occurs at the end of the day, when the terminal cost function is taken account.

The final results of applying the dynamic programming successive approximations algorithm of Chapter 4 to this problem are shown in Table 5.9. The improvement in total benefit was from 362.5 to 401.3 units. Note that the optimum policy is very different from the initial policy. Nine iterations were required to obtain convergence; these calculations required approximately 30 seconds on the Burroughs 5500 computer. Because of the linearity of the performance criterion, it was possible to compute the true optimum solution by linear programming; for this problem the successive approximations solution was the same as the true optimum solution.

For several other cases, substantially the same results as quoted here were obtained. The extension to more complex performance criteria and constraints represents little difficulty. Stochastic

Table 5.8. Initial Policy

k	$x_1(k)$	$x_2(k)$	$x_3(k)$	$x_4(k)$	$u_1(k)$	$u_2(k)$	$u_3(k)$	$u_4(k)$
0	5	5	5	5	2	3	3	5
1	5	5	5	5	2	3	3	5
2	5	5	5	5	2	3	3	5
3	5	5	5	5	2	3	3	5
4	5	5	5	5	2	3	3	5
5	5	5	5	5	2	3	3	5
6	5	5	5	5	2	3	3	5
7	5	5	5	5	2	3	3	5
8	5	5	5	5	2	3	3	5
9	5	5	5	5	2	3	3	5
10	5	5	5	5	2	3	3	5
11	5	5	5	5	2	3	3	3
12	5	5	5	7				

Total Benefit = 362.5

Table 5.9. Optimum Policy

k	$x_1(k)$	$x_2(k)$	$x_3(k)$	$x_4(k)$	$u_1(k)$	$u_2(k)$	$u_3(k)$	$u_4(k)$
0	5	5	5	5	1	4	0	0
1	6	4	8	7	0	1	0	2
2	8	5	10	5	0	2	4	7
3	10	7	8	1	2	0	4	7
4	10	10	4	0	3	3	4	7
5	9	10	3	0	3	4	4	7
6	8	9	3	0	3	4	4	7
7	7	8	3	0	3	4	4	7
8	6	7	3	0	3	4	4	7
9	5	6	3	0	3	4	4	7
10	4	5	3	0	3	4	4	0
11	3	4	3	7	0	2	0	0
12	5	5	5	7				

Total Benefit = 401.3

inflows can be explicitly taken into account by employing the ideas presented in Chapter 3.

5.13 Consider a traveling salesman who wishes to start at a given city (call it city 1) and visit N-1 other cities once, and only once, finishing his route back at city 1. Given the cost t_{ij} of going from city i to city j, use dynamic programming to formulate a feasible computational approach to the solution of this problem.

Solution: First of all, we note that at each point of the four it is necessary to know all the cities visited so far. Thus, it is not sufficient, as in the process discussed in the text, to keep track only of the current position.

Let the state of the process be described by the vector

$$p = [i_1, i_2, \ldots, i_s] ,$$

where $[i_1,i_2,\ldots,i_s]$ denotes the cities already visited starting from city 1, with i_s being the current point. Note that the order in which $i_1,i_2,\ldots,i_s$ appears is irrelevant. Once we are at i_s, it suffices to know that we have visited $i_1,i_2,\ldots,i_{s-1}$ in some order. This feature is essential for computational purposes.

The principle of optimality yields the functional equation

$$I(i_1,i_2,\ldots,i_s) = \min_j \left[t_{i_s j} + I(i_1,i_2,\ldots,i_s,j)\right] ,$$

where

$I(i_1,i_2,\ldots,i_s)$ = the cost required to optimally tour the remaining points and return to city 1.

When $i_s = N$, and $i_1,i_2,\ldots,i_{s-1}$ represent $2,3,\ldots,N-1$ in some order, we have

$$I(i_1,i_2,\ldots,i_s) = t_{i_s 1} .$$

---

* 5.14 A system has been designed to perform a mission in the face of a possible (random) component failure. If the system functions without the failure, the system equation is

$$x(k+1) = x(k) + u(k) .$$

If the failure occurs, the decision mechanism becomes inoperative, and the system equation becomes

$$x(k+1) = x(k) .$$

At each stage k, $k = 0,1,2$, it is known whether or not the component has failed. If the failure has occurred, the mission controller has a choice of repairing the failure, at a cost of one unit, or of proceeding with the system in the failed mode. Once a repair has been made, the system will not fail again over the remainder of the mission. If the system is working, the controller can also make a pre-emptive repair that will prevent any future failures, again at a cost of one unit.

If the performance criterion is

$$J = \sum_{k=0}^{2} [x^2(k) + u^2(k)] + [\text{one unit if repair is made}] \ ,$$

if the probability of component failure at each stage is 1/2, and if the system is constrained to operate in the region

$$0 \leq x \leq 2 \ ,$$

$$-2 \leq u \leq 0 \ ,$$

find the optimum decision and repair policy and minimum expected cost function, taking into account the information available at each stage as to whether the system is working, failed, or repaired, and the flexibility in the decision-making process of repairing the component.

(HINT: Let the state vector of the system be either (x,W), (x,F), or (x,R), where W = system working normally, F = component failed, R = component has failed but been repaired. If the state is (x,F) or (x,W), let the decision vector be (u,A) or (u,B), where A = repair the system and B = do <u>not</u> repair the system.)

<u>Solution</u>: We first evaluate I(x,W,3), I(x,F,3) and I(x,R,3). In all cases the first decision variable is clearly $\hat{u} = 0$. If the state is either I(x,F,3) or I(x,W,3), no repair needs to be made. Therefore,

$$I(x,R,3) = 0 \ ,$$
$$\hat{u}(x,R,3) = 0 \ ,$$

$$I(x,F,3) = 0 \ ,$$
$$\hat{u}(x,F,3) = (0,B) \ ,$$

$$I(x,W,3) = 0 \ ,$$
$$\hat{u}(x,W,3) = (0,B) \ .$$

Next, we proceed to stage 2. Clearly, for the repaired system,

$$I(x,R,2) = x^2 ,$$

$$\hat{u}(x,R,2) = 0 .$$

If the system has failed, we must consider the possibility of repairing. However, since the optimum decision for the repaired system is the same decision that results from the unrepaired system, namely $\hat{u} = 0$, and since the repair costs one unit, we see immediately that

$$I(x,F,2) = x^2 ,$$

$$\hat{u}(x,F,2) = (0,B) .$$

Finally, for the working system, we must again decide whether or not to repair. Also, we must note that we have 0.5 probability of having the component fail in the next stage. However, since

$$I(x,F,3) = I(x,W,3) = I(x+u,R,3) = 0 ,$$

the extra cost of the repair is not justified, and we have again

$$I(x,W,2) = x^2 ,$$

$$\hat{u}(x,W,2) = (0,B) .$$

Now we proceed to stage 1. First, we see that for the repaired case,

$$I(x,R,1) = \frac{3}{2} x^2 ,$$

$$\hat{u}(x,R,1) = -\frac{1}{2} x .$$

For the failed state, we must decide whether or not to repair. We have

$$I(x,F,1) = \operatorname*{Min}_{A,B} \begin{bmatrix} A: & \operatorname*{Min}_u \, [x^2 + u^2 + 1 + I(x+u,R,2)] \\ B: & x^2 + I(x,F,2) \end{bmatrix} .$$

Substituting for $I(x,R,2)$ and $I(x,F,2)$, we have

$$I(x,F,1) = \min_{A,B} \begin{bmatrix} A: & \min_u [x^2 + u^2 + 1 + (x+u)^2] \\ B: & 2x^2 \end{bmatrix} .$$

Clearly, for alternative A, $\hat{u}(x,F,1) = -0.5x$, $I(x,F,1) = 1.5x^2 + 1$. We then have to determine whether $2x^2$ or $1.5x^2 + 1$ is smaller. We obtain

$$I(x,F,1) = \begin{cases} \frac{3}{2} x^2 + 1 , & x \geq \sqrt{2} , \\ 2x^2 , & x \leq \sqrt{2} , \end{cases}$$

$$\hat{u}(x,F,1) = \begin{cases} (-0.5x, A) , & x \geq \sqrt{2} , \\ (0, B) , & x \leq \sqrt{2} . \end{cases}$$

Finally, for the case where the system is working, we have the recursive equation

$$I(x,W,1) = \min_{A,B} \begin{bmatrix} A: & \min_u [x^2 + u^2 + 1 + I(x+u,R,2)] \\ B: & \min_u [x^2 + u^2 + \frac{1}{2} I(x+u,W,2) \\ & \qquad + \frac{1}{2} I(x+u,F,2)] \end{bmatrix} .$$

We note again that $I(x,R,2) = I(x,F,2) = I(x,W,2) = x^2$. Therefore, no repair is justified, and we obtain

$$I(x,W,1) = \frac{3}{2} x^2 ,$$

$$\hat{u}(x,W,1) = (-0.5x, B) .$$

We at last proceed to stage 0. For the repaired case,

$$I(x,R,0) = 1.6x^2 ,$$

$$\hat{u}(x,R,0) = -0.6x^2 .$$

If the system has failed, we have

$$I(x,F,0) = \operatorname*{Min}_{A,B} \begin{bmatrix} A: & \operatorname*{Min}_{u} [x^2 + u^2 + 1 + I(x+u,R,1)] \\ B: & [x^2 + I(x,F,1)] \end{bmatrix}.$$

For the repair option, $I(x+u,R,1) = \frac{3}{2}(x+u)^2$. For the no repair option, $I(x,F,1)$ takes on one value in one region and another value in the other. Let us assume $0 \leq x \leq \sqrt{2}$ and try $I(x,F,1) = 2x^2$. We have

$$I(x,F,0) = \operatorname*{Min}_{A,B} \begin{bmatrix} A: & \operatorname*{Min}_{u} [x^2 + u^2 + 1 + 1.5(x+u)^2] \\ B: & 3x^2 \end{bmatrix}$$

$$= \operatorname*{Min}_{A,B} \begin{bmatrix} A: & 1.6x^2 + 1 \\ B: & 3x^2 \end{bmatrix}.$$

Clearly,

$$I(x,F,0) = \begin{cases} 1.6x^2 + 1 , & x \geq \sqrt{5/7} , \\ 3x^2 , & x \leq \sqrt{5/7} , \end{cases}$$

$$\hat{u}(x,F,0) = \begin{cases} (-0.6x,R) , & x \geq \sqrt{5/7} , \\ (0,F) , & x \leq \sqrt{5/7} . \end{cases}$$

We note that the basic assumption required at stage 1, $x \leq \sqrt{2}$, is satisfied if $x \leq \sqrt{5/7}$. Therefore, the above analysis holds and we do not need to repeat the analysis using the function $I(x,F,1) = \frac{3}{2} x^2 + 1$.

Finally, we consider the system to be working at stage 0. We have

$$I(x,W,0) = \operatorname*{Min}_{A,B} \begin{bmatrix} A: & \operatorname*{Min}_{u} [x^2 + u^2 + 1 + I(x,R,1)] \\ B: & \operatorname*{Min}_{u} [x^2 + u^2 + \frac{1}{2} I(x+u,W,1) \\ & \quad + \frac{1}{2} I(x+u,F,1)] \end{bmatrix}.$$

In order to determine $I(x+u,F,1)$, we again have the problem of deciding into which region $x+u$ falls in for stage 1. Let us again assume $0 \leq x+u \leq \sqrt{2}$ and use $I(x+u,F,1) = 2x^2$. We obtain

$$I(x,W,u) = \underset{A,B}{\text{Min}} \begin{bmatrix} A: & 1.6x^2 + 1 \\ B: & \underset{u}{\text{Min}}\ [x^2 + u^2 + \frac{3}{4}(x+u)^2 + (x+u)^2] \end{bmatrix}$$

$$= \underset{A,B}{\text{Min}} \begin{bmatrix} A: & x^2 + 1 \\ B: & x^2 + \left(-\frac{7}{11}x\right)^2 + \frac{7}{4}\left(\frac{4}{11}x\right)^2 \end{bmatrix}$$

$$= \underset{A,B}{\text{Min}} \begin{bmatrix} A: & 1.6x^2 + 1 \\ B: & x^2 \end{bmatrix} .$$

We see that for $0 \leq x \leq 2$,

$$x^2 = 1.636x^2 < 1.6x^2 + 1 .$$

Therefore, for the region of interest, we do not make the preemptive repair. Thus,

$$I(x,W,0) = \frac{11}{7}x^2 ,$$

$$\hat{u}(x,W,0) = \left(-\frac{7}{11}x, B\right) .$$

As a final check, we verify that for $0 \leq x \leq 2$, $0 \leq x+\hat{u}(x,W,0) \leq \sqrt{2}$. We see that this requires $0 \leq \frac{4}{11}x \leq \sqrt{2}$, which is always satisfied for $0 \leq x \leq 2$. The above result is thus correct and we again do not have to examine the case $x+u \geq \sqrt{2}$.

SUPPLEMENTARY PROBLEMS

5.1 Assume there are N different stages of production involved in turning out a particular item. The probability that the item is processed correctly at stage i is p_i.

Assume that k machines may be used to increase the accuracy of any particular stage so that if one machine is added to stage i, p_i changes to p_{i_1}, if two machines are added then p_i becomes p_{i_2}, and so on.

(a) How should we distribute the k machines to maximize the overall accuracy of the process?

(b) What if at most m_i machines may be added to stage i?

(c) What if a machine at stage i costs d_i dollars and we have at most d dollars to spend?

(d) What if a machine at stage i requires h_i operators and at most h men are available?

5.2 If a mistake found at point i requires t_i amount of time and a cost c_i to remedy, and the cost of turning out a defective item is z, how much money should be spent on checking out the equipment and how should it be allocated? (Take into account labor costs, machine costs, and the costs of a defective item.)

5.3 Assume we have a warehouse with fixed capacity and an initial stock of a certain item which is subject to known seasonal price and cost variations. We wish to determine the optimal pattern of purchasing, storage, and sales.

Let B be the fixed warehouse capacity, A the initial stock. For the ith period, let

c_i = cost per unit,

p_i = selling price per unit,

x_i = amount bought,

y_i = amount sold.

The constraints are:

(a) the stock on hand at the end of the ith period cannot exceed B, the warehouse capacity;

(b) the amount sold in period i cannot exceed the amount available at the end of period (i-1);

(c) the amounts purchased or sold in any period are non-negative.

Determine the policy which maximizes the overall profit from an N-stage process.

5.4 Consider the following transportation problem. We have sources $S_1, S_2, \ldots, S_M$ and terminals $T_1, T_2, \ldots, T_N$. Each source has a quantity x_i of resources which must be transported to various terminals such that the total quantity arriving at T_j fulfills an a priori demand y_j. Assume $\sum_i x_i = \sum_j y_j$. Given the distances d_{ij} between the sources and terminals, and assuming the cost of shipping a unit quantity of resources between S_i and T_j is d_{ij}, determine the routing which minimizes the total cost of supplying the demands.

5.5 Consider a three-terminal communication system with message loads at each of the terminals for the other terminals. Let r_{ij} denote the maximum number of messages that can be sent from T_i to T_j in unit time, and consider two cases:

first, where there is no interference between signals going from T_i to T_j and those going from T_j to T_i and, second, where the total traffic in both directions cannot exceed r_{ij}.

Let x_{ij}, $i,j = 1,2,3$, $i \neq j$, denote the quantity of messages at T_i with ultimate destination T_j, and assume that a unit time is consumed transmitting a message from any T_i to any T_j. Denote by $I_n(x_{ij})$ the maximum quantity of messages that can be delivered in n time units and derive a recurrence relation for $\{I_n(x_{ij})\}$.

5.6 A newspaper delivers papers to a number of newstands. Assuming that the distribution of sales at each of these stands is known, and assuming that a certain quantity of unsold papers may be returned, suitably discounted, how many papers should be published, and how should they be distributed?

5.7 Suppose we have N different activities in which to invest capital. Let $g_i(x_i)$ be the return from investing x_i in activity i. Given an initial quantity of capital x, we are required to invest in at most k activities so as to minimize the total return. Denote the maximum return by $I_{n,k}(x)$. Derive a recurrence relation for $I_{N,k}(x)$.

5.8 Suppose that we have a mechanism consisting of N interacting parts. Let part i have weight w_i and size s_i, and assume that we know the probability distribution for the length of time that any particular part will function without a replacement due to a breakdown. Assume also that we know the time and cost required for replacement, and the cost of a breakdown. Assuming that there are weight and size limitations on the total quantity of spare parts we are allowed to stock, how do we stock to minimize:

(a) the expected time lost due to breakdowns,

(b) the expected costs of breakdowns,

(c) a given function of time and cost,

(d) the probability that the time lost due to breakdowns will exceed T,

(e) the probability that the cost due to breakdown will exceed C?

5.9 A company is planning its production schedule for the next five months. Its product is produced at the rate of one batch per month. However, only two batch sizes are possible: 500 items and 1,000 items. The 500-item batch costs $500 to produce, while the 1,000-item batch costs $800. Monthly demands, which must be met, are 1,500; 1,000; 500; 500; 1,500 for months 1 through 5, respectively. At the end of each month, an inventory charge of $0.20 per item is charged against every item in stock at that time. The inventory level at the beginning of month 1 is 1,500, and the desired inventory at the end of month 5 is zero. What production schedule minimizes the sum of inventory and production costs?

5.10 The chess board shown below is taken from the Fischer-Spassky World Championship Match. Recall that a Knight can move one square horizontal or vertical and one square diagonal, but cannot occupy a square already occupied by another piece. Find the minimum number of moves for the White Knight

				WK		WR	
WP	WKt						WP
			WP	WQ			
		WP	BP	BB		WB	
						BKt	
BP					BP		
	BP			BP		BP	
		BR	BR			BK	

(WKt) in the upper left of the figure to capture the Black King (BK) in the lower right of the figure (assuming no other pieces move). Also, find all paths that achieve this minimum number of moves. Finally (Optional — only for chess fans), show how White can checkmate in five moves.

5.11 Consider the problem of extending the maximal marginal return algorithm to the case of non-concave return functions. (HINT: consider the concave hull of the set of functions $\{\ell_i(x_i)\}$. Recall the concave hull Ω of a set of functions $\{p_i(t)\}$ is defined as

$$\Omega = \left\{ f(t) \; : \; f(t) = -\sum_{i=1}^{n} \beta_i p_i(t), \quad \beta_i \geq 0, \quad \beta_1 + \ldots + \beta_{n=1} \right\} .$$

REFERENCES

<u>Allocation Problems</u>: [B-31], [B-115], [B-116], [E-10], [F-14], [K-20], [L-15], [M-3], [N-3], [P-7], [P-16], [S-16], [W-14], [W-27].

<u>Scheduling Problems</u>: [B-11], [F-16], [G-7], [H-13], [H-28], [J-7], [M-18], [N-1], [N-5], [R-6], [R-14], [R-21], [Y-3].

<u>Inventory Problems</u>: [A-19], [B-7], [D-16], [D-17], [E-8], [G-9], [H-2], [R-24].

<u>Routing Problems</u>: [B-24], [B-44], [B-89], [D-14], [G-13], [P-9], [S-8].

CHAPTER 6

TWO CASE STUDIES

INTRODUCTION

Our principal objective in this two volume set has been to present the methodology of dynamic programming in a form which would be accessible to practicing system analysts, as well as to students. Although the subject has many fascinating, and highly nontrivial mathematical features which deserve careful analysis, there can be little doubt that significant applications is "where the action is." For this reason, we have attempted to present as diverse a collection of worked examples as our experience, interests, and patience would allow. However, it must be confessed that the majority of examples presented in the previous chapters were either academic, to illustrate a theoretical point, or semi-realistic, to keep the presentation to a reasonable length.

In this chapter, we conclude our exposition of dynamic programming by exhibiting the complete details of two real-life problems in which dynamic programming techniques entered the analysis in a fundamental way. Our objective is not only to substantiate the implicit claims that dynamic programming provides a <u>practical</u> problem-solving tool, but also to illustrate (in painstaking detail) the precise steps taken to translate these words into workable programs.

The two problems we present both involve optimal planning and regulation of complex industrial processes; one example being regulation of an electrical power generation complex, the other the planning of a nuclear fuel processing plant. To complicate issues, in one case we must determine regulation policies involving three different time scales, while in the second problem there are a number of nonstandard constraints that must be satisfied. In both problems, high dimensionality precludes use of standard computational approaches. Thus, almost all of the techniques and subterfuges discussed in earlier chapters must be employed to bring the solution of these problems into the realm of feasibility.

CASE STUDY I: DISPATCHING AND OPERATIONS PLANNING FOR AN ELECTRIC UTILITY WITH MULTIPLE TYPES OF GENERATION

As the demand served by electric utilities has grown, there has been a constant requirement for installing new sources of generation. Many of the newer sources are quite different in character from conventional hydro or steam plants. Pumped storage units are an example of such a source. The increasing utilization of interchange energy has effectively provided another new type of source. The program of dispatching power from alternative sources is thus becoming more difficult, not only because of the larger number of plants, but also because of the more complex characteristics of the new plants.

This example describes a dynamic programming successive approximation algorithm for the optimization of power generation schedules for utilities participating in a coordination agreement. The power sources available for scheduling can include:

(1) conventional hydro plants,
(2) thermal generation plants,
(3) pumped-storage units,
(4) diversity interchange contracts with outside utilities.

The objective of the optimization is to schedule hourly generation of these sources such that generation costs are minimized over the entire year.

The algorithm is implemented in three separate but coupled computer programs. These programs perform three optimizations:

(1) monthly optimization over one year,

(2) daily optimization over one month,

(3) hourly optimization over 48 hours.

The monthly optimization, daily optimization, and hourly optimization programs form an optimization procedural hierarchy. Each program in turn addresses power generation scheduling with a shorter time horizon but on a more detailed and refined system model. The optimal long-term policy and long-term optimal costs of a higher-ordered program are transferred as input to the program of adjacent rank in order that the short-term optimization explicitly considers long-term effects of current decisions.

The monthly and daily optimization programs provide support to the hourly optimization program in the hourly dispatching function. Each program is of direct use in its own right for operations planning studies. The monthly program, for example, has been used for budget planning and maintenance scheduling applications and for evaluating alternate proposals of diversity interchange agreements.

The optimal schedules are determined by a dynamic programming successive approximation computational procedure. The convexity of the performance function is sufficient to ensure monotonic convergence to a globally optimum solution.

In the next three sections the monthly, daily, and hourly optimizations are discussed. The next section describes the interaction between the programs. In the following section results of applying the programs to a medium-size utility system are presented. In the final section conclusions are drawn.

THE MONTHLY OPTIMIZATION PROBLEM

Problem Formulation

In the highest-level program, energy is scheduled in monthly blocks over a one-year period. In general, the combined system may consist of:

(1) n_1 diversity interchange contracts with outside utilities,
(2) n_2 hydro plants,
(3) n_3 steam plants,
(4) n_4 pumped storage units.

The objective is to allocate monthly blocks of generation from the above sources such that the total cost of thermal generation is minimized over one year.

Formulation of the Dynamic Optimization Problem

One state variable will be required for each source where energy can be stored. There will be n_1 states to keep track of the energy balances remaining on the n_1 diversity contracts. Since these contracts require payment for all contracted energy, it can be assumed without loss of generality that all contracted compare the quantities

During the return period, the outside utilities schedule the energy deliveries to their system. The return schedules are not control variables of the system and merely represent an additional demand which the system generation must satisfy.

Contracts providing for energy purchases at specified price, for which the total draw is to be determined by the optimization, can be included with minor modifications. The dynamics of a diversity interchange contract can be described by

$$XC_j(k+1) = XC_j(k) - UC_j(k) , \quad j = 1,\ldots,n_1 , \quad k = 1,\ldots,13 , \tag{6.1}$$

where $XC_j(k)$ = energy balance remaining on contract j at beginning of month k,

$UC_j(k)$ = energy scheduled for month k from contract j.

There will be n_2 state variables to keep track of the water flow through the n_2 hydro plants. The reservoir dynamics are described by

$$XH_j(k+1) = XH_j(k) + IN_j(k) + \sum_{i=1}^{n_2} \delta_{ij} UH_i(k) - UH_j(k) \ ,$$

$$j = 1,\ldots,n_2 \ , \quad k = 1,\ldots,13 \ , \qquad (6.2)$$

where $XH_j(k)$ = water storage level at Reservoir j during month k,

$IN_j(k)$ = water inflow forecast,

$UH_j(k)$ = water outflow scheduled from Reservoir j during month k,

$$\delta_{ij} = \begin{cases} 1, & \text{if Reservoir } i \text{ is directly upstream from Reservoir } j, \\ 0, & \text{otherwise.} \end{cases}$$

Each steam plant j has an hourly heating requirement H (MMBTU) which is a convex function of the power P_j generated. The heating requirement multiplied by the fuel cost f_j ($/MMBTU) of plant j determines the hourly operating cost C_j of the plant

$$C_j(P_j) = H(P_j)\, f_j \ . \qquad (6.3)$$

Maximum and minimum generating limits are specified for each plant. The total thermal generation capacity $\overline{S}$ is the sum of the individual plant capacities $\overline{P}_j$

$$\overline{S} = \sum_{j=1}^{n_3} \overline{P}_j \ . \qquad (6.4)$$

The pumped storage reservoir will always be returned to the same level at the end of each month. Since no energy can be stored during one month for application during a later month, no state variables are required. The cost of generation at the pumped storage is determined by computing the cost of thermal generation required to refill the pumped storage reservoir. The cost is computed on the basis that pumping will be done when the load and, hence, the cost of generating the energy required for pumping are minimum. An optimal balance between thermal generation and pump storage generation is obtained for each month in the following manner. The entire load is first assumed to be satisfied by thermal and pumped storage generation only. Where the available thermal capacity in conjunction with the generation capacity (MW) of the pumped storage unit is insufficient to satisfy the load, a high cost fictitious steam plant generates the slack. The limited generation system consisting only of thermal plants and pumped storage units is optimally dispatched in a suboptimization. In the monthly optimization the pumped storage unit generation constraints MW are explicitly considered, but the reservoir constraints are not. The reservoir constraints are simplified to a requirement that all energy generated will be restored by pumping during the month. The suboptimization determines which portion of the load should be satisfied by thermal generation and which by pump storage generation (or by a mix of the two) in order to satisfy the load at least cost. The suboptimization determines the optimal incremental cost for dispatching the load duration curve in the absence of hydro generation or contracted energy deliveries. This procedure is a generalization of the well known techniques for optimally allocating generation requirements among thermal plants alone. The incremental costs based upon optimal use of the thermal pump storage subsystems are then used to determine the optimal long-term utilization of hydro and contracted energy.

The goal of the monthly optimization is thereby reduced to scheduling hydro generation and contracted energy deliveries such that the thermal generation cost (including refilling pump storage reservoir) is minimized. By scheduling generation from these sources, the thermal generation requirement is reduced by an amount r(k) during month k. The replacement energy r(k) is given by

$$r(k) = \sum_{j=1}^{n_1} UC_j(k) + \sum_{j=1}^{n_2} cv[X_j(k)]\, UH_j(k) , \qquad (6.5)$$

where $cv(U_j,X_j)$ = conversion factor (includes hydro efficiency and elevation head effects) for transforming potential energy of stored water into generation energy.

The monthly optimal thermal generation cost RC*(k) will be shown in the next section to depend upon r(k). The cost will be based upon optimal utilization of the monthly energy allotments during the month taking into account the actual distribution of load over one-month intervals. With the aid of the optimal replacement thermal energy cost function RC*[r(k)], the monthly optimization problem can be reduced to a formulation suitable for solution by standard dynamic programming techniques.

The following data are required by the dynamic programming algorithms:

(1) Monthly water inflow forecasts $IN_j(k)$; $j = 1,\ldots,n_2$; $k = 1,\ldots,12$;

(2) Optimal replacement thermal energy cost functions RC*[r(k)];

(3) Conversion factors cv(U,X) for transforming potential energy stored in reservoir into generation energy.

The system equations become

$$XC_j(k+1) = XC_j(k) - UC_j(k) , \qquad j = 1,\ldots,n_1 ,$$

$$XH_j(k+1) = XH_j(k) + \sum_{j=1}^{n_2} \delta_{ij} UH_i(k) - UH_j(k) , \quad j = 1,\ldots,n_2 . \qquad (6.6)$$

The operating constraints are described by

$$\underline{UC}_j \leq UC_j(k) \leq \overline{UC}_j ,$$

$$\underline{UH}_j \leq UH_j(k) \leq \overline{UH}_j ,$$

$$\underline{XC}_j \leq XC_j(k) \leq \overline{XC}_j ,$$

$$\underline{XH}_j \leq XH_j(k) \leq \overline{XH}_j ,$$

$$r(k) = \sum_{j=1}^{n_1} UC(k) + \sum_{j=1}^{n_2} cv[X_j(k)]\ UH_j(k) . \tag{6.7}$$

The initial and final conditions are

$$XC_j(1) = XC_j(t_o) , \quad j = 1,\ldots,n_1 ,$$

$$XH_j(1) = XH_j(t_o) , \quad j = 1,\ldots,n_2 ,$$

$$XC_j(13) = XC_j(t_o) , \quad j = 1,\ldots,n_1 ,$$

$$XH_j(13) = XH_j(t_o) , \quad j = 1,\ldots,n_2 . \tag{6.8}$$

The performance criterion is

$$J = \sum_{k=1}^{12} RC^*[r(k)] . \tag{6.9}$$

The technique of dynamic programming with successive approximation was employed to minimize the performance criterion. The performance function yet to be explicitly derived can be shown to be a convex function of the state and control variables. This property is sufficient to guarantee monotonic convergence to the optimal solution.

Calculation of the Performance Criterion by a Static Optimization

In order to solve the dynamic optimization problem of scheduling hydro and diversity interchange energy sources monthly over a one-year period, it is necessary to determine the optimal replacement thermal energy cost function, $RC^*[r(k)]$, by a static optimization.

The static optimization problem is:

Given:

(1) A load distribution $\gamma_k(\xi)$ characterizing the amount of time the load persists at a given level for month k.*

(2) Hourly operating cost functions $C_j(P_j)$ for producing P_j megawatts at Plant j.

(3) Generation and pumping capacity (MW) of pump storage unit.

(4) $r(k)$ megawatt hours of hydro and diversity interchange energy to be scheduled.

(5) W megawatts of hydro and diversity interchange capacity.

Find: The minimum cost of thermal generation required to satisfy load $\gamma_k(\xi)$.

The solution to this static optimization proceeds in two phases. The suboptimization of optimally allocating generation among n_3 thermal generation plants and a pump storage unit is considered. That result is then used to solve the problem of making maximum use of replacement energy $r(k)$ during the month in order to obtain $RC^*[r(k)]$.

In order to obtain maximum benefit from a specified amount of hydro and diversity interchange energy, it is necessary to consider in detail the distribution of the load over the month as well as the hourly cost $C^*(S)$ of optimally dispatching S megawatts of thermal generation. The demand for each month is characterized by a load distribution function $\gamma_k(\xi)$. This function describes the distribution of the load over the month interval (see Fig. 6.1).

---

* $\gamma_k(\xi)$ is the load duration curve scaled by the peak system load.

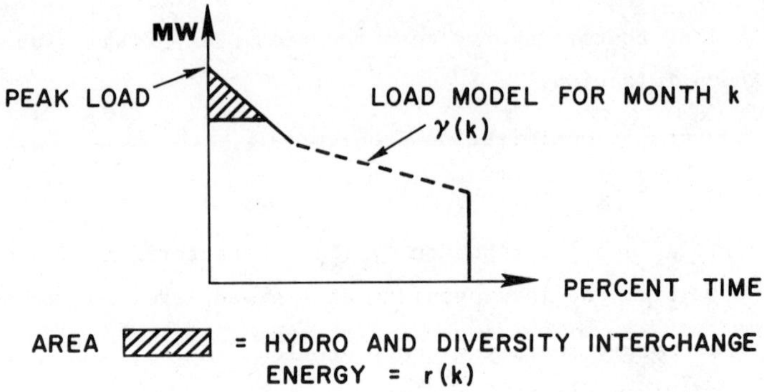

Figure 6.1. Optimal Use of Hydro Energy

The abcissa ξ is the percentage of time that the load $\gamma_k(\xi)$ is exceeded.

The demand $\gamma_k(\xi)$ must be satisfied by the thermal generation and the hydro and diversity interchange sources. The hydro and diversity interchange energy can be used to replace thermal generation in meeting the load requirement. In order to achieve maximum benefit from this replacement energy (i.e., obtain the greatest reduction in steam cost savings) the most costly thermal generation is replaced first. Because the cost of thermal generation increases monotonically with generation, the optimal utilization of a block of hydro and diversity interchange energy within month k is to peak shave as shown in Figure 6.1.

Because of the physical limitation on the amount of power the hydro and diversity interchange sources can produce, it is not always possible to replace all of the most expensive remaining thermal generation energy. Instead, the most expensive thermal generation of the feasible region is replaced as shown in Figure 6.2.

The value of the replaced thermal generation is then the difference between the minimum* cost of dispatching the original

*Dispatching thermal generation at minimum cost neglecting transmission losses is known as optimal bus-bar dispatching.

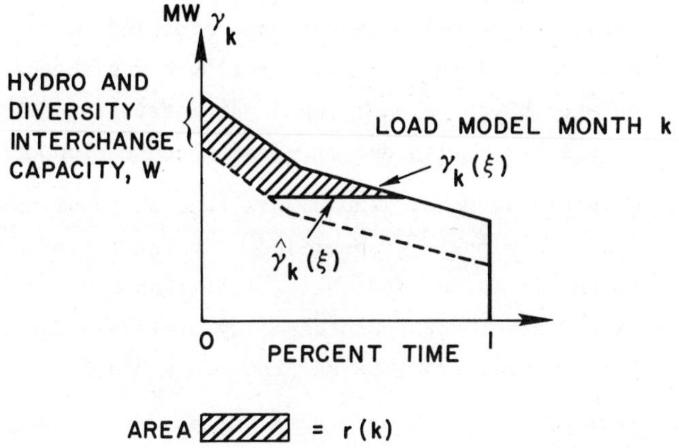

Figure 6.2. Optimal Use of Hydro Energy and Diversity Interchange Considering Capacity Limitations

generation requirement γ_k and the minimum cost of dispatching the reduced thermal generation requirement $\hat{\gamma}_k$ after optimal peak shaving. The maximum savings of replaced thermal generation is given by:

$$RC^*[r(k)] = -h_k \int_0^1 \gamma_k(\xi)\; C^*(\xi)\; d\xi + h_k \int_0^1 \gamma_k(\xi) \cdot C^*(\xi)\; d\xi\ , \tag{6.10}$$

where γ_k is a function of γ_k, r(k), and W.

Inputs to the Monthly Optimization Program

An attempt was made to minimize the amount of input data required in order to facilitate use of the program. For each month the following data is required as input:

(1) Load Data: Three parameters are used to describe the two-piece linear approximation to the load duration

curve. External sales as nonconforming loads which were not included in the load duration curve forecasts are entered by three additional parameters: total energy, peak accentuation and off-peak accentuation of the sale.

(2) Maintenance Data: The plants that are available for scheduling must be specified. In the optimization algorithms it is assumed that all available plants are on-line. Also, the hydro capacity available for peak shaving is specified from the maintenance schedule.

(3) Regional Fuel Prices: Fuel prices are entered for each plant or plant region.

(4) Water Inflow Forecasts: Monthly water inflow forecasts into each reservoir system can be entered whenever the forecast differs from the internally stored forecasts based on historical data.

The beginning reservoir elevations and contract energy balances are additional input.

DAILY OPTIMIZATION PROBLEM

Problem Formulation

The daily optimization program accepts as input the monthly allotments of diversity exchange, hydro, and pump storage generation energies. It further subdivides the monthly schedules into eight-hour blocks. A detailed load model (piecewise constant over eight hours) is constructed based on maximum load, minimum load, and total daily MWH requirements. Because the load model is piecewise constant, a static optimization is no longer required to determine the optimal cost of replacement energy. A state variable is introduced to account for level of the pumped storage reservoir. No assumption is made regarding operation of the pump storage cycle. The algorithm will determine the optimal operation such that all capacity and reservoir constraints are satisfied. Spinning reserve constraints can also be imposed upon the optimal schedule. Except as noted, the formulation is identical to the monthly optimization problem.

Inputs to the Daily Optimization Problem

The monthly program determines the long-term costs of optimal operation to the end of the year that would result from having a given reservoir storage or energy balance of a diversity contract at the end of a specified month. These long-term costs and the associated energy levels are inputs to the daily program which are balanced against the short-term daily operating costs in order to derive optimal daily schedules. For each day the following input data is also required:

(1) Load Data: Peak load, off-peak load, and total energy requirement are specified for each day.

(2) Plant Commitment Schedule: The plant commitment schedule may be given on an eight-hour basis corresponding to work shifts.

(3) Regional Fuel Prices: Fuel prices are entered for each plant or plant region.

(4) Water Inflow Resources: Inflow rates can be entered for each day or a single average inflow rate for the entire period can be entered.

HOURLY OPTIMIZATION PROBLEM

Problem Formulation

The hourly optimization program accepts as input the daily allotment of diversity exchange, hydro, and pumped storage generation (or pumping) energies for a 48-hour period. It further subdivides these 48-hour allotment totals on an hour-by-hour basis for dispatcher use. The nominal 48-hour allotments are not necessarily followed in the hourly optimization. The long-term operating cost data which is available from the daily program enables the hourly optimization program to perform a tradeoff between the short-term savings of increased hydro generation versus the long-term cost incurred by not using the hydro energy later as planned by the daily program. The formulation is identical to the daily optimization problem except for one minor refinement. Transport delays reflecting the time for water released at an upper reservoir to reach lower reservoirs are modeled where the effect is significant.

Inputs to the Hourly Optimization Program

Target reservoir elevations, target diversity exchange energy balances, and associated long-term optimal operating costs are handed down from the daily optimization. All inputs for the daily optimization required on a daily or eight-hour basis are now required on an hourly basis. The load forecast input is given on an hourly basis. The inflow forecast for each reservoir can be given as a single number representing average hourly inflow over the 48-hour period. The times during which plants are unavailable for scheduling is required input. This optimization procedure establishes a unit commitment schedule from this data and determines the corresponding optimal generation schedule.

COUPLING BETWEEN HIGH-LEVEL OPTIMIZATION AND LOWER-LEVEL OPTIMIZATION

The higher-level optimization performs its allocation of generation energies under the assumption that optimal use will be made of each generation block by the lower-level optimizations. For example, all production costing of monthly generation allocation alternatives was based upon optimal hourly dispatching of power. Performing the optimizations in this manner ensures that the decomposition of the higher-level allocations by the lower-level optimizations will result in minimum cost over the entire year.

The lower-level optimization program selects as a "target value" for the end of its optimization interval the values of reservoir levels and contract withdrawals along the optimal trajectory obtained by the higher-level optimization. In addition, the dynamic programming feedback solution provides minimum costs for optimal operation to the end of the higher-level optimization interval for all admissible end states of the lower-level optimization interval. This additional cost information can be used to specify terminal costs for final reservoir and contract balances in the lower-level optimization.

The advantage of providing such cost coupling is that the lower-level program becomes self-adapting to variations unforeseen at the time the higher-level optimization was performed. The lower-level program will compensate for a change by weighting the longer-term future optimal operating costs (terminal cost data) against current generation costs in order to determine the optimal generation schedule and corresponding target deviation warranted by the change.

DISCUSSION OF RESULTS

The combined system contains two conventional hydro plants on the same river and one pumped storage unit. The remaining generating capacity is provided by 16 thermal plants and 2 diversity interchange contracts. The scheduling is performed by coordinated use of the monthly optimization program, the daily optimization program, and the hourly optimization program.

The output of the monthly optimization program is shown in Table 6.1. The amount of generation scheduled from each source in thousands of megawatt hours is given in the columns on the left. The total system energy requirements (demand + pumping energy) for each month is displayed in the final column. Not shown but also printed out are the target elevations and dynamic programming optimal return cost information. The program requires approximately 30 seconds on a UNIVAC 1108 for a single run over 12 months.

Table 6.1. Monthly Optimization

MONTH	PUMPED GEN	STORAGE PUMP	HYDRO 1 GEN	HYDRO 2 GEN	DIV EXC C_1	DIV EXC C_2	STEAM GEN	SYSTEM ENERGY REQUIREMENTS
1	0	0	27	14	0	27	553	620
2	0	0	52	26	0	26	491	595
3	0	0	60	30	0	12	434	536
4	0	0	59	30	0	12	369	470
5	6	9	9	5	0	16	477	513
6	0	0	37	18	51	16	484	605
7	0	0	39	19	75	28	548	708
8	0	0	29	14	92	28	575	739
9	0	0	2	1	44	20	462	529
10	0	0	56	28	0	12	421	518
11	0	0	59	30	0	12	446	546
12	0	0	61	30	0	24	523	638
TOTAL	6	9	488	243	262	233	5,785	7,017

The output of the hourly optimization is shown in Table 6.2. All generations are in MWH. Hourly steam generation costs are displayed in dollars and average cost in dollars/MWH. A single run of the hourly program over 48 hours requires approximately one minute on the UNIVAC 1108. A single run of the daily program over one month requires about the same time.

The operating cost of the hourly optimal schedule was compared to the cost of actual operations for several 48-hour periods. The boundary conditions of all states were constrained during the optimization to those which resulted from actual operations. Even with this restriction, thermal generation cost savings greater than one percent per day were realized.

Based on analysis of the results of all three programs, it appears that by coupling them in the manner previously described, combined hourly, daily, and monthly savings of about one and one-half to two percent in thermal generation costs can be obtained over the entire year.

CONCLUSIONS

The major point of this study is that long-term optimal dispatching of power can be achieved using dynamic programming algorithms. The hierarchy of monthly, daily, and hourly optimization programs is used routinely by utility companies to support their dispatching functions. In addition, the monthly and daily programs have proved to be valuable in their own right for operations planning, budgeting, and contract evaluation studies.

The programs are very decidedly cost-effective. Based on the running times quoted earlier and using an exceptionally high charge of $20/minute for time on a UNIVAC 1108, the total cost of running all three programs for dispatching purposes is less than $10,000 per year. By comparison, a medium-size utility will spend $10,000,000 to $15,000,000 per year on thermal generation costs;

Table 6.2. Hourly Optimization

HOUR	PUMPED GEN	STORAGE PUMP	HYDRO 1 GEN	HYDRO 2 GEN	DIV EXC H_1	DIV EXC H_2	STEAM GEN	SYSTEM REQUIREMENT
7	0	0	99	0	147	0	605	850
8	0	0	99	0	220	0	616	935
9	0	0	99	0	220	0	762	1,081
10	0	0	99	0	220	0	859	1,178
11	0	0	99	0	220	77	883	1,278
12	0	0	99	27	220	134	882	1,362
13	0	0	99	63	220	134	889	1,405
14	0	0	99	108	220	134	951	1,512
15	0	0	99	108	220	134	976	1,537
16	0	0	99	108	220	134	985	1,545
17	0	0	99	108	220	134	1,007	1,568
18	0	0	99	108	220	134	997	1,557
19	0	0	99	108	220	134	968	1,529
20	0	0	99	108	220	134	902	1,463
21	0	0	99	81	220	134	886	1,419
22	0	0	99	36	220	134	886	1,375
23	0	0	99	9	220	96	843	1,266
24	0	0	99	0	147	19	850	1,115
1	0	0	99	0	73	0	830	1,001
2	0	0	99	0	0	0	843	942
3	0	0	99	0	0	0	807	906
4	0	0	99	0	0	0	759	858
5	0	0	99	0	0	0	746	845
6	0	0	99	0	73	0	679	850
TOTAL	0	0	2,369	972	3,960	1,665	20,410	29,376

a 1-1/2 percent savings would then amount to \$150,000 to \$225,000 per year.

The amount of savings realized by the use of the programs is conservatively estimated to be one and one-half to two percent of the thermal generation costs. These savings are indicative of what can be obtained in a complex system such as the one considered here.

Further savings could be obtained by adding features such as optimum management of the fuel supply network for the thermal plants, on-line analysis of the sale of economy energy, and consideration of transmission losses. Preliminary analysis indicates that these and other features can be added to the programs without a substantial increase in computer time.

It is important to note that these programs do not replace the human dispatcher. Instead, the results serve as an operating guide to the dispatcher and provide him with data he could not otherwise easily obtain. The programs have been tailored to provide the maximum useful information while requiring as little manual input data as possible. As a result of using this program, the dispatcher has a better idea of how to maximize the performance of the system, and he has more time to consider other aspects of system operation.

Finally, the applicability of the programs is not restricted to special systems. They can readily be extended to any system characterized by two or more of the following sources of generalization:

(1) pumped storage units,
(2) conventional hydro plants,
(3) thermal plants,
(4) coordinated operation with other utilities,
(5) diversity interchange contracts.

CASE STUDY II: A NUCLEAR FUEL REPROCESSING PLANT PLANNING PROBLEM

A. INTRODUCTION

The purpose of this example is to provide a study of the feasibility of applying advanced dynamic programming computational procedures to planning the installation of nuclear fuel reprocessing plants.

The general nature of the problem and the details of the sample problem are described in the next section. In Section C a dynamic programming formulation is developed. The standard dynamic programming computational procedure is presented in Section D, and the difficulties that arise in applying it to the problem are examined.

In Sections E and F advanced dynamic programming procedures that can be applied to this problem are discussed; Section E describes general procedures, while Section F develops procedures based on the structure of the problem.

In Section G specific algorithms based on these procedures are applied to the sample problem, and computational results are presented. In Section H the results obtained for the sample problem are summarized, and the feasibility of solving even more extensive problems is evaluated.

B. DESCRIPTION OF THE PROBLEM

One of the critical steps in the nuclear power reactor fuel cycle is the reprocessing of spent fuel. The requirement for plants to perform the reprocessing is presently rather small, but it is expected to grow rapidly in the next decades. The spent fuel can be assumed to arise from a number of discrete sources, each corresponding to one geographical region. Similarly, plants can be built at a number of discrete sites, each corresponding to geographical regions.

The construction costs for these plants are very high, but a pronounced economy of scale is realized from building large plants. A still-developing technology plus political and environmental factors limit the maximum size of plant that can be built in any year. It is permissible to store spent fuel and defer plant construction; however, the cost of storing the spent fuel is substantial. Shipping spent fuel to a distant plant with excess capacity is also permitted; however, the additional cost over shipping to a local plant is again significant.

The planning problem is to determine that schedule for plant construction (by year, site, and size), fuel reprocessing, fuel storage, and fuel shipping which results in the minimum discounted cost over the planning period.

Data for a sample problem having two sources and two construction sites is summarized in Table 6.3. The planning horizon is taken to be 15 years. The amount of spent fuel originating at each source for each year is shown in Columns 1 and 2. The maximum size plant that can be built each year is shown in Column 3. The capacity of a newly built plant is assumed to be immediately available. However, each plant has a finite lifetime of ten years; for example, the capacity of a plant built in Year 3 would be available only in Years 3 through 12 inclusive. Multiple plants can exist at a site in a given year by having been constructed there at different times. No correction is made for the capital cost of unused plant life which extends beyond the planning horizon.

The plant capacity cost is given by (9×10^6) (capacity in metric tons/years)$^{0.35}$. The reprocessing cost is \$500/metric ton. The cost of storing fuel for one year is \$10,600/metric ton. No spent fuel may remain in storage at the end of the planning period. The round-trip shipping costs are \$1,625/metric ton from Source 1 to Site 1; \$2,000/metric ton from Source 2 to Site 2; and \$4,060/metric ton from either Source 1 to Site 2 or Source 2 to Site 1.

Table 6.3. First Sample Problem — Parameters

	ANNUAL METRIC TONS SPENT FUEL AT SOURCE		PLANT SIZE LIMIT
Year, k	1	2	METRIC TONS/YEAR
1	94	0	260
2	228	86	520
3	370	285	780
4	529	455	1,040
5	781	684	1,300
6	1,204	1,090	1,560
7	1,694	1,589	1,820
8	2,079	2,002	2,080
9	2,694	2,674	2,080
10	3,525	3,602	2,080
11	4,358	4,633	2,080
12	4,860	5,400	2,080
13	5,370	6,090	2,080
14	5,910	6,770	2,080
15	6,510	7,640	2,080

Plant Capital Cost = ($\$9 \times 10^6$) (metric tons/year capacity)$^{0.35}$

Fuel Reprocessing Cost = \$500/metric ton

Storage Cost = \$10,600/metric ton

Round-Trip Shipping Costs: Source 1 - Site 1 = \$1,625/metric ton
Source 2 - Site 2 = \$2,000/metric ton
Source 1 - Site 2, Source 2 - Site 1 = \$4,060/metric ton

Interest Rate = 12%/year

Plant Life = 10 years

Discounting to present worth is obtained using an annual interest rate of 12 percent.

The size of plant that can be built is constrained to be a multiple of 260 metric tons/year. Two sample problems are considered. In the first, 8 allowable plant sizes are 260, 520, 760, 1,040, 1,300, 1,560, 1,820, and 2,080 metric tons/year. In the second, 4 allowable sizes are 260, 520, 1,040, and 2,080 metric tons/year.

C. DYNAMIC PROGRAMMING FORMULATION

The dynamic programming formulation of this problem proceeds as follows. First, the index k is taken to represent time in years. The state variables are taken to be

$x_1(k)$ = amount of capacity at Site 1 in Year k,

$x_2(k)$ = amount of capacity at Site 2 in Year k,

$x_3(k)$ = amount of spent fuel stored at Source 1 in Year k,

$x_4(k)$ = amount of spent fuel stored at Source 2 in Year k.

The decision variables are taken to be

$u_1(k)$ = amount of capacity installed at Site 1 in Year k,

$u_2(k)$ = amount of capacity installed at Site 2 in Year k,

$u_3(k)$ = amount of fuel from Source 1 processed at Site 1,

$u_4(k)$ = amount of fuel from Source 2 processed at Site 2,

$u_5(k)$ = amount of fuel from Source 2 processed at Site 1,

$u_6(k)$ = amount of fuel from Source 1 processed at Site 2.

The first two system equations describe the evolution of the total capacity at the sites:

$$x_1(k+1) = x_1(k) + u_1(k+1) - u_1(k-9) , \tag{6.11}$$

$$x_2(k+1) = x_2(k) + u_2(k+1) - u_2(k-9) . \tag{6.12}$$

The term $u_i(k+1)$ is the capacity added in the current year, while $u_i(k-9)$ is the capacity built ten years ago whose lifetime has just expired. To keep Eqs. (6.11) and (6.12) formally correct, $u_1(k)$ and $u_2(k)$ are defined to be zero for $k \leq 0$.

The last two system equations are storage balances at the two sources:

$$x_3(k+1) = x_3(k) + d_1(k+1) - u_3(k+1) - u_6(k+1) , \qquad (6.13)$$

$$x_4(k+1) = x_4(k) + d_2(k+1) - u_4(k+1) - u_5(k+1) . \qquad (6.14)$$

The terms $d_1(k+1)$ and $d_2(k+1)$ are the spent fuel requirements arising in Year (k+1) at Sites 1 and 2, respectively; these quantities are taken from Table 6.3. The terms $u_3(k+1)$ and $u_6(k+1)$ represent fuel from Source 1 processed at Sites 1 and 2, respectively, while $u_4(k+1)$ and $u_5(k+1)$ represent fuel from Source 2 processed at Sites 2 and 1, respectively.

The performance criterion becomes

$$J = \sum_{k=1}^{15} \frac{1}{(1.12)^k} \Big[9\times10^6 \{ [u_1(k)]^{0.35} + [u_2(k)]^{0.35} \} + 500[u_1(k) + u_2(k) + u_3(k) + u_4(k)] + 10{,}600[x_3(k) + x_4(k)] + \{1{,}625u_3(k) + 2{,}000u_4(k) + 4{,}060[u_5(k) + u_6(k)]\} \Big] . \qquad (6.15)$$

The factor $\frac{1}{(1.12)^k}$ discounts the costs to present worth. The first term inside the large brackets is the plant construction cost; the second term in the fuel processing cost; the third term is the storage cost; and the fourth term is the shipping cost. The summation is over the 15-year planning horizon.

There are a number of constraints on the problem. The size of plant that can be built in any year is bounded by

$$u_1(k) \leq c(k) \ , \tag{6.16}$$

$$u_2(k) \leq c(k) \ , \tag{6.17}$$

$$k = 1,2,\ldots,15 \ ,$$

where $c(k)$ is the plant size limit at Year k as obtained from Table 6.3. The plant capacity $u_1(k)$ and $u_2(k)$ are also constrained to be positive or zero.

The amount of fuel in storage at the final time must be zero

$$x_3(15) = 0 \ , \tag{6.18}$$

$$x_4(15) = 0 \ . \tag{6.19}$$

In addition, the fuel stored at earlier times, $x_3(k)$ and $x_4(k)$, $k=1,2,\ldots,14$, must be positive or zero.

The amount of fuel processed at the sites is bounded by the capacity at the site:

$$u_3(k) + u_6(k) \leq x_1(k) \ , \tag{6.20}$$

$$u_4(k) + u_5(k) \leq x_2(k) \ , \tag{6.21}$$

$$k = 1,2,\ldots,15 \ .$$

In addition, all the variables $u_3(k)$, $u_4(k)$, $u_5(k)$, and $u_6(k)$, must be positive or zero.

At the initial time there is no capacity at either site and no fuel in storage at either source. The initial condition is thus

$$x_1(0) = x_2(0) = x_3(0) = x_4(0) = 0 \ . \tag{6.22}$$

These basic relations can be summarized as follows. The system equations become

$$x(k+1) = f[x(k), u(k+1), u(k-9), k] , \quad (6.23)$$

where the components of f are defined by Eqs. (6.11) through (6.14).

The performance criterion is written

$$J = \sum_{k=1}^{15} L[x(k), u(k), k] , \quad (6.24)$$

where L is defined by Eq. (6.15).

The constraints can be summarized as

$$x \, \varepsilon \, X(k) , \quad (6.25)$$

$$u \, \varepsilon \, U(x,k) , \quad (6.26)$$

where the sets $X(k)$ and $U(x,k)$ are defined by Eqs. (6.16)-(6.21) and where all components of $x(k)$ and $u(k)$ are restricted to positive or zero values. Finally, the initial condition, Eq. (6.22), can be written as

$$x(0) = 0 . \quad (6.27)$$

The problem can now be formulated as follows:

Find: The sequence of decisions $u(1), u(9), \ldots, u(15)$, that minimizes the performance criterion in Eq. (6.24), subject to the system equations Eq. (6.23), the constraints of Eqs. (6.25) and (6.26), and the initial condition in Eq. (6.27).

D. *THE STRAIGHTFORWARD DYNAMIC PROGRAMMING SOLUTION*

The problem formulated in the previous section can be solved, at least in principle, by a straightforward dynamic programming procedure. If the finite lifetime of the plants is ignored for the moment, the system equations become*

$$x(k+1) = f[x(k), u(k+1), k] . \tag{6.28}$$

The minimum cost function, $I(x,k)$, can then be defined as the minimum cost achievable starting at given present state x and present time k and going to the end of the planning horizon. Formally

$$I(x,k) = \min_{u(k+1),\ldots,u(15)} \left\{ \sum_{j=k+1}^{15} L[x(j),u(j),j] \right\} , \tag{6.29}$$

where $x(k) = x$ becomes the initial condition and where the minimization is subject to the system equation, Eq. (6.28) and the constraints, Eqs. (6.25) and (6.26). The basic dynamic programming equation is then obtained as

$$I(x,k) = \min_{u(k+1)} \Big\{ L[f(x,u(k+1),k), u(k+1), k] + I\{f[x, u(k+1), k], k+1\} \Big\} . \tag{6.30}$$

This equation determines the optimum decision at each state and time, denoted as $\hat{u}(x,k)$, as that value of u for which the minimum in Eq. (6.30) is attained. The optimum sequence of decision from the initial state $x(0) = 0$ can easily be recovered from this information.

The solution of Eq. (6.30) is obtained by iterating backward in time. Because no credit is given for unexpired capacity

---

*The system equations are customarily written as

$$x(k+1) = f[x(k),u(k),k] .$$

However, the nature of Eqs. (6.11)-(6.14) dictates that they appear in the above form. The more familiar form could be obtained by defining $v(k)$ to be $u(k+1)$ and transforming the equations, but later discussions are clearer if Eq. (6.28) is carried through as it stands.

beyond the planning horizon, the starting condition for these iterations is

$$I(x,15) = 0 . \tag{6.31}$$

The iterative equation is solved by quantizing the state variables to a discrete set of values and solving Eq. (6.30) at these discrete points. The minimization in Eq. (6.30) is performed by quantizing the decision variables to a discrete set of values, evaluating the expression in brackets for each discrete decision and choosing the minimum by a direct search. The quantization in both state and decision variables is determined by the allowed plant sizes. The various aspects of this procedure have already been covered in detail.

Unfortunately, there are two difficulties with applying the standard methods to the above nuclear fuel reprocessing plant planning problem. The first is the familiar "curse of dimensionality" — the exponential growth of computer memory and time requirements with the number of state variables. The sample problem has four state variables; for a reasonable grid size in the quantization, the sample problem is solvable on a large machine. However, for the actual problem in which there are 8 regions and 16 state variables, there is little hope of a straightforward solution, even on an advanced machine.

The second difficulty is related to the finite lifetime of the plants. The presence of the terms $u_1(t-9)$ and $u_2(t-9)$ in Eqs. (6.11) and (6.12), respectively, means that these equations do not have the Markovian property required of dynamic programming system equations. This situation can be remedied in a number of ways, but all of them require defining a substantial number of additional state variables. The resulting computational requirements are excessive for the sample problem and completely unreasonable for the actual problem.

In the next section some of the procedures from Chapter 4 that are appropriate to the nuclear fuel reprocessing plant planning problem are reviewed.

E. *ADVANCED DYNAMIC PROGRAMMING COMPUTATIONAL PROCEDURES*

As indicated in Section D, there are serious computational difficulties in solving this problem with the straightforward dynamic programming procedure. However, there are a number of advanced dynamic programming procedures that are capable of reducing the computational difficulties to the point where a solution to actual problems can be obtained. In this section, the procedures that have been found to be of most promise are summarized.

Forward Dynamic Programming

The most severe difficulty cited in Section D is the additional state variables required to account for the finite lifetime of the plants and the resulting terms $u_1(t-9)$ and $u_2(t-9)$ in Eqs. (6.11) and (6.12). A procedure that eliminates the need for these state variables is based upon forward dynamic programming as discussed in Chapter 4 of Part I.

The iterative equation for forward dynamic programming is similar to Eq. (6.30). The minimum cost function is defined somewhat differently; instead of being the minimum cost that can be achieved starting at the present state and going to the end of the process, it is the minimum cost that can be achieved starting in the initial state and arriving at the present state at the present time. Writing this minimum cost as $I'(x,t)$ to distinguish it from the minimum cost function for the procedure of Section D, the definition becomes

$$I'(x,k) = \min_{u(0),\ldots,u(k-1)} \left\{ \sum_{j=0}^{k-1} L[x(j),u(j),j] \right\}, \qquad (6.32)$$

where

$$f[x(k-1),\ u(k),\ k-1] = x\ . \tag{6.33}$$

Note that in this definition x is the last state along a trajectory, rather than the initial state, as in Eq. (6.29). The quantity $h[x, u(k), k]$ is now defined by

$$f[h[x,\ u(k),\ k],\ u(k),\ k-1] = x\ . \tag{6.34}$$

The iterative equation can then be derived as

$$I(x,k) = \min_{u(k-1)} \Big\{ L[h[x,\ u(k),\ k]\ u(k),\ k-1] + I'[h[x,\ u(k),\ k],\ k-1] \Big\}\ . \tag{6.35}$$

Note that this iterative procedure determines $I(x,k)$ in terms of $I(x,k-1)$; thus, the calculations do indeed go forward in time rather than backward. The initial condition to start the iteration is

$$I'(x,0) = 0\ , \tag{6.36}$$

i.e., no cost is incurred before the planning interval begins. Note also that the optimum decision which minimizes Eq. (6.35) tells us what decision should have been applied at the last time increment to result in state $x(k) = x$, rather than which decision to apply at x. This decision is denoted as $\hat{u}'(x,k)$, where the prime is again used to distinguish this quantity from the corresponding quantity in the backward procedure.

One way of carrying out the computations is to use a procedure exactly analogous to that described in Section D: at a given quantized x and k try all possible quantized decisions $u(k-1)$; find the corresponding previous state $x(k-1)$; evaluate the quantity inside brackets in Eq. (6.35), using interpolation procedures if necessary, and pick the optimum decision and minimum cost. This procedure has some deficiencies. In the first place, there is no a priori way of ensuring that a given decision $u(k-1)$ is in fact an admissible decision at $x(k-1) = h[x,u(k-1),k]$. Second, it may be difficult to compute the inverse functional h.

A procedure which overcomes these difficulties and offers other advantages as well is the following: at each quantized state $x(k-1)$ where $I'[x(k-1), k-1]$ has been computed, apply all the quantized admissible decisions; for each corresponding next state, $x(k) = f[x(k-1),u(k-1),k-1]$, check to see if this state has been the next state for any decision applied at previous values of $x(k-1)$; if it has not previously been a next state, then store the quantity in braces in Eq. (6.35) as the tentative minimum cost at that point; if it has, then compare the quantity in braces in Eq. (6.35) with the tentative minimum cost already computed at that point, and if this quantity is less than the tentative minimum cost, replace the value stored there. This procedure continues until the quantized admissible decisions have been applied at every quantized state $x(k-1)$. The tentative minimum costs and optimum decisions at each $x(k) = f[x(k-1),u(k),k-1]$ are then the true minimum costs and optimum decisions at these points.

If the values of the ten previous capacity installations are now stored along with the minimum cost function $I'(x,k)$, it is possible to eliminate the plants at the end of their lifetime by using Eqs. (6.11) and (6.12). Because there are only a few discrete plant sizes, these ten values can be coded and stored in one or two words of memory.

Successive Approximations

The philosophy behind this method is to break up a problem with several decision variables into subproblems with a single decision variable each. For a problem with an equal number of decision and state variables, the subproblems have only one state variables, two of the decision variables can be eliminated by off-line optimization of $L[x(k),u(k),k]$ in Eq. (6.15) with respect to $u_5(k)$ and $u_6(k)$. The procedure for doing this is indicated as $\{x^{(0)}(k)\}$.

Next, one of the n state variables, say x_{i_1}, is selected. A dynamic programming problem is then solved in which the sequence of all other states, namely $\{x_i^{(0)}(k)\}$, $i \neq i_1$, is held fixed. The problem thus has only one state variable. On the other hand, there are still n decision variables. However, the constraint that the sequence of (n-1) of the states must remain fixed imposes (n-1) equality constraints on the n decision variables so that effectively there is only one control variable as well. The performance criterion remains the same as in Eq. (6.24), and the constraints are as in Eqs. (6.25) and (6.26). The solution to this optimization problem is a new control sequence, $\{u^{(1)}(k)\}$, and a new trajectory $\{x^{(1)}(k)\}$.

Next, a different state variable, say x_{i_2}, is selected and the above optimization procedure repeated. This continues until every state variable has been selected for optimization at least once. Termination occurs when the same control sequence and trajectory are obtained for optimization with respect to any of the state variables.

The advantage of this procedure is that the solution of an n-dimensional dynamic programming problem is reduced to the solution of one-dimensional problems. Thus, the computational requirements, both time and storage, increase linearly with n rather than exponentially. The computational savings in high-dimensional problems can be truly astronomical.

N-Best Trajectories

In the general reprocessing plant planning problem it may not be possible to prove that either of the two procedures discussed in the previous sections converge to the absolute optimum.

This may also be true for some of the methods suggested in the next section. In all these cases the reason a global optimum cannot be guaranteed is that some of the state variables, e.g., those associated with the finite plant lifetime in the forward dynamic programming procedure of Section F, are suppressed. Nevertheless, it is expected that one or a combination of these procedures will come close to the global optimum.

A procedure that is often very effective in finding the global optimum in discrete problems of this type is computation of the N-best trajectories. The central idea of this procedure is to compute not just the single best decision at each state and stage, but the N best decisions. If computing only the best policy at each state and stage yields a close approximation to the optimum solution, then computing the N best decisions, for N sufficiently large, will obtain the true optimum with high confidence.

The N best decisions at a given state and stage can be computed in much the same manner as the single optimal decision is computed in Eq. (6.30). If $I^{(N)}(x,k)$ is defined as the cost corresponding to the N<u>th</u> best policy from state $\underline{x}$, stage k, then the iterative equation can be written as

$$I^{(N)}(x,k) = \min_{u(k+1),N'}^{(N)} \{L[f(x,u(k+1),k),\ u(k+1),\ k] + I^{(N')}[f(x,\ u(k+1),\ k),\ k+1]\}\ , \qquad (6.37)$$

$$N' = 1,2,\ldots,\overline{N}\ ,$$
$$N = 1,2,\ldots,\overline{N}\ ,$$

where N' = index for N'-best cost at stage $(k+1)$,

N = index for N-best cost at stage k,

$\overline{N}$ = largest value of N that is to be considered,

$\min_{u(k+1),N'}^{(N)}$ = <u>o</u>perator that denotes the determination of the N-best values of the argument, where a search is made over $u(k+1)$, the decision, and N', the index of the N'-best values of $I[f(x,u(k+1), k), k+1]$.

The operator $\min^{(N)}$ can be implemented in a number of ways. The most straightforward procedure is to evaluate $L[f(x,u(k+1),k), u,k] + I^{(N')}[f(x,u)k+1),k),k+1]$ for all $u(k+1) \varepsilon U$ and for all $N' = 1,2,\ldots,\bar{N}$, and then to retain the $\bar{N}$-best values, in order, as $I^{(N)}(x,k)$, $N = 1,2,\ldots,\bar{N}$.

F. *PROCEDURES BASED ON THE STRUCTURE OF THE PROBLEM*

The procedures described in Section E can be applied to any dynamic programming problem. However, it is also possible to exploit the particular structure of the problem to reduce computational requirements even further. In this section the most promising procedures for taking advantage of the problem structure are discussed.

Minimum Inventory Policy

Two of the four state variables in this problem are associated with the accumulated inventory at each source. The procedure described in this section takes advantage of the relative cost of processing, shipping and storing fuel to eliminate the storage state variables. The storage corresponding to the state variables for capacity is instead calculated by applying a set of rules called the minimum inventory policy.

From Table 6.3 it is seen that the processing and round-trip shipping cost for one metric ton of fuel is \$2,125 for fuel originating at Source 1 and processed at Site 1, \$2,500 for fuel originating at Site 2 and processed at Site 2, and \$4,560 for fuel originating at Source 1 and processed at Site 2 or originating at Source 2 and processed at Site 1. The cost for storing one metric ton at either source is \$10,600.

Based on a comparison of these costs, it is seen that at either source there is a specific set of rules for handling the fuel. This set of rules, which will be called the minimum inventory policy, can be summarized as:

(1) Process as much fuel as possible at the local site (Site 1 for Source 1, Site 2 for Source 2). If all the fuel at this source is then processed, stop; otherwise, continue to step (2).

(2) Process as much fuel as possible with the unused capacity of the remote site. If all fuel at the source is now processed, stop; otherwise, continue to step (3).

(3) Store all fuel remaining at the source.

Processing fuel at the local site rather than at a remote site is clearly preferred because it avoids incurring the additional round-trip shipping cost. Processing fuel at a remote site is preferred to storing fuel because the cost of storage is so high; even if the stored fuel could be processed at no charge in the next time interval, it would still be less expensive to process fuel now at a remote site than to save it for a more favorable shipping situation.

This ordering can be implemented within the forward dynamic programming procedure to eliminate the storage state variables x_3 and x_4. Assume that for the present capacity state characterized by $x_1(k)$ and $x_2(k)$, values of the inventories $x_3(k)$ and $x_4(k)$ are known. Then, for an arbitrary next capacity state $x_1(k+1)$, $x_2(k+1)$, the corresponding inventories $x_3(k+1)$ and $x_4(k+1)$ can be specified by applying the minimum inventory policy to Eqs. (6.13) and (6.14). This can be seen by examining these equations, reproduced here for convenience:

$$x_3(k+1) = x_3(k) + d_1(k+1) - u_3(k+1) - u_6(k+1) , \qquad (6.13)$$

$$x_4(k+1) = x_4(k) + d_2(k+1) - u_4(k+1) - u_5(k+1) . \qquad (6.14)$$

First, $u_3(k+1)$ is set equal to $[x_3(k) + d_1(k+1)]$. If this value is less than $x_1(k+1)$, then $x_3(k+1)$ is set to zero, as is $u_6(k+1)$. A similar analysis is made for $u_4(k+1)$.

If $[x_3(k) + d_1(k+1)]$ is larger than $x_1(k+1)$, a check is made to see if there is any unused capacity at Site 2; i.e., to see if $[x_4(k) + d_2(k+1)]$ is less than $x_2(k+1)$. If it is less, then as much as possible of the excess fuel from Site 1, $[x_3(k) + d_1(k+1)]$, is processed there, and $u_6(k+1)$ is set to this value. Any remaining fuel is stored at Site 1, and $x_3(k+1)$ is set to this value. A similar analysis is performed if $[x_4(k) + d_2(k+1)]$ is larger than $x_2(k+1)$.

This procedure, in addition to computing $x_3(k+1)$ and $x_4(k+1)$, determines $u_3(k+1)$, $u_4(k+1)$, $u_5(k+1)$, and $u_6(k+1)$. It is thus possible to evaluate the single-stage cost function, L, in Eq. (6.15).

The forward dynamic programming procedure in Section E can now be performed using only the capacity state variables x_1 and x_2, provided that the corresponding inventories x_3 and x_4 are computed by the minimum inventory policy and carried forward along with the minimum cost function and optimum decision function.

In order to start the procedure it is necessary to know the inventories corresponding to all possible initial states. In this problem there is initially zero inventory at both sources.

The increase in computational requirements over a standard forward dynamic programming procedure applied to a problem with two stage variables is very slight. The additional calculations of the minimum inventory policy are mostly simple logic decisions. The additional data storage needed for the inventories for each state can be obtained by packing the data into one word.

The minimum inventory policy cannot be implemented directly in a backward dynamic programming procedure because of the ambiguity that arises when the inventory is zero at the next stage.

If this occurs, there is a whole range of inventories at the present state for which the minimum inventory policy will obtain the zero values.

It is conceptually possible that the forward procedure will not determine the global optimum. Even if the finite plant lifetime is ignored, the procedure does not use a full state-space description. Instead, only one inventory is propagated forward for each capacity state. It may happen that, for some intermediate state along the true optimal trajectory, an inventory other than the one corresponding to the true optimum is selected on the basis of costs incurred up to that point. It can be argued that this is unlikely, and it has not occurred in simple examples where this could be checked. Nevertheless, attempts to find conditions under which it can be proven that this will not happen have thus far been unsuccessful.

Lumped Site and Source

Another approach to reducing the computational requirements is to consider the two sources and two sites as one equivalent source and one equivalent site. The number of state variables is thus reduced from 4 to 2. The high cost of storage ($10,600/metric ton) relative to the difference in shipping cost between a local and a remote site ($2,435/metric ton at most) makes this "lumped approximation" a reasonable one.

In this procedure the two system equations for capacity, Eqs. (6.11) and (6.12), are replaced by the single equation

$$x_1(k+1) = x_1(k) + u_1(k+1) + u_2(k+1) \,, \tag{6.38}$$

where $x_1(k)$ = total capacity at both sites at Year k,

$u_1(k)+u_2(k)$ = total amount of capacity added at both sites during Year k.

It is permissible to build two plants in any one year; this is explicitly noted by carrying both u_1 and u_2 in Eq. (6.38). However, the plant size limitation of Eqs. (6.16) and (6.17) are still imposed.

The two system equations for inventory, Eqs. (6.13) and (6.14), are replaced by the single equation

$$x_2(k+1) = x_2(k) + d_1(k+1) + d_2(k+1) - u_3(k+1) , \qquad (6.39)$$

where $x_2(k)$ = total inventory at both sources in Year k,

$d_1(k)+d_2(k)$ = total demand for fuel to be processed from both sources,

$u_3(k)$ = total amount of fuel processed at both sites.

The total demand $(d_1 + d_2)$ is taken by adding the two individual demands in Table 6.3. The amount of fuel processed is limited by the total amount of capacity available:

$$u_3(k) \leq x_1(k) . \qquad (6.40)$$

The cost function becomes

$$J = \sum_{k=1}^{15} \frac{1}{(1.12)^k} \Big\{ 9\times10^6 [(u_1(k))^{0.35} + (u_2(k))^{0.35}] + 500[u_3(k)] + s(k)[u_3(k)] + 10{,}600[x_2(k)] \Big\} , \qquad (6.41)$$

where the variable $s(k)$ is the assumed shipping cost at stage k. This variable can be set at one of the local site round-trip shipping costs, the remote-site round-trip shipping cost, or any value in between. Computational experience indicates that for the numerical values considered here the solution is not very sensitive to $s(k)$.

The solution to the problem specified in Eqs. (6.38)-(6.41) determines the size and timing of plant construction, but it does not specify where the plants should be built. The location can be determined either by further analysis, by simulation of all alternatives, or by some heuristic rule, such as building alternately at the two sites.

Replacement Cost Penalty Function

As stated in Section D, the major difficulty in a dynamic programming solution to this problem is the finite lifetime of the plants. One approach to overcoming this difficulty is to use a penalty function to account for the fact that every plant must be replaced at the end of its lifetime. This can be implemented by revising the single-stage cost function $L[\underline{x}(k),\underline{u}(k),k]$ as follows:

$$L'[x(k),u(k),k] = L[x(k),u(k),k] + \varphi[u_1(k),u_2(k),k] , \quad k = 1,\ldots,5 ,$$

$$L'[x(k),u(k),k] = L[x(k),u(k),k] , \quad k = 6,\ldots,10 ,$$

$$L'[x(k),u(k),k] = L[x(k),u(k),k] - \varphi[u_1(k-10),u_2(k-10), k-10] , \quad k = 11,\ldots,15 , \tag{6.42}$$

where $L'[x(k),u(k),k]$ = revised single-stage cost functions $k = 1,\ldots,15$,

$\varphi[u_1(k),u_2(k),k]$ = replacement cost penalty function $k = 1,\ldots,5$.

The original problem is then solved using L' rather than L. The penalty function adds an anticipated replacement cost for plants constructed during the first five years and then removes this cost ten years later when the plant's lifetime actually ends. At this latter time the plant's capacity is removed from the system, and the optimization must determine the actual least cost for replacing the removed capacity.

A number of penalty functions were tried in this problem. The first penalty function computes the cost of replacing the expired capacity at year k+10 on the assumption that some capacity would be added at each site during that year to meet the expanding demand regardless of any removal of plant capacity. Therefore, it appears appropriate to compute an average replacement cost on the assumption that these plants are replaced as part of building one of the largest allowed plants. The average replacement cost was selected to be

$$9\times 10^6 \cdot \frac{\{[2908]^{0.35} - [2080 - c(k)]^{0.35}\}}{c(k)}, \tag{6.43}$$

where $c(k)$ is the maximum capacity that can be installed in each region in year k (see Table 6.3). The maximum capacity that can be built during year $k+10$ is 2080. The present-worth replacement cost $\varphi_1(k)$ becomes

$$\varphi_1[u_1(k), u_2(k), k] = \frac{9\times 10^6}{1.12^{k+10}} \left\{\frac{[2080]^{0.35} - [2080 - c(k)]^{0.35}}{c(k)}\right\} \cdot [u_1(k) + u_2(k)]. \tag{6.44}$$

Because φ_1 is linear in the control variables, it is actually a Lagrange multiplier, as discussed in earlier chapters.

The second penalty function reflects the replacement cost directly:

$$\varphi_2[u_1(k), u_2(k), k] = \frac{9\ 10^6}{1.12^{k+10}} \{[u_1(k)]^{0.35} + [u_2(k)]^{0.35}\}. \tag{6.45}$$

These two penalty functions performed better in computational experiments than any other choices tried.

It can be shown that there does exist a penalty function such that optimizing the sum of the penalty function and the original criterion using the system equations without the finite lifetime yields the same solution as optimizing the original criterion using the system equations with the finite lifetime. However, no general method for determining the correct penalty function is known.

G. *COMPUTATIONAL RESULTS*

All of the computational procedures described in Sections E and F were applied to the sample problems of Section B. The most effective solution methods proved to be combinations of these procedures. In this section some of the computational results on the two cases of the sample problem are presented. In Case 1 plant sizes are 260, 520, 780, 1040, 1300, 1560, 1820, and 2080 metric tons/year, while in Case 2 plant sizes are 260, 520, 1040, and 2080 metric tons/year. A tabular summary of these results is shown in Tables 6.4 and 6.5, respectively.

In all of these programs no attempt was made to minimize the core memory required. About 53,000 words of usable core was available in the UNIVAC 1108 system configuration. The amount of core used in each program was the maximum consistent with efficient program execution. As a result, the storage requirements given for the programs are quite conservative, while the running times are close to the fastest possible.

The Program DP1 - Minimum Inventory Policy and Forward Dynamic Programming

One of the basic computer programs developed for this problem, called DP1, combines the minimum inventory policy (Section F) with the forward dynamic programming procedure (Section E). As described in Section F, the program has two state variables — one for the capacity at each site — while the inventory corresponding to each state is evaluated by the minimum inventory policy.

Table 6.4. Construction Policies for Solutions to Case 1

Under each solution entry in first column is plant size built at site 1, while entry in second column is plant size built at site 2. Allowed plant sizes are 260, 520, 780, 1040, 1300, 1560, 1820, 2080.

SOLUTION		A		B		C		D		E		F	
ALGORITHM UTILIZED		• DP1 • DP1 + 1 • 3 Best DP1 + 1 (1st Solution) • DP2 Starting With Site 2 • DP3 + 1 • 3 Best DP3 + 1 (1st Solution)		• 3 Best DP1 + 1 (2nd Solution)		• DP1 + 2 • 3 Best DP1 + 1 (3rd Solution)		• DP2 Starting With Site 1		• DP3		• 3 Best DP1	
YEAR	MAXIMUM PLANT SIZE IN EACH SITE												
1	260	0	0	0	0	0	0	0	0	0	0	0	0
2	520	0	0	0	0	0	0	0	0	0	0	0	0
3	780	780	0	780	0	780	0	780	0	780	0	780	0
4	1040	0	0	0	0	0	0	0	0	0	0	0	0
5	1300	0	0	0	0	0	0	0	0	0	1300	0	0
6	1560	0	1560	0	1560	1560	1560	0	1560	0	0	1560	1560
7	1820	1820	0	1820	0	0	0	1820	0	1820	0	0	0
8	2080	0	0	0	0	0	0	0	0	0	0	0	0
9	2080	0	2080	0	2080	0	2080	0	2080	0	2080	0	2080
10	2080	0	0	0	0	2080	0	2080	0	0	0	2080	0
11	2080	2080	2080	2080	2080	0	2080	0	0	2080	2080	0	2080
12	2080	0	0	0	0	0	0	0	2080	0	0	0	0
13	2080	2080	0	0	2080	2080	0	2080	0	2080	0	0	0
14	2080	0	2080	2080	0	0	2080	0	2080	0	2080	0	2080
15	2080	520	0	520	0	780	0	520	0	0	2080	730	0
TOTAL COST IN $ MILLION		459.7		460.4		460.6		461.0		461.0		461.5	

Table 6.5. Construction Policies for Solutions to Case 2

SOLUTION		A'		B'		C'	
ALGORITHM UTILIZED		• DP 1 • 3 Best DP1 + 1 • DP2 Starting With Site 1 or Site 2 • DP3 + 1		• 3 Best DP1 + 1 (2nd Solution)		• 3 Best DP1 + 1 (3rd Solution)	
YEAR	MAXIMUM PLANT SIZE IN EACH SITE						
1	260	0	0	0	0	0	0
2	520	0	0	0	0	0	0
3	520	0	0	0	0	0	0
4	1040	1040	1040	1040	1040	1040	1040
5	1040	0	0	0	0	0	0
6	1040	0	0	0	0	0	0
7	1040	0	0	0	0	0	0
8	2080	2080	2080	2080	2080	2080	2080
9	2080	0	0	0	0	0	0
10	2080	0	0	0	0	0	0
11	2080	2080	2080	2080	2080	2080	2080
12	2080	0	0	0	0	0	0
13	2080	0	2080	2080	0	0	2080
14	2080	2080	0	0	2080	2080	0
15	2080	0	2080	0	2080	2080	0
TOTAL COST IN $ MILLION		470.1		470.4		470.5	

The capabilities for computing the N-best trajectories (Section E) and for adding replacement cost penalty functions (Section F) are available as optional features.

The solution obtained by DP1 on Case 1 is shown as Solution A in Table 6.4. A more detailed presentation of this solution and the corresponding costs is given in Table 6.6 and Figure 6.3. Solution A has the lowest cost, $\$459.7\times10^6$, of any solution obtained to this problem, and analysis of these results strongly indicates that it is the true optimum. This solution required approximately 90 seconds on a UNIVAC 1108 and utilized 40,000 words of core memory.

The program was run with the two replacement penalty costs φ_1 and φ_2. With φ_1 Solution A was again obtained; while with φ_2 Solution C, which was a cost of $\$460.6\times10^6$, was found. Computer time and storage requirement were almost identical with those of the unmodified DP1.

For the N-best trajectories calculation with $N = 3$, the best solution obtained was Solution F, which has a cost of $\$461.5\times10^6$. It is at first surprising that a better solution is obtained with only one trajectory carried forward than with a procedure that propagates three trajectories. However, this apparent discrepancy can be explained in a manner that adds insight into the nature of the reprocessing plant planning problem.

It has been previously pointed out that DP1 does not contain a complete state description for the problem. First, the minimum inventory policy propagates forward only one inventory for each capacity state, and as a result the true optimum solution may not be obtained (see Section F). More important, the procedure does not carry state variables for taking into account the finite plant lifetimes (see Section E).

When the three best trajectories are carried forward, it is a priori more likely that the true solution will be obtained than

Table 6.6. Detailed Analysis of Solution A, The Optimal Solution to Case 1, Plant Sizes 260, 520, 780, 1,040, 1,300, 1,560, 1,820, 2,080

	CAPACITY		PLANTS INSTALLED		INVENTORY		DEMAND		SHIPPING FROM SOURCE I TO SOURCE J				TOTAL PROCESSED	
YEAR	SITE 1	SITE 2	SITE 1	SITE 2	SRCE 1	SRCE 2	SRCE 1	SRCE 2	SRCE 1 SITE 1	SRCE 1 SITE 2	SRCE 2 SITE 1	SRCE 2 SITE 2	SITE 1	SITE 2
1	0.	0.	0.	0.	94.	0.	94.	0.	0.	0.	0.	0.	0.	0.
2	0.	0.	0.	0.	322.	96.	228.	86.	0.	0.	0.	0.	0.	0.
3	780.	0.	780.	0.	0.	283.	370.	285.	692.	0.	88.	0.	780.	0.
4	780.	0.	0.	0.	0.	487.	529.	455.	529.	0.	251.	0.	780.	0.
5	780.	0.	0.	0.	1.	1171.	781.	684.	780.	0.	0.	0.	780.	0.
6	780.	1560.	0.	1560.	425.	701.	1204.	1090.	780.	0.	0.	1560.	780.	1560.
7	2600.	1560.	1820.	0.	0.	249.	1694.	1589.	2119.	0.	481.	1560.	2600.	1560.
8	2600.	1560.	0.	0.	0.	170.	2079.	2002.	2079.	0.	521.	1560.	2600.	1560.
9	2600.	3640.	0.	2080.	0.	0.	2694.	2674.	2600.	94.	0.	2844.	2600.	2938.
10	2600.	3640.	0.	0.	887.	0.	3525.	3602.	2600.	38.	0.	3602.	2600.	3640.
11	4680.	5720.	2080.	2080.	0.	0.	4358.	4633.	4680.	565.	0.	4633.	4680.	5198.
12	4680.	5720.	0.	0.	0.	0.	4860.	5400.	4680.	180.	0.	5400.	4680.	5580.
13	5980.	5720.	2080.	0.	0.	0.	5370.	6090.	5370.	0.	370.	5720.	5740.	5720.
14	5980.	7800.	0.	2080.	0.	0.	5910.	6770.	5910.	0.	0.	6770.	5910.	6770.
15	6500.	7800.	520.	0.	0.	0.	6510.	7640.	6500.	10.	0.	7640.	6500.	7650.

	BUILDING COST		INVENTORY COST		PROCESSING COSTS AT SITE J FROM SOURCE I				SHIPPING COSTS FROM SOURCE I TO SITE J				TOTAL
YEAR	SITE 1	SITE 2	SOURCE 1	SOURCE 2	SRCE 1 SITE 1	SRCE 1 SITE 2	SRCE 2 SITE 1	SRCE 2 SITE 2	SRCE 1 SITE 1	SRCE 1 SITE 2	SRCE 2 SITE 1	SRCE 2 SITE 2	COST
1	0.0000	0.0000	8.8964+05	0.0000	0.000	0.000	0.000	0.000	0.000	0.000	0.000	0.000	8.89643+05
2	0.0000	0.0000	2.7210+06	7.2672+05	0.000	0.000	0.000	0.000	0.000	0.000	0.000	0.000	4.33735+06
3	6.5890+07	0.0000	0.0000	2.1352+06	2.463+05	0.000	3.132+04	0.000	8.004+05	0.000	2.543+05	0.000	7.36951+07
4	0.0000	0.0000	0.0000	3.2807+06	1.681+05	0.000	7.976+04	0.000	5.443+05	0.000	6.476+05	0.000	7.84176+07
5	0.0000	0.0000	6.0147+03	7.0432+06	2.213+05	0.000	0.000	0.000	7.192+05	0.000	0.000	0.000	8.64074+07
6	0.0000	5.9776+07	2.2824+06	3.7646+06	1.976+05	0.000	0.000	3.952+05	6.422+05	0.000	0.000	1.581+06	1.55046+08
7	5.6330+07	0.0000	0.0000	1.1939+06	4.793+05	0.000	1.088+05	3.528+05	1.558+06	0.000	8.834+05	1.411+06	2.17363+08
8	0.0000	0.0000	0.0000	7.2780+05	4.198+05	0.000	1.052+05	3.150+05	1.364+06	0.000	8.543+05	1.260+06	2.22410+08
9	0.0000	4.7055+07	0.0000	0.0000	4.698+05	1.695+04	0.000	5.128+05	1.524+06	1.376+05	0.000	2.051+06	2.74176+08
10	0.0000	0.0000	3.0273+06	0.0000	4.186+05	6.117+03	0.000	5.799+05	1.360+06	4.967+04	0.000	2.319+06	2.81937+08
11	3.7512+07	3.7512+07	0.0000	0.0000	6.727+05	8.121+04	0.000	6.659+05	2.186+06	6.594+05	0.000	2.664+06	3.63890+08
12	0.0000	0.0000	0.0000	0.0000	6.006+05	2.310+04	0.000	6.930+05	1.952+06	1.876+05	0.000	2.772+06	3.70118+08
13	2.9904+07	0.0000	0.0000	0.0000	6.153+05	0.000	4.240+04	6.554+05	2.000+06	0.000	3.443+05	2.622+06	4.06301+08
14	0.0000	2.6700+07	0.0000	0.0000	6.047+05	0.000	0.000	6.926+05	1.945+06	0.000	0.000	2.771+06	4.39034+08
15	1.4675+07	0.0000	0.0000	0.0000	5.938+05	9.135+02	0.000	6.979+05	1.930+06	7.417+03	0.000	2.792+06	4.59730+08

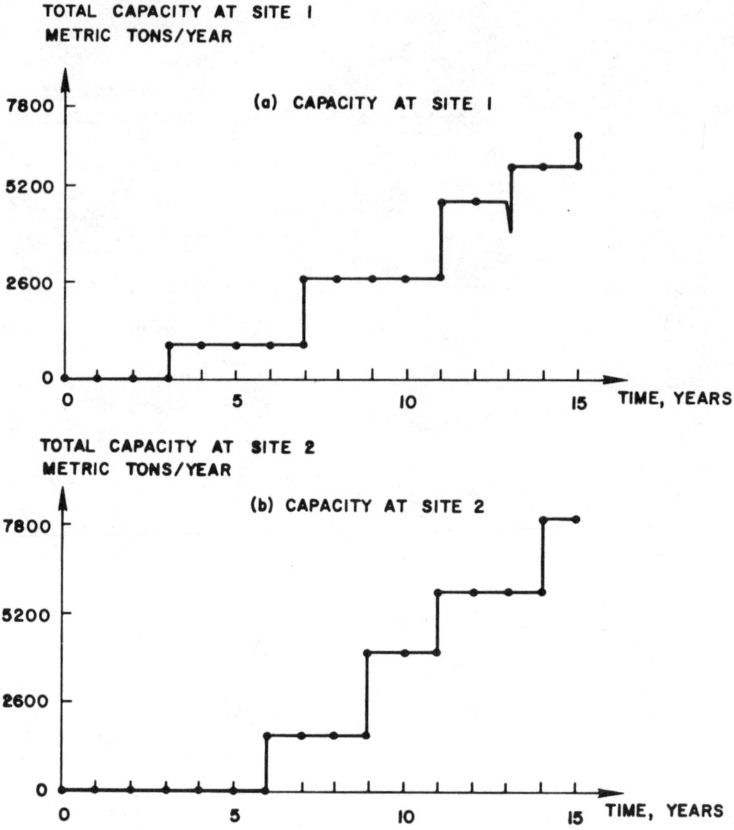

Figure 6.3. Total Capacity as a Function of Time for Solution A to Case 1

with the unmodified DP1. However, in view of the incomplete state space, it may happen that the second- and third-best trajectories carried forward later replace the true optimal trajectory at some intermediate state. This effect actually occurred in this particular case. The optimal trajectory of Solution A was propagated until Stage 13. At this time, the 780-metric-ton-per-year plant built in Year 3 reached the end of its lifetime, and consequently the total capacity at Site 1 decreased by this amount. For the capacity state along the optimal trajectory at $k = 13$, namely $x_1 = 5980$, $x_2 = 5720$, three trajectories were found that had lower cost than the value along the optimal trajectory. All three could be traced back to second- and third-best trajectories at earlier states. In all three cases, these trajectories involved building plants at $k = 4$ and/or $k = 5$. The main reason that these costs were less at $k = 13$ than the cost along the optimal trajectory is that these plants still had unused lifetime, while the optimal trajectory had built an additional 780 metric tons/year of capacity that was just lost. Naturally, these trajectories lost the capacity installed at $k = 4$ and $k = 5$ in the next two time increments, and as a result none of them propagated to the end of the planning interval. Nevertheless, at $k = 13$ they still managed to eliminate the true optimal trajectory from further consideration.

This analysis shows the effect of not giving credit for unused capacity at the end of the planning interval. This aspect of the problem formulation makes it very desirable to build plants whose lifetime barely extends to the end of the planning interval. It is interesting to note that the six best solutions obtained to this problem, Solutions A to F in Table 6.6, all build plants in the Years 6 and 7, the earliest two years where the ten-year lifetime will reach Year 15, the end of the planning interval. In future studies of this problem it would be desirable to reconsider this aspect of the formulation.

The running time for DP1 with the N-best trajectories procedure for $N = 3$ was about 3 minutes on the UNIVAC 1108, less than 3 times the time for DP1 without this procedure. The entire usable core storage of the machine, 53,000 locations, was utilized.

To verify that the finite plant lifetime causes the discrepancy just discussed, DP1 with the N-best trajectories calculations for $N = 3$ was rerun using the replacement cost penalty function φ_1. This penalty function enabled the procedure to make some correction for the finite lifetime. The three best solutions obtained from the procedure were Solutions A, B and C of Table 6.4, which are the best three solutions obtained by all methods. The cost of these solutions are $\$459.7\times10^6$, $\$460.4\times10^6$, and $\$460.6\times10^6$, respectively.

The computer time and storage requirements were approximately the same with φ_1 as without it.

The program DP1 with no additional features was also run in Case 2 (plant size in either 260, 520, 1040, or 2080 metric tons/year). Solution A' in Table 6.5, which has a cost of $\$470.1\times10^6$, was found. Again, this solution proved to be the best solution found with any method. The solution is presented in more detail in Table 6.7 and Figure 6.4. The computer time required was 40 seconds on the UNIVAC 1108, less than for DP1 on Case 1, while the storage requirements were 40,000 locations, the same as Case 1.

Finally, the program was run using DP1 with the N-best trajectories procedure for $N = 3$ and the replacement cost penalty function φ_1. The three best solutions are Solutions A', B', and C' of Table 6.5, the three best solutions found by any method. The costs corresponding to these solutions are $\$470.1\times10^6$, $\$470.4\times10^6$, and $\$470.5\times10^6$, respectively. The computer time required was just over 2 minutes, on the UNIVAC 1108, somewhat less than the time for the same program on Case 1. The entire usable storage capacity of the computer, 53,000 locations, was again utilized.

The Program DP2 - Minimum Inventory Policy, Forward Dynamic Programming and Successive Approximations

Another basic program, called DP2, combines the minimum inventory policy (Section F), forward dynamic programming (Section E), and successive approximations (Section E).

The program has the same two state variables as DP1. However, only one state variable is changed at a time, as described in

Table 6.7. Detailed Analysis of Solution A', the Optimum Solution to Case 2, Plant Sizes 260, 520, 1,040, 2,080

YEAR	CAPACITY SITE 1	CAPACITY SITE 2	PLANTS INSTALLED SITE 1	PLANTS INSTALLED SITE 2	INVENTORY SRCE 1	INVENTORY SRCE 2	DEMAND SRCE 1	DEMAND SRCE 2	SHIPPING FROM SOURCE I TO SOURCE J SRCE 1 SITE 1	SRCE 1 SITE 2	SRCE 2 SITE 1	SRCE 2 SITE 2	TOTAL PROCESSED SITE 1	TOTAL PROCESSED SITE 2
1	0.	0.	0.	0.	94.	0.	94.	0.	0.	0.	0.	0.	0.	0.
2	0.	0.	0.	0.	322.	86.	228.	86.	0.	0.	0.	0.	0.	0.
3	0.	0.	0.	0.	692.	371.	370.	285.	C.	0.	0.	0.	0.	0.
4	1040.	1040.	1040.	1040.	0.	0.	529.	455.	1040.	181.	0.	826.	1040.	1007.
5	1040.	1040.	0.	0.	0.	0.	781.	684.	781.	0.	0.	684.	781.	684.
6	1040.	1040.	0.	0.	164.	50.	1204.	1090.	1040.	0.	0.	1040.	1040.	1040.
7	1040.	1040.	0.	0.	818.	599.	1694.	1589.	1040.	0.	0.	1040.	1040.	1040.
8	3120.	3120.	2080.	2080.	0.	0.	2079.	2002.	2897.	0.	0.	2601.	2897.	2601.
9	3120.	3120.	0.	0.	0.	0.	2694.	2674.	2694.	0.	0.	2674.	2694.	2674.
10	3120.	3120.	0.	0.	405.	482.	3525.	3602.	3120.	0.	0.	3120.	3120.	3120.
11	5200.	5200.	2080.	2080.	0.	0.	4358.	4633.	4763.	0.	0.	5115.	4763.	5115.
12	5200.	5200.	0.	0.	0.	0.	4860.	5400.	4860.	0.	200.	5200.	5060.	5200.
13	5200.	7280.	0.	2080.	0.	0.	5370.	6090.	5200.	170.	0.	6090.	5200.	6260.
14	6240.	6240.	2080.	0.	0.	200.	5910.	6770.	5910.	0.	330.	6240.	6240.	6240.
15	6240.	8320.	0.	2080.	0.	0.	6510.	7640.	6240.	270.	0.	7840.	6240.	8110.

YEAR	BUILDING COST SITE 1	BUILDING COST SITE 2	INVENTORY COST SOURCE 1	INVENTORY COST SOURCE 2	PROCESSING COSTS AT SITE J FROM SOURCE I SRCE 1 SITE 1	SRCE 1 SITE 2	SRCE 2 SITE 1	SRCE 2 SITE 2	SHIPPING COSTS FROM SOURCE I TO SITE J SRCE 1 SITE 1	SRCE 1 SITE 2	SRCE 2 SITE 1	SRCE 2 SITE 2	TOTAL COST
1	0.0000	0.0000	8.8964+05	0.0000	0.000	0.000	0.000	0.000	0.000	0.000	0.000	0.000	8.89643+05
2	0.0000	0.0000	2.7210+06	7.2672+05	0.000	0.000	0.000	0.000	0.000	0.000	0.000	0.000	4.33735+06
3	0.0000	0.0000	5.2211+06	2.7991+06	0.000	0.000	0.000	0.000	0.000	0.000	0.000	0.000	1.23575+07
4	6.5063+07	6.5063+07	0.0000	0.0000	3.305+05	5.751+04	0.000	2.625+05	1.074+06	4.670+05	0.000	1.050+06	1.45724+08
5	0.0000	0.0000	0.0000	0.0000	2.216+05	0.000	0.000	1.941+05	7.201+05	0.000	0.000	7.762+05	1.47636+08
6	0.0000	0.0000	8.8073+05	2.6851+05	2.634+05	0.000	0.000	2.634+05	8.562+05	0.000	0.000	1.054+06	1.51222+08
7	0.0000	0.0000	3.9222+06	2.8721+06	2.352+05	0.000	0.000	2.352+05	7.645+05	0.000	0.000	9.409+05	1.60193+08
8	5.2701+07	5.2701+07	0.0000	0.0000	5.850+05	0.000	0.000	5.253+05	1.901+06	0.000	0.000	2.101+06	2.70708+08
9	0.0000	0.0000	0.0000	0.0000	4.857+05	0.000	0.000	4.821+05	1.579+06	0.000	0.000	1.929+06	2.75183+08
10	0.0000	0.0000	1.3822+06	1.6450+06	5.023+05	0.000	0.000	5.023+05	1.632+06	0.000	0.000	2.009+06	2.82856+08
11	3.7512+07	3.7512+07	0.0009	0.0000	6.846+05	0.000	0.000	7.352+05	2.225+06	0.000	0.000	2.941+06	3.64465+08
12	0.0000	0.0000	0.0000	0.0000	6.237+05	0.000	2.567+04	6.674+05	2.027+06	0.000	2.084+05	2.669+06	3.70687+08
13	0.0000	2.9904+07	0.0000	0.0000	5.959+05	1.948+04	0.000	6.978+05	1.937+06	1.582+05	0.000	2.791+06	4.06790+08
14	2.6700+07	0.0000	0.0000	4.3379+05	6.047+05	0.000	3.376+04	6.384+05	1.965+06	0.000	2.741+05	2.554+06	4.39994+08
15	0.0000	2.3839+07	0.0000	0.0000	5.700+05	2.466+04	0.000	7.162+05	1.853+06	2.003+05	0.000	2.865+06	4.70061+08

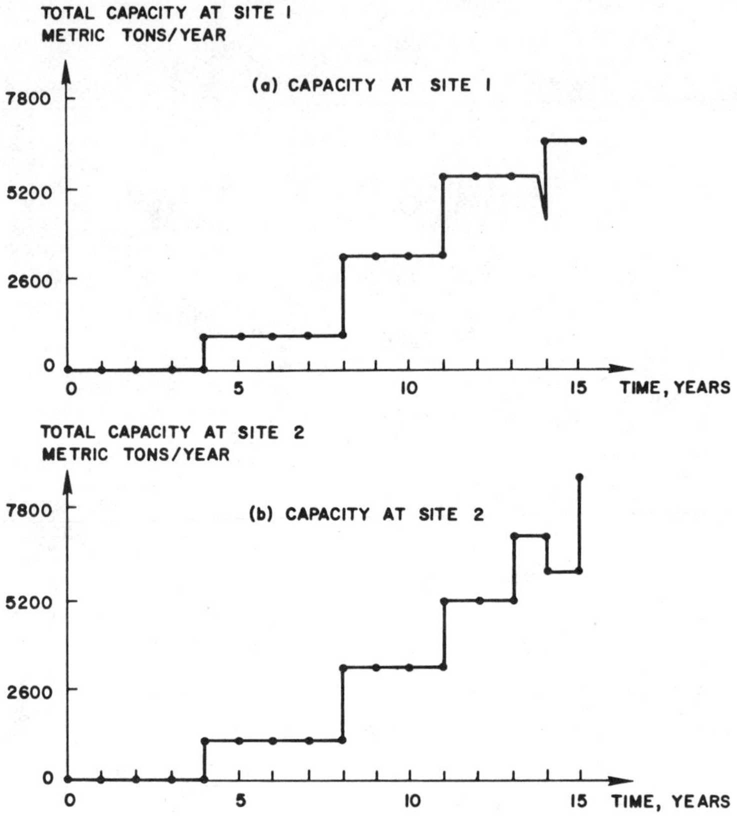

Figure 6.4. Total Capacity as a Function of Time for Solution A' to Case 2

Section E. To start the procedure a nominal building schedule at both sites must be known. This is obtained in two steps. First, the optimal building schedule for one of the sites is computed assuming that there is no capacity at the other site and no demand at the remote source. A building schedule at the other site is then obtained, holding the schedule at the first site fixed and allowing demand at both sources. The normal iterations then begin at the first site.

This program DP2 was applied to Case 1. When the initialization was performed starting at Site 2, Solution A was obtained. This solution has a cost of $\$459.7\times10^6$, the lowest cost obtained with any of the procedures. When the initialization was performed starting at Site 1, Solution D was obtained. This solution has a cost of $\$461.0\times10^6$. Each iteration required approximately 4 seconds on a UNIVAC 1108 and 40,000 words of core memory. All solutions were obtained within five iterations. No apparent reduction in core memory requirements occurred due to the fact that now no packing of information into a single word was necessary. The storage released by the successive approximations technique was utilized to speed the program execution. With packing as in DP1, less than 8,000 words of storage would have been required.

This program was also applied to Case 2. With initialization starting at either site, Solution A' was obtained. This solution has a cost of $\$470.1\times10^6$, which is the least cost obtained in this problem by any of the procedures. Computer time and storage requirements were essentially the same as for Case 1.

The Program DP3 - Lumped Site and Source, Minimum Inventory Policy and Forward Dynamic Programming

A third program developed for this problem, DP3, is based on the lumped-site, lumped-source procedure and the minimum inventory policy of Section F, and the forward dynamic programming procedure of Section E.

This program has one state variable, the total capacity at both sites. The inventory corresponding to a given total capacity is computed by the minimum inventory policy. The shipping cost at each stage k, $s(k)$, is a program input. The output of the program is the size and timing (but not the location) of plant construction. Two possible assignments of plants to sites are then considered; both assign plants alternately to sites, one starting at Site 1 and one starting at Site 2. These two assignments, as well as other promising assignments, are then put into a detailed simulation; this simulation determines the actual processing, shipping, and storage policy by use of the minimum inventory policy and calculates the corresponding cost. The complete solution is then selected by taking the assignment that results in least cost.

The program DP3 was applied to Case 1. Several assignments of the plants chosen for construction by this program to the two sites were then simulated. The two assignments based on alternate assignment to sites were simulated, as well as three other slightly modified assignments. Solution E in Table 6.4 has a cost of $\$461.0\times10^6$, the least cost of all of the modified assignments. The major flaw in this solution is that it builds a plant of capacity 1300 metric tons/year in Year 5; it is thus necessary to replace this amount of capacity in Year 15 when the plant lifetime expires. The total running time for both DP3 and the simulations was 6 seconds on the UNIVAC 1108. The core storage requirements were 7,000 words for DP3 and 3,000 for the simulation program.

When the penalty function φ_1 was added to DP3, one of the alternate plant assignments yielded Solution A, the best solution obtained by any method. This penalty function enabled DP3 to account for the fact that the plant built in Year 5 would have to be replaced and hence to avoid this decision. The computational requirements for the program with the penalty function were essentially the same as without it.

Addition of the N-best trajectories calculation with $N = 3$ to DP3 with the penalty function φ_1 again resulted in Solution A'. Total computation time increased to 20 seconds, and core storage requirements increased to 20,000 words.

Application of DP3 (with penalty function φ_1) to Case 2 resulted in Solution A', which has a cost of $\$470.1\times10^6$, the least cost obtained in this case by any procedure. Computational requirements were the same as for Case 1.

H. SUMMARY AND CONCLUSIONS

All three programs, DP1, DP2 and DP3, obtained the optimum answer for both Case 1 and Case 2. Computation time and storage requirements were very reasonable in all cases. DP1 required about 90 seconds on a UNIVAC 1108 and approximately 40,000 words of core storage; DP2 required about 20 seconds on the UNIVAC 1108 and 40,000 words of core storage; and DP3 required about 5 seconds on the UNIVAC 1108 and approximately 7,000 words of core storage.

The three programs are not restricted to the problem formulation of Section B. The functions for the building, processing, transportation, and inventory cost can all be changed. In particular, a different function for economy of scale in building cost can be used, as well as nonlinear functions for any of the processing, transportation, and inventory costs.

The finite lifetime of the plants can be accounted for as it is in the present program or in a number of other ways. Adding a function to give credit for unexpired capacity at the end of the planning interval is not only desirable from the standpoint of completeness but, if handled as indicated in Section G, would also reduce the analytical problems caused by a finite plant lifetime.

The program DP1 cannot easily be extended to eight sources and eight sites, which is the size of the model used in previous studies for planning the installation of reprocessing plants for the entire United States. This problem requires at least eight state variables, one for each site, and the exponential growth of the computational requirements at DP1 with the number of state variables limits its extendibility to at most four or five sites and sources.

Program DP2, on the other hand, could easily be extended to an eight-source, eight-site problem. Again, eight state variables are required. However, in this program computing time and core storage increase approximately linearly with the number of state variables. The only calculations that grow faster than linearly with the number of state variables are calculations related to the extension of the minimum inventory policy procedure. In a problem with a large number of sites and sources, it is necessary to solve at each time increment a linear programming problem to determine the shipping and processing policy corresponding to a given set of installed plants. The size of this linear programming problem is still relatively small for an eight-source, eight-site problem; if shipping is allowed only between a given subset of source-site pairs, then the problem is even smaller. Thus, it is felt that for a problem of this size the linear programming solution will not add substantially to the total computational requirements. The largest size of problem that can be solved with DP2 is difficult to determine, but it appears at this time that it is somewhat larger than the eight-source, eight-site case.

Procedure DP3 can likewise be easily extended to the eight-source, eight-site problem. For DP3, there is only one state variable no matter how many sources and sites there are. However, a problem of assigning the plants to sites becomes substantially more difficult in a large problem.

REFERENCES

[A-1] Aczel, J., Lectures on Functional Equations and Their Applications, Academic Press, New York, 1966.

[A-2] Altshuler, S., "Variational Principles for the Wave Equation Function Theory," Phys. Rev., 109, 1958, pp. 1830-1836.

[A-3] Angel, E., "Dynamic Programming and Partial Differential Equations," USCEE-265, Univ. of Southern California, Los Angeles, March 1968.

[A-4] Angel, E., "A Building Block Technique for Elliptic Boundary-Value Problems Over Irregular Regions," USCEE-285, Univ. of Southern California, Los Angeles, June 1968.

[A-5] Angel, E., "Discrete Invariant Imbedding and Elliptic Boundary-Value Problems Over Irregular Regions," USCEE-279, Univ. of Southern California, Los Angeles, June 1968.

[A-6] Angel, E., "Noniterative Solutions of Nonlinear Elliptic Equations," USCEE-325, Univ. of Southern California, Los Angeles, October 1968.

[A-7] Angel, E., "Invariant Imbedding and Three-Dimensional Potential Problems," USCEE-325, Univ. of Southern California, Los Angeles, January 1969.

[A-8] Angel, E., "Inverse Boundary-Value Problems: Elliptic Equations," USCEE-343, Univ. of Southern California, Los Angeles, April 1969.

[A-9] Aoki, M., "On the Approximation of Trajectories and Its Application to Control Theory Optimization Problems," JMAA, 9, 1964, pp. 23-41.

[A-10] Aoki, M., "On Optimal and Suboptimal Policies in Control Systems," Chapter 1 in Advances in Control Systems, Vol. I; Theory and Applications (C. T. Leondes, ed.), Academic Press, New York, 1964.

[A-11] Aoki, M., "Optimal Bayesian and Min-Max Controls of a Class of Stochastic and Adaptive Dynamic Systems," Preprints, pp. II-21-II-31, paper presented at IFAC Tokyo Symposium on Systems Engineering for Control System Design, Tokyo, Japan, September 1965.

[A-12] Aoki, M., "Control of Large Dynamic Systems by Aggregation," IEEE Trans. Autom. Control, June 1968.

[A-13] Aris, R., G.L. Nemhauser, and D.J. Wilde, "Optimization of Multistage Cyclic and Branching Systems by Serial Procedures," A.I.Ch.E.J., 10, 1964, p. 913.

[A-14] Aris, R., Discrete Dynamic Programming, Blaisdell, New York, 1964.

[A-15] Arnold, R.F., D.L. Richards, "Monotone Reduction Algorithms," in Recent Mathematical Advances in Operations Research, No. 129, Univ. of Michigan, Ann Arbor, Summer 1964.

[A-16] Aronsson, G., "Minimization Problems for the Functional sup F(x,f(x),f'(x))-II," Ark. Mat., 6, 1966, pp. 409-431.

[A-17] Arrow, K.D., T.E. Harris and J. Marshak, "Optimal Inventory Policy," Econometrica, Vol. 19, 1951, pp. 250-272.

[A-18] Arrow, K.J., S. Karlin and H. Scarf, Studies in the Mathematical Theory of Inventory and Production, Stanford University Press, Stanford, California, 1958.

[A-19] Arrow, K.J., "Applications of Control Theory to Economic Growth," AMS Lec. Appl. Math, 12, pp. 85-119, 1968.

[A-20] Ash, M., Optimal Shutdown Control of Nuclear Reactors, Academic Press, New York, 1966.

[A-21] Athans, M. and P.L. Falb, Optimal Control, McGraw-Hill, New York, 1966.

[A-22] Avriel, M. and D.J. Wilde, "Golden Block Search for the Maximum of a Unimodal Function," Manag. Sci., 14, 1968, pp. 307-319.

[A-23] Azen, S., Higher Order Approximation to the Computational Solution of Partial Differential Equations, The RAND Corp., RM-3917, 1964.

[A-24] Azen, S.P., Successive Approximations by Quadratic Fitting As Applied to Optimization Problems, The RAND Corp., RM-3301-PR, 1966.

[B-1] Babushka, I. and S.L. Sobolev, "Optimization of Numerical Processes" (in Russian), Appl. Math., 10, 1965, pp. 96-129.

[B-2] Bailey, P.B. and L. Shampine, Nonlinear Two-Point Boundary-Value Problems, Academic Press, New York, 1969.

[B-3] Baldwin, J.F. and J.S. Sims-Williams," An On-Line Control Scheme Using a Successive Approximations in Policy Space Approach," J. Math. Anal. Appl., 22, 1968, pp. 523-536.

[B-4] Banks. H.T., "Variational Problems Involving Functional Differential Equations," J. SIAM Contr., 7, 1969, pp. 1-17.

[B-5] Beckenbach, E.F. and R. Bellman, Inequalities, Springer, Berlin, 1961.

[B-6] Beckwith, R.E., Analytic and Computational Aspects of Dynamic Programming Processes of High Dimension, JPL, California Institute of Technology, Pasadena, 1959, pp. 1-125.

[B-7] Bellman, R., I. Glicksberg and O. Gross, "On the Optimal Inventory Equation," Management Science, Vol. 2, 1955, pp. 83-104.

[B-8] R. Bellman, "Some Functional Equations in the Theory of Dynamic Programming - I: Functions of Points and Point Transformations," Trans. Amer. Math. Soc., 80, 1955, pp. 51-71.

[B-9] Bellman, R., I. Glicksberg and O. Gross, "On the Bang-Bang Control Problem," Q. Appl. Math., Vol. 14, 1956, pp. 11-18.

[B-10] Bellman, R., "A Variational Problem with Constraints in Dynamic Programming," J. Soc. Indus. Appl. Math., 4, 1956, pp. 48-61.

[B-11] Bellman, R., "Mathematical Aspects of Scheduling Theory," J. Soc. Ind. Appl. Math, 4, 1956, pp. 168-205.

[B-12] Bellman, R., "Dynamic Programming and Its Application to Variational Problems in Mathematical Economics," Proc. Symp. Calculus of Variations and Appli., April 1956, pp. 115-138.

[B-13] Bellman, R., I. Glicksberg and O. Gross, "Some Nonclassical Problems in the Calculus of Variations," Proc. Amer. Math. Soc., 7, 1956, pp. 87-94.

[B-14] Bellman, R., "On the Computational Solution of Linear Programming Problems Involving Almost-Block Diagonal Matrices," Management Sci., 1957, pp. 403-406.

[B-15] Bellman, R., Dynamic Programming, Princeton University Press, New Jersey, 1957.

[B-16] Bellman, R., "Terminal Control, Time Lags, and Dynamic Pro-Gramming," Proc. Nat. Acad. Sci. USA, 43, 1957, pp. 927-930.

[B-17] Bellman, R., "Functional Equations in the Theory of Dynamic Programming - VII: A Partial Differential Equation for the Fredholm Resolvent," Proc. Amer. Math. Soc., 8, 1957, pp. 435-440.

[B-18] Bellman, R., "Functional Equations in the Theory of Dynamic Programming - VI: A Direct Convergence Proof," Ann. Math., 65, 1957, pp. 215-223.

[B-19] Bellman, R., "On Monotone Convergence to Solutions of $u' = g(u,t)$," Proc. Amer. Math. Soc., 8, 1957, pp. 1007-1009.

[B-20] Bellman, R., "On a Class of Variational Problems," Quart. Appl. Math., 14, 1957, pp. 353-39.

[B-21] Bellman, R. and H. Osborn, "Dynamic Programming and the Variation of Green's Functions," J. Math. Mech., 7, 1958, pp. 81-86.

[B-22] Bellman, R., I. Glicksberg and O. Gross, Some Aspects of the Mathematical Theory of Control Processes, The RAND Corp., R-313, 1958.

[B-23] Bellman, R., "On a Routing Problem," Quart. Appl. Math., 16, 1958, pp. 87-90.

[B-24] Bellman, R., I. Cherry, and G.M. Wing, "A Note on the Numerical Integration of a Class of Nonlinear Hyperbolic Equations," Quart. Appl. Math., 16, 1958, pp. 181-183.

[B-25] Bellman, R., "Functional Equations in the Theory of Dynamic Programming - XI: Limit Theorems," Rend. Circ. Mat. Palermo, 8, 1959, pp. 1-3.

[B-26] Bellman, R., M. Ash, and R. Kalaba, "On Control of Reactor Shutdown Involving Minimal Xenon Poisoning," Nuclear Sci. Eng., 6, 1959, pp. 152-156.

[B-27] Bellman, R., J. Jacquez and R. Kalaba, "Some Mathematical Aspects of Chemotherapy - I: One-Organ Models," Bull. Math. Biophys., 22, 1960, pp. 181-198.

[B-28] Bellman, R., J. Jacquez and R. Kalaba, "The Distribution of a Drug in the Body," Bull. Math. Biophys., 22, 1960, pp. 309-322.

[B-29] Bellman, R. and P. Brock, "On the Concepts of a Problem and Problem-Solving," Amer. Math. Monthly, 67, 1960, pp. 119-134.

[B-30] Bellman, R., "Combinatorial Processes and Dynamic Programming," Proc. Ninth Symp. Appl. Math., American Mathematical Society, 1960.

[B-31] Bellman, R. and R. Kalaba, "On k-th Best Policies," J. Soc. Indus. Appl. Math., 8, 1960, pp. 582-588.

[B-32] Bellman, R., Adaptive Control Processes, Princeton University Press, Princeton, New Jersey, 1961.

[B-33] Bellman, R., "Successive Approximations and Computer Storage Problems in Ordinary Differential Equations," Comm. Assoc. Comput. Machinery, 4, 1961, pp. 222-223.

[B-34] Bellman, R., "On the Reduction of Dimensionality for Classes of Dynamic Programming Processes," J. Math. Anal. Appl., 3, 1961, pp. 358-360.

[B-35] Bellman, R. and R. Kalaba, "Reduction of Dimensionality, Dynamic Programming and Control Processes," J. Basic. Eng.,

[B-36] Bellman, R. and W. Karush, "On a New Functional Transform in Analysis: The Maximum Transform," Bull. Amer. Math. Soc., 67, 1961, pp. 501-503.

[B-37] Bellman, R., "On the Computational Solution of Differential-Difference Equations," J. Math. Anal. Appl., 2, 1961, pp. 108-110.

[B-38] Bellman, R., "On the Reduction of Dimensionality for Classes of Dynamic Programming Processes," J. Math. Anal. Appl., 23, 1961, pp. 358-360.

[B-39] Bellman, R., "On the Approximation of Curves by Line Segments Using Dynamic Programming," Comm. Assoc. Comput. Machinery, 4, 1961, pp. 284.

[B-40] Bellman, R. and J.M. Richardson, "A Note on an Inverse Problem in Mathematical Physics," Quart. Appl. Math., 19, 1961, pp. 269-271.

[B-41] Bellman, R. and S. Dreyfus, Applied Dynamic Programming, Princeton University Press, Princeton, New York, 1962.

[B-42] Bellman, R., ed., Proc. Symp. Appl. Math. in Mathematical Problems in the Biological Sciences, Vol. 14, American Mathematical Society, 1962.

[B-43] Bellman, R. and R. Kalaba, "Dynamic Programming Applied to Control Processes Governed by General Functional Equations," Proc. Nat. Acad. Sci. USA, 48, 1962, pp. 1735-1737.

[B-44] Bellman, R., "Dynamic Programming Treatment of the Traveling Salesman Problem," J. Assoc. Comput. Mach., 9, 1962, pp. 61-63.

[B-45] Bellman, R. and W. Karush, "On the Maximum Transform and Semigroups of Transformations," Bull. Amer. Math. Soc., 68, 1962, pp. 516-518.

[B-46] Bellman, R., "Quasilinearization and Upper and Lower Bounds for Variational Problems," Quart. Appl. Math., 19, 1962, pp. 349-350.

[B-47] Bellman, R., R. Kalaba and B. Kotkin, "On a New Approach to the Computational Solution of Partial Differential Equations," Proc. Nat. Acad. Sci. USA, 48, 1962, pp. 1325-1327.

[B-48] Bellman, R., "Some Questions Concerning Difference Approximations to Partial Differential Equations," Bull. UMI, 17, 1962, pp. 118-190.

[B-49] Bellman, R., "From Chemotherapy to Computers to Trajectories," Math. Prob. Biol. Sci., 1962, pp. 225-232.

[B-50] Bellman, R. and W. Karush, "On the Maximum Transform," J. Math. Anal. Appl., 6, 1963, pp. 67-74.

[B-51] Bellman, R. and W. Karush, "Functional Equations in the Theory of Dynamic Programming - XII: An Application of the Maximum Transform," J. Math. Anal. Appl., 6, 1963, pp. 155-157.

[B-52] Bellman, R., "A Note on Dynamic Programming and Perturbation Theory," Nonlinear Vibration Problems, 1963, pp. 242-244.

[B-53] Bellman, R. and R. Kalaba, "Invariant Imbedding and the Integration of Hamilton's Equations," Rend. Circ. Mat. Palermo, 12, 1963, pp. 1-11.

[B-54] Bellman, R. and R. Kalaba, "A Note on Hamilton's Equations and Invariant Imbedding," Quart. Appl. Math., 21, 1963, pp. 166-168.

[B-55] Bellman, R. and R. Kalaba, "A Note on Nonlinear Summability Techniques in Invariant Imbedding," J. Math. Appl., 6, 1963, pp. 465-472.

[B-56] Bellman, R. and R. Kalaba, "An Inverse Problem in Dynamic Programming and Automatic Control," J. Math. Anal. Appl., 7, 1963, pp. 322-325.

[B-57] Bellman, R., "Mathematical Model-Making as an Adaptive Control Process," Mathematical Optimization Techniques, Univ. of California Press, 1963, pp. 333-339.

[B-58] Bellman, R. and R. Karush, "Dynamic Programming: A Bibliography of Theory and Application," The RAND Corp. Memo RM-3951-PR, 1964.

[B-59] Bellman, R., "Dynamic Programming and Markovian Decision Processes with Particular Applications to Baseball and Chess," Applied Combinatorial Mathematics, Wiley, New York, 1964, pp. 221-236.

[B-60] Bellman, R. and R. Bucy, "Asymptotic Control Theory," J. SIAM Cont., 2, 1964, pp. 11-18.

[B-61] Bellman, R. and R. Kalaba (eds.), Mathematical Trends in Control Theory, Dover, New York, 1964.

[B-62] Bellman, R., B. Gluss and R. Roth, "On the Identification of Systems and the Unscrambling of Data: Some Problems Suggested by Neurophysiology," Proc. Nat. Acad. Sci. USA, 52, 1964, pp. 1239-1249.

[B-63] Bellman, R., "Control Theory," Sci. Amer., Sept. 1964, pp. 186-200.

[B-64] Bellman, R. and R. Kalaba, "Dynamic Programming, Invariant Imbedding and Quasilinearization: Comparisons and Interconnections," Computing Methods in Optimization Problems, (A. Balakrishnan and L. Neustadt eds.), Academic Press, 1965.

[B-65] Bellman, R. and R. Kalaba, Dynamic Programming and Modern Control Theory, Academic Press, New York, 1965.

[B-66] Bellman, R. and T. Brown, "Projective Metrics in Dynamic Programming," Bull. Amer. Math. Soc., 1965, pp. 773-775.

[B-67] Bellman, R. and R. Kalaba, Quasilinearization and Nonlinear Boundary-Value Problems, American Elsevier Publishing Co., Inc., New York, 1965.

[B-68] Bellman, R. and R. Kalaba, "On a New Approach to the Numerical Solution of a Class of Partial Differential-Integral Equations of Transport Theory," Proc. Nat. Acad. Sci. USA, 54, 1965, pp. 1293-1296.

[B-69] Bellman, R. and R.S. Lehman, "Invariant Imbedding, Particle Interation and Conservation Relations," J. Math. Anal. Appl., 10, 1965, pp. 112-122.

[B-70] Bellman, R. and K.L. Cooke, "Existence and Uniqueness Theorems in Invariant Imbedding - II: Convergence of a New Difference Algorithm," J. Math. Anal. Appl., 12, 1965, pp. 247-253.

[B-71] Bellman, R., H. Kagwada and R. Kalaba, "Nonlinear Extrapolation and Two-Point Boundary-Value Problems," Comm. A.C.M., 1965, pp. 511-512.

[B-72] Bellman, R. and K.L. Cooke, "On the Computational Solution of a Class of Functional Differential Equations," J. Math. Anal. Appl., 12, 1965, pp. 495-500.

[B-73] Bellman, R., "Young's Inequality and the Problem of the Optimum Transversal Contours," Theory of Optimum Aerodynamics Shape (A. Milele, ed.), Academic Press, New York, 1965, pp. 315-317.

[B-74] Bellman, R., "Dynamic Programming, Generalized States and Switching Systems," J. Math. Anal. Appl., 12, 1965, pp. 360-363.

[B-75] Bellman, R., "Segmental Differential Approximation and the 'Black Box' Problem," J. Math. Anal. Appl., 12, 1965, pp. 91-104.

[B-76] Bellman, R., "Functional Equations," Handbook of Mathematical Psychology, Vol. III, Wiley, New York, 1965.

[B-77] Bellman, R., "An Application of Dynamic Programming to the Coloring of Maps," ICC Bull., 4, 1965, pp. 3-6.

[B-78] Bellman, R., J. Buel and R. Kalaba, "Mathematical Experimentation in Time-Lag Modulation," Comm. Assoc. Comput. Mach., 9, 1966, p. 752.

[B-79] Bellman, R., "On Analogues of Poincare-Lyapunov Theory for Multipoint Boundary-Value Problems - I," J. Math. Anal. Appl., 14, 1966, pp. 522-526.

[B-80] Bellman, R., R. Kalaba and J. Lockett, Numerical Inversion of the Laplace Transform, Elsevier, New York, 1966.

[B-81] Bellman, R. and R. Roth, "Segmental Differential Approximation and Biological Systems: An Analysis of a Metabolic Process," J. Theor. Biol., 11, 1966, pp. 168-176.

[B-82] Bellman, R., H. Kagiwada and R. Kalaba, "Dynamic Programming and an Inverse Problem in Neutron Transport Theory," Computing, 2, 1967, pp. 5-16.

[B-83] Bellman, R., "Dynamic Programming: A Reluctant Theory," New Methods of Thought and Procedure (F. Zwicky and A. Wilson, eds.), Springer, Berlin, 1967.

[B-84] Bellman, R., "On the Validity of Truncation for Infinite Systems of Ordinary Differential Equations Associated with Nonlinear Partial Differential Equations," Math & Phys. Sci., 1, 1967, pp. 95-100.

[B-85] Bellman, "Stratification and the Control of Large Systems with Application to Chess and Checkers," Inform. Sci., 1, 1968, pp. 7-21.

[B-86] Bellman, R., "Some Inequalities for the Square Root of a Positive Definite Matrix," Linear Algebra Its Appl., 1, 1968, pp. 321-324.

[B-87] Bellman, R., Introduction to the Mathematical Theory of Control Processes, Vol. I: Linear Equations and Quadratic Criteria, Academic Press, New York, 1968.

[B-88] Bellman, R., "A New Type of Approximation Leading to Reduction of Dimensionality in Control Processes," JMAA, 27, 1969, pp. 454-459.

[B-89] Bellman, R. and K.L. Cooke, "The Konigsberg Bridges Problem Generalized," JMAA, 25, 1969, pp. 1-7.

[B-90] Bellman, R., Dynamic Programming and Problem-Solving, TR-69-2, Univ. of Southern California, 1969.

[B-91] Bellman, R., Introduction to Matrix Analysis, 2nd ed., McGraw-Hill, New York, 1970.

[B-92] Bellman, R., Methods of Nonlinear Analysis, Academic Press, New York, 1970.

[B-93] Bellman, R., K.L. Cooke and J. Lockett, Algorithms, Graphs and Computers, Academic Press, New York, 1970.

[B-94] Bellman, R., "Functional Equations in the Theory of Dynamic Programming - XV: Layered Functionals and Partial Differential Equations," JMAA, to appear.

[B-95] Bellman, R. and Roth, R., "A Technique for the Analysis of a Broad Class of Biological Systems," Bionics Symp., pp. 725-737.

[B-96] Bellman, R., "A Note on Asymptotic Control Theory," JAAA, to appear.

[B-97] Bellman, R., "On Analogues of Poincare-Lyapunov Theory for Multipoint Boundary-value Problems-Correction," JAAA, to appear.

[B-98] Bellman, R. and J. Casti, "Differential Quadrature and Long-Term Integration," JMAA, 34, 1971, pp. 235-236.

[B-99] Bellman, R., "A Function is a Mapping - Plus a Class of Algorithms," Inform. Sci., to appear.

[B-100] Bellman, R., "Dynamic Programming and Inverse Optimal Problems in Mathematical Economics," JMAA, to appear.

[B-101] Berkowitz, L.D., "Variational Methods in Problems of Control and Programming, JMAA, 1961, pp. 145-169.

[B-102] Berkowitz, L.D., "A Variational Approach to Differential Games," Study No. 52, Annals of Mathematics, Princeton University Press, Princeton, New Jersey, 1964.

[B-103] Bertele, U. and F. Brioschi, "A New Algorithm for the Solution of the Secondary Optimization Problem in Nonserial Dynamic Programming," JMAA, 27, 1969.

[B-104] Bertele, U. and F. Brioschi, Parameterization in Nonserial Dynamic Programming, Academic Press, New York, 1969.

[B-105] Bharucha-Reid, A.T., Elements of the Theory of Markov Processes and Their Applications, McGraw-Hill, New York, 1960.

[B-106] Blackwell, D. and M.A. Girschick, Theory of Games and Statistical Decisions, Wiley, 1954.

[B-107] Bliss, G.A., Calculus of Variations, Univ. of Chicago Press, Chicago, 1925.

[B-108] Boltianskii, V.G., Mathematical Methods of Optimal Control, Izdat, Nauka, Moscow, 1961; English Trans. by Interscience (Wiley), New York, 1962.

[B-109] Boltianskii, V.G., "Sufficient Conditions for Optimality and the Justification of the Dynamic Programming Method," Jour. SIAM Cont., 4, 1966, pp. 326-361.

[B-110] Borch, K., "Economic Objectives and Decision Problems," IEEE Trans. Sys. Sci. and Cybernet., Sept. 1968, pp. 266-270.

[B-111] Brown, T.A. and R.E. Strauch, "Dynamic Programming in Multiplicative Lattices," JMAA, 12, 1965, pp. 365-370.

[B-112] Breakwell, J.V., J.L. Speyer and E.A. Bryson, "Optimization and Control of Nonlinear Systems Using the Second Variation," J. SIAM Cont., Ser. A., Vol, 1, No. 2, March 1963, pp. 193-223.

[B-113] Budak, B.M., E.M. Berkovich, and E.N. Soloveva, "Difference Approximations in Optimal Control Problems," J. SIAM Cont., 7, 1969, pp. 18-31.

[B-114] Burkov, V.N. and S.E. Lovetskii, "Methods for the Solution of External Problems of Combinatorial Type (review) Autom. and Remote Control, No. 11, 1968, pp. 1785-1806.

[B-115] Burt, O. and C. Jarris, Jr., "Appointment of the U.S. House of Representatives: A Minimum Range, Integer Solution, Allocation Problem," Operations Research, 11, 1963, pp. 648-652.

[B-116] Burt, O., "Optimal Resource Use Over Time with an Application to Ground Water," Management Science, 11, 1964, pp. 80-93.

[B-117] Butkowsky, A.G., A.I. Egonov and K.A. Jurie, "Optimal Control of Distributed Systems (A Survey of Soviet Publications)," J. SIAM Cont., 6, 1968, pp. 437-476.

[B-118] Butkowsky, A.G., Distributed Control Processes, Elsevier, New York, 1970.

[B-119] Butler, T. and A.V. Martin, "On a Method of Courant for Minimizing Functionals," J. Math. & Phys., 41, 1962, pp. 291-299.

[B-120] Butz, "Convergence with Hilbert's Space Filling Curve," J. Comp. Sys. Sci., 3, 1969, pp. 128-146.

[C-1] Calogero, F., Variable Phase Approach to Potential Scattering, Academic Press, New York, 1970.

[C-2] Canon, M.D., C.D. Cullum and E. Polak, Theory of Optimal Control and Mathematical Programming, McGraw-Hill, New York, 1970.

[C-3] Caratheodory, C., Calculus of Variations and Partial Differential Equations of the First Order - I: Partial Differential Equations of the First Order, Holden-Day, San Francisco.

[C-4] Cartaino, S. and S. Dreyfus, "Application of Dynamic Programming to the Minimum Time-to-Climb Problem," Aero. Engr. Rev., Vol. 16, No. 6, June 1957, pp. 74-77.

[C-5] Charnes, A. and W.W. Cooper, Management Models and Industrial Applications of Linear Programming, Vols. I and II, John Wiley & Sons, New York, 1961.

[C-6] Chazan, D., "Profit Functions and Optimal Control: An Alternate Description of the Control Problem," JMAA, 21, 1968, pp. 169-205.

[C-7] Chernoff, H. and L.E. Moses, Elementary Decision Theory, John Wiley & Sons, New York, 1959.

[C-8] Chu, W.W., Optimal Adaptive Search, Tech. Rep. 6252-1, Stanford Electronics Laboratories, 1966.

[C-9] Churchman, C.W., R.L. Ackoff, and E.L. Arnoff, Introduction to Operations Research, John Wiley & Sons, New York, 1957.

[C-10] Clement, M.F., "Modele Categorique de processes de decisions, theoreme de Bellman, Applications," C. R. Acad. Sci. Paris, 268, 1969, pp. 1-4.

[C-11] Collatz, L., The Numerical Treatment of Differential Equations, Springer, Berlin, 1966.

[C-12] Collins, D.C., "Reduction of Dimensionality in Dynamic Programming Via the Method of Diagonal Decomposition," JMAA, 1969.

[C-13] Collins, D.C. and A. Lew, "Dimensional Approximation in Dynamic Programming by Structural Decomposition," JMAA, 1970.

[C-14] Conti, R., "Recent Trends in the Theory of Boundary Value Problems for Ordinary Differential Equations," Boll. Unione Mat. Ital., 22(3), 1967, pp. 135-178.

[C-15] Courant, R. and D. Hilbert, Methods of Mathematical Physics, Interscience Publishers, New York, 1958.

[C-16] Conway, E.D. and E. Hopf, "Hamilton's Theory and Generalized Solutions of the Hamilton-Jacobi Equation," J. Math. Mech., 13, 1964, pp. 939-986.

[C-17] Cox, D.R. and H.D. Miller, The Theory of Stochastic Processes, Wiley, New York, 1965.

[C-18] Cullum, J., "Discrete Approximations to Continuous Optimal Control," J. SIAM Cont., 7, 1969, pp. 37-49.

[D-1] Dantzig, G.B., Linear Programming and Extensions, Princeton University Press, New Jersey, 1963.

[D-2] Denardo, E.V., "Contraction Mappinds in the Theory Underlying Dynamic Programming," SIAM Review, 9, pp. 165-177, April 1967.

[D-4] Denardo, E.V. and L.G. Mitten, "Elements of Sequential Decision Processes," J. Ind. Engr., 18, 1967, pp. 106-112.

[D-5] Doesoer, C.A., "Pontriagin's Maximum Principle and the Principle of Optimality," J. Franklin Inst., 27, 1961, pp. 361-367.

[D-6] Dijkstra, E.W., "A Note on Two Problems in Connection with Graphs," Numerische Mathematick, 1, 1959.

[D-7] Dobell, A.R., "Some Characteristic Features of Optimal Control Problems in Economic Theory," IEEE Trans. Auto. Cont., AC-14, 1969, pp. 39-47.

[D-8] Dorato, P., A Selected Bibliography and Summary of Articles on Automatic Control, Res. Rep. PIBMRI-1196-63, Polytechnic Institute, Microwave Research Institute, Jan. 1963.

[D-9] Dreyfus, S., "Computational Aspects of Dynamic Programming," Operations Research, 5, 1957, pp. 409-415.

[D-10] Dreyfus, S. and M. Freimer, A New Approach to the Duality Theory of Mathematical Programming, The RAND Corp., P-2334, 1961.

[D-11] Dreyfus, S., "The Numerical Solution of Variational Problems," JMAA, 5, 1962, pp. 30-45.

[D-12] Dreyfus, S.E., Dynamic Programming and the Calculus of Variations, Academic Press, 1965.

[D-13] Dreyfus, S.E., "The Numerical Solution of Nonlinear Optimal Control Problems," Numerical Solutions of Nonlinear Differential Equations, Wiley, New York, 1966, pp. 97-113.

[D-14] Dreyfus, S., An Appraisal of Some Shortest-Path Algorithms, The RAND Corp., RM-5433-PR, Oct. 1967.

[D-15] Dreyfus, S.E., "Dynamic Programming and the Hamilton-Jacobi Method of Classical Mechanics," J. Opt. Theory & Appl., 2, 1968, pp. 15-26.

[D-16] Dvoretsky, A., J. Kiefer and J. Wolfowitz, "The Inventory Problem, I: Case of Known Distributions of Demand," Econometrica, 20, 1952, pp. 450-455.

[D-17] Dvoretsky, A. J. Kiefer and J. Wolfowitz, "The Inventory Problem, II: Case of Unknown Distributions of Demand," Econometrica, 20, 1952, pp. 456-466.

[E-1] J. E. Eckles, "Optimum Maintenance with Incomplete Information," Opns. Res., 16, Sept. 1968, pp. 1058-1067.

[E-2] D. O. Ellis, An Abstract Setting for the Notion of Dynamic Programming," The RAND Corp, P-783, 1955.

[E-3] S. E. Elmaghraby, "The Concept of 'State' in Discrete Dynamic Programming," JMAA, 43, 1973, pp. 642-693.

[E-4] S. E. Elmaghraby and V. Y. Bawle, "Optimization of Batch Ordering Under Deterministic Variable Demand," Man. Sci.: Theory, 18, pp. 508-517, May 1972.

[E-5] L. E. El'sgol'c, "Variational Problems with a Delayed Argument," Amer. Math. Soc. Trans., Ser. 2, 16, 1960.

[E-6] D. Erlenkotter, "Sequencing Expansion Projects," Opns. Res., 21, March 1973, pp. 542-553.

[E-7] A. O. Esogbur and B. R. Marks, "Nonserial Dynamic Programming-A Survey," Operational Research Quarterly, 25, 1974, pp. 253-265.

[E-8] R. V. Evans, "Sales and Restocking Policies in a Single Item Inventory System," Man. Sci.: Theory, 14, pp. 463-472, March 1968.

[E-9] R. V. Evans, "Optimality of (s,S) Policies when Set-Up Costs Vary," Opsearch, 6, March 1969, pp. 25-46.

[E-10] H. Everett, "Generalized Lagrange Multiplier Method for Solving Problems of Optimum Allocation of Resource," Operations Research, 11, 399-417, 1963.

[F-1] C. H. Falkner, "Jointly Optimal Deterministic Inventory and Replacement Policies," Man. Sci.: Theory, 16, May 1970, pp. 622-635.

[F-2] L. T. Fan and C. S. Wang, The Discrete Maximum Principle: A Study of Multistage Systems Optimization, John Wiley and Sons, New York, 1964.

[F-3] L. T. Fan et al., "A Sequential Union of the Maximum Principle and Dynamic Programming," J. Electron. Control, XVII, 1964, pp. 593-600.

[F-4] A. A. Feldbaum, "Dual Control Theory-I," Automation and Remote Control, Vol. 21, No. 9, April 1961, pp. 874-880.

[F-5] D. Feldman, "Contributions to the Two-Armed Bandit Problem," Ann. Math. Stat., Vol. 33, 1962, pp. 847-856.

[F-6] W. Feller, "On the Integral Equation of Renewal Theory," Ann. Math. Stat., 21, 1941, pp. 243-267.

[F-7] W. Fenchel, "On Conjugate Convex Functions," Can. J. Math., 1, 1949, pp. 73-77.

[F-8] F. S. Fine and S. G. Bankoff, "Second Variational Methods for Optimization of Discrete Systems," Industrial & Engineering Chemistry Fundamentals, 6, May 1967, pp. 293-299.

[F-9] B. A. Finlayson and L. G. Scriven, "On the Search for Variational Principles," Int. J. Heat Mass Transfer, 10, 1967, pp. 799-821.

[F-10] S. I. Firstman, B. Gluss, "Optimum Search Routines for Automatic Fault Location," Ops. Res., Vol. 8, 1960, pp. 512-523.

[F-11] C. E. Flagle, W. H. Huggins, R. H. Roy, Operations Research and Systems Engineering, The Johns Hopkins Press, Baltimore, 1960.

[F-12] I. Fleischer and H. Young, "The Differential Form of the Recursive Equation for Partial Optima," J. Math. Anal. Appl., 9, 1964, pp. 294-302.

[F-13] I. Fleischer and A. Kooharian, "Optimization and Recursion," J. Soc. Indus. Appl. Math., 12, 1964, pp. 186-188.

[F-14] W. H. Fleming, A Resource Allocation Problem in Continuous Form, The RAND Corp., RM-1430, 1955.

[F-15] W. H. Fleming, Discrete Approximations to Some Differential Games, The RAND Corp., RM-1526, 1955.

[F-16] M. Florian, M. Klein, "Deterministic Production Planning and Concave Costs and Capacity Constraints," Man. Sci.: Theory, 18, Sept. 1971, pp. 12-20.

[F-17] L. R. Ford, Jr., and D. R. Fulkerson, "Constructing Maximal Dynamic Flows from Static Flows," Oper. Res., 5, 1948, pp. 419-433.

[F-18] G. E. Forsythe, "Today's Computational Methods of Linear Algebra," SIAM Rev., 9, 1967, pp. 489-515.

[F-19] R. Fortet, "Properties of Transition Mappings in Dynamic Programming," (in French), METRA, 2, 1963, pp. 79-97.

[F-20] B. L. Fox, "Finite State Approximations to Denumerable-State Dynamic Programs," J. Math. Analysis and Applications, 34, 1971, pp. 665-670.

[F-21] B. L. Fox, "Discretizing Dynamic Programs," J. Opt. Theory and Appl., 11, 1973, pp. 228-234.

[F-22] J. N. Franklin, Well-Posed Stochastic Extensions of Ill-Posed Linear Problems, Tech. Rep. 135, California Institute of Technology, Pasadena, 1969.

[F-23] M. Freimer, M. R. Rao, and M. H. Weingartner, "Note on Municipal Bond Coupon Schedules with Limitations on the Number of Coupons," Man. Sci., 19, Dec. 1972, pp. 379-380.

[F-24] K. O. Friedrichs, "Ein Verfahren der Variationsrechung," Nachr. Ges. Cottingen, 1929, pp. 13-20.

[F-25] B. L. Fry, Network-Type Management Control Systems Bibliography, The RAND Corp., RM-3047-PR, 1963.

[F-26] K. S. Fu, Sequential Methods in Pattern Recognition and Machine Learning, Academic Press, New York, 1968.

[F-27] A. T. Fuller, "Bibliography of Optimum Nonlinear Control of Deterministic and Stochastic Systems," J. Electron. Control, First Series, 13, 1962, pp. 589-611.

[F-28] A. T. Fuller, "Optimization of Some Nonlinear Control Systems by Means of Bellman's Equation and Dimensional Analysis," Int. J. Control, 3, 1966, pp. 359-394.

[F-29] A. T. Fuller, "Linear Control of Nonlinear Systems," Int. J. Control, 5, 1967, pp. 197-243.

[G-1] F. Gantmacher, Theory of Matrices, Chelsea Publishing Co., New York, 1959, Vols. I and II.

[G-2] J. H. George and W. G. Sutton, "Application of Liapunov Theory to Boundary-Value Problems," JMAA to appear.

[G-3] S. B. Gershwin, "On the Higher Derivatives of Bellman's Equation," J. Math. Anal. Appl., 27, 1969.

[G-4] P. A. Gilmore, "Structuring of Parallel Algorithms," J. Assoc. Comp. Mach., 15, 1968, pp. 176-192.

[G-5] P. C. Gilmore and R. E. Gomory, "Multi-stage Cutting Stock Problems of Two or More Dimensions," Opns, Res., 13, 1965, pp. 94-120.

[G-6] I. V. Girsanov, "Certain Relations Between the Bellman and Krotov Functions for Dynamic Programming Problems," J. SIAM Control 7, 1969, pp. 64-67.

[G-7] C. R. Glassey, "Minimum Change-Over Scheduling of Several Products of One Machine," Opns. Res., 16, March 1968, pp. 342-352.

[G-8] B. Gluss, "An Optimum Policy for Detecting a Fault in a Complex System," Opns. Res., Vol. 7, 1959, pp. 468-477.

[G-9] B. Gluss, "An Optimal Inventory Solution for Some Specific Demand Distributions," Naval Research Logistics Quarterly, Vol. 7, 1960, pp. 45-48.

[G-10] B. Gluss, "Lease Squares Fitting of Planes to Surfaces Using Dynamic Programming," Comm. Assoc. Computing Machinery, 6, 1963, pp. 172-175.

[G-11] B. Gluss, "An Alternative Method for Continuous Line Segment Curve-Fitting," Inform. Control, 7, 1964, pp. 200-206.

[G-12] B. Gluss, "Generalizations of the Two-Armed Bandit Problem," Ph.D. Thesis, Electrical Engineering Dept., Univ. of Calif., Berkeley, 1965.

[G-13] R. E. Gomory, "The Traveling Salesman Problem," Proc. IBM Sci., Comput. Symp. Combinatorial Probl., 1964.

[G-14] R. E. Gomory, "On the Relation Between Integer and Non-Integer Solutions to Linear Programs," Proc. Nat. Acad. Sci., 53, 260-265.

[G-15] R. H. Gonzalez, "Solution of the Traveling Salesman Problem by Dynamic Programming on the Hypercube," Tech. Rep. No. 18, O.R. Center, M.I.T., 1962.

[G-16] D. Gorman and J. Zaborsky, "Functional Representation of Nonlinear Systems, Interpolation and Lagrange Expansion for Functionals," ASME Trans., 1966.

[G-17] R. L. Graves and P. Wolfe (eds.) Recent Advances in Mathematical Programming, McGraw-Hill, New York, 1963.

[G-18] H. J. Greenberg, "Dynamic Programming with Linear Uncertainty," Opns. Res., 16, pp. 1100-1114, Nov. 1968.

[G-19] D. E. Greenspan, "Approximate Solution of Initial Value Problems for Ordinary Differential Equations by Boundary-Value Techniques," J. Math. Phys. Sci., 1, 1967, pp. 261-274.

[G-20] Z. Grihches, "Distributed Lags: A Survey," Econometrica, 35, 1967, pp. 16-49.

[G-21] R. C. Grinold, "A Generalized Discrete Dynamic Programming Model," Man. Sci., 20, pp. 1092-1103, March 1974.

[G-22] J. J. G. Guignabodet, "Marjoration des Errerus de Quantification dans les Calcules de Programmation Dynamique," Compt. Rend. 255, 1962, pp. 828-830.

[G-23] P. I. Gulyayev, "The Significance of Dynamic Programming and Theory of Games for Physiology," (In Russian with English Summary), Colloq. Nervous System, 3, LGU Leningrad, 1962, pp. 177-189.

[G-24] T. L. Gunckel, "Optimum Design of Sampled-Data Systems with Random Parameters," S.E.L.T.R. No. 2101-2, Stanford Electronics Laboratory, Stanford, Calif.

[H-1] G. Hadley, Linear Programming, Addison-Wesley, Reading, Massachusetts, 1962.

[H-2] G. Hadley and T. Whitin, Analysis of Inventory Systems, Prentice-Hall, 1963.

[H-3] G. Hadley, Nonlinear and Dynamic Programming, Addison-Wesley, Reading, Mass., 1964.

[H-4] Hahn, Susan G., "On the Optimal Cutting of Defective Sheets," Opns. Res., 16, pp. 1100-1114, Dec. 1968.

[H-5] J. K. Hale and J. P. LaSalle, "Differential Equations: Linearity vs. Nonlinearity," SIAM Rev., 5, 1963, pp. 249-272.

[H-6] H. Halkin, "The Principle of Optimal Evolution," in Nonlinear Differential Equations and Nonlinear Mechanics, Academic Press, New York, 1963, pp. 284-302.

[H-7] R. W. Hamming, Numerical Methods for Scientists and Engineers, McGraw-Hill Book Co., Inc., New York, 1962.

[H-8] H. Hancock, Theory of Maxima and Minima, Dover, N.Y., 1960.

[H-9] H. Happ, The Theory of Network Diakoptics, Academic Press, New York, 1970.

[H-10] B. K. Harrison, "A Discussion of Some Mathematical Techniques Used in Kron's Method of Tearing," J. Soc. Indus. Appl. Math., 11, 1963.

[H-11] N. A. J. Hastings, "Some Notes on Dynamic Programming and Replacement," Operat. Res. Quart., 19, pp. 453-464, Dec. 1968.

[H-12] W. H. Hausman and J. L. Thomas, "Inventory Control with Probabilistic Demand and Periodic Withdrawals," Man. Sci: Theory, 18, pp. 265-275, Jan. 1975.

[H-13] M. Held and R. M. Karp, "A Dynamic Programming Approach to Sequencing Problems," J. Soc. Indust. and Appl. Math., 10, 1962, pp. 196-210.

[H-14] M. Held, R. M. Karp, and R. Shareshian, "Assembly-Line Balancing-Dynamic Programming with Precedence Constraints," Opns. Res., 11, 1962, pp. 442-459.

[H-15] R. Hermann, Differential Geometry and the Calculus of Variations, Academic Press, New York, 1968.

[H-16] H. Hermes and J. P. LaSalle, Functional Analysis and Time Optimal Control, New York, 1969.

[H-17] M. Hestenes, Calculus of Variations and Optional Control Theory, Wiley, New York, 1966.

[H-18] T. J. Higgins, "A Resume of the Development and Literature of Nonlinear Control-System Theory," Trans. ASME, April 1957, pp. 445-453.

[H-19] F. B. Hildebrand, Introduction to Numerical Analysis, McGraw-Hill Book Co., Inc., New York, 1956.

[H-20] E. Hille and R. Phillips, Functional Analysis and Semigroups, Amer. Math. Soc. Colloq. Vol. XXXI, 1948.

[H-21] S. Hinatsuka and A. Ichikawa, "Optimal Control of System with Transportation Lags," IEEE Trans. Autom. Control., AC-14, 1969, pp. 237-247.

[H-22] Yu-Chi Ho and R. C. K. Lee, "A Bayesian Approach to Problems in Stochastic Estimation and Control," 1964 JACC, pp. 382-387.

[H-23] R. A. Howard, Dynamic Programming and Markov Processes, John Wiley and Sons, New York, 1960.

[H-24] R. A. Howard, "Information Value Theory," IEEE Trans. Sys. Sci. Cyber., Vol. SSC-2, No. 1, August 1966, pp. 22-26.

[H-25] R. A. Howard, Dynamic Probabilistic Systems, 2, John Wiley and Sons, 1970.

[H-26] A. Hordijk and H. Tijms, "Convergence Results and Approximations for Optimal (s,S) Policies," Man. Sci., 20, July 1974, pp. 1432-1438.

[H-28] W. Hyman and L. Gordon, "Commercial Airlines Scheduling Technique," Transportation Research, 2, pp. 23-29, March 1968.

[I-1] T. Ikebe and T. Kato, "Applications of Variational Methods to the Thomas-Fermi Equation," J. Phys. Soc. Japan, 12, 1957, pp. 201-203.

[I-2] E. Ingall, "Optimal Continuous Review Policies for Two Product Inventory Systems with Joint Setup Costs," Man. Sci.: Theory, 15, pp. 278-283, Jan. 1969.

[I-3] P. Ivanescu, "Dynamic Programming with Bivalent Variables," No. 7261, Mat. Vesnik 3(19), 1966, pp. 87-99.

[J-1] D. H. Jacobson and D. Q. Mayne, Differential Dynamic Programming, Elsevier, New York, 1970.

[J-2] D. L. Jaquette, "A Discrete-Time Population-Control Model with Setup Cost," Opns, Res., 22, pp. 298-304, March 1974.

[J-3] P. A. Jensen, "Optimization of Series-Parallel-Series Networks," Opns. Res., 18, pp. 471-482, May 1970.

[J-4] P. A. Jensen, "Optimum Network Partitioning," Opns. Res., 19, pp. 916-932, May 1971.

[J-5] C. D. Johnson, "Optimal Control with Chebyschev Minimax Performance Index," ASME Trans. J. Basic Eng., June 1967.

[J-6] E. L. Johnson, "Optimality and Computation of (s,S) Policies in the Multi-Item Infinite Horizon Inventory Problem," Man. Sci., 13, March 1967, pp. 475-491.

[J-7] S. Johnson, "Optimal Two- and Three-Stage Production Schedules with Setup Times Included," Nav. Res. Log. Quart., 1, 61-68, 1954.

[J-8] P. D. Joseph and T. J. Tou, "On Linear Control Theory," Trans. AIEE, Pt. II, Vol. 80, No. 56, pp. 193-196, Sept. 1961.

[J-9] E. I. Jury and T. Pavlidis, "A Literature Survey of Biocontrol Systems," IEEE Trans. on Autom. Control, AC-8, No. 3, July 1963, pp. 210-217.

[K-1] S. J. Kahne, "Feasible Control Computations Using Dynamic Programming," Proc. Third IFAC Congress, London, June 1966, Paper No. 18G, Session 18.

[K-2] R. Kalaba, "On Nonlinear Differential Equations, the Maximum Operation and Monotone Convergence," J. Math. Mech., 8, 1959, pp. 519-574.

[K-3] R. E. Kalman, "On the General Theory of Control Systems," Proc. 1st IFAC Congress, Moscow, USSR, pp. 481-492, Butterworths Press, London, 1960.

[K-4] R. E. Kalman and R. W. Koepcke, "Optimal Synthesis of Linear Sampling Control Systems Using Generalized Performance Indexes," Trans. ASME, Vol. 80, No. 6, Nov. 1958, pp. 1820-1838.

[K-5] R. E. Kalman and R. S. Bucy, "New Results in Linear Filtering and Prediction Theory," Trans. ASME J. Basic Engr., Vol. 83, Series D, No. 1, pp. 95-108, March 1961.

[K-6] R. E. Kalman, "Towards a Theory of Difficulty of Computation in Optimal Control," Proc. IBM SCI. Symp. Control Theory Appl., 1964.

[K-7] R. E. Kalman, "Algebraic Aspects of the Theory of Dynamical Systems," in Differential Equations and Dynamical Systems (J. K. Hale and J. P. LaSalle, eds.) Academic Press, New York, 1967, pp. 133-146.

[K-8] R. E. Kalman, "New Developments in Systems Theory Relevant to Biology, Systems Theory and Biology," Proc. III Sys. Symp. at Case Institute of Technology, Springer, Berlin, 1968.

[K-9] S. Karlin, Duality in Dynamic Programming, The RAND Corp. RM-971, 1952.

[K-10] S. Karlin, "The Structure of Dynamic Programming Models," Naval Res. Logs. Quart., 2, 1955, pp. 285-294.

[K-11] R. M. Karp and M. Held, "Finite-State Processes and Dynamic Programming," SIAM J. Appl. Math., 15, 3, 1967.

[K-12] A. Kaufmann, Graphs, Dynamic Programming and Finite Games, Academic Press, New York, 1967.

[K-13] A. Kaufmann and R. Cruon, Dynamic Programming: Sequential Scientific Management, Academic Press, New York, 1967.

[K-14] W. G. Keckler, R. E. Larson, "Dynamic Programming for the Pre-Launch Calculation," Preliminary Report, Contract DA-01-021-AMC-90006(y), SRI Project 5811, Stanford Research Institute, Menlo Park, Calif., July 1967.

[K-15] W. G. Keckler, "Optimization About a Nominal Trajectory Via Dynamic Programming," Engineer's Thesis, Department of Electrical Engineering, Stanford, Calif., Sept. 1967.

[K-16] E. B. Keeler, "Projective Metrics and Economic Growth Models," The RAND Corp., RM-6153-PR, Sept. 1969.

[K-17] H. B. Keller, "On the Pointwise Convergence of the Discrete-ordinate Method," J. Soc. Indus. Appl. Math., 9, 1960, pp. 560-567.

[K-18] J. Kelly, "A New Interpretation of Information Rate," Bell System Technical Journal, Vol. 35, 1956, pp. 917-926.

[K-19] J. G. Kemeny and J. L. Snell, Finite Markov Chains, D. Van Nostrand, Princeton, New Jersey, 1959.

[K-20] J. D. Kettelle, Jr., "Lease-Cost Allocation of Reliability Investment," Opns. Res., 10, 1962, pp. 249-265.

[K-21] M. A. Khvedelidze, "Investigation of the Control Processes in Living Organisms and Ways of Developing New Cybernetic Systems," (In Russian) Tr. Inst. Kibernet., AN GRUZ SSR, 1, 1963, pp. 169-190.

[K-22] D. L. Kleinman, T. Fortmann and M. Athans, "On the Design of Linear Systems with Piecewise-Constant Feedback Gains," IEEE Trans. Automat. Control AC-13, pp. 354-361.

[K-23] D. E. Knuth, The Art of Computer Programming, Volume I: Fundamental Algorithms, Addison-Wesley, Reading, Mass., 1968.

[K-24] P. V. Kokotovic and P. Sannuti, "Singular Perturbation Method for Reducing the Model Order in Optimal Control Design," IEEE Trans. Automat. Control, AC-13, 1968, pp. 377-383.

[K-25] R. E. Kopp, "Pontryagin Maximum Principle," in Optimization Techniques (G. Leitmann, ed.), Academic Press, New York, 1962.

[K-26] A. Korsak, Perturbed Optimal Control Problems, Dept. of Math. Univ. of Calif., Berkeley, 1966.

[K-27] A. J. Korsak and R. E. Larson, "A Dynamic Programming Successive Approximations Technique with Convergence Proofs, Part I: Description of the Method and Applications, and Part II: Convergence Proofs," Automatica, Vol. 6, pp. 245-260, March 1970.

[K-28] D. J. Kuck, "Illiac IV Software and Application Programming," IEEE Trans. Computers, C-17, 1968, pp. 758-770.

[K-29] H. W. Huhn and A. W. Tucker, "Nonlinear Programming," in Proc. of the Second Berkeley Symp. on Math. Statistics and Probability, 481-490, Univ. of Calif. Press, 1951.

[K-30] H. W. Kuhn and A. W. Tucker (eds.), "Contributions to the Theory of Games II," Annals of Mathematical Studies, 24, Princeton University Press, Princeton, New Jersey, 1953.

[K-31] H. J. Kushner, Stochastic Stability and Control, Academic Press, New York, 1967.

[K-32] J. D. R. Kramer, Jr., "On Control of Linear Systems with Time Labs," Inform. and Control, 3, 1960, pp. 299-326.

[K-33] N. N. Krasoviskii, "On the Analytical Construction of an Optimal Control in Systems with Time Lags," J. Appl. Math. Mech., 26, 1962, pp. 50-67.

[K-34] H. G. L. Karuse, Astrorelativity, NASA Tech. Rep. TR R-188, Jan. 1964.

[K-35] P. Krolak and L. Cooper, "An Extension of Fibonaccian Search to Several Variables," Comm ACM, 6, 1963, pp. 369-641.

[K-36] E. Kreindler, "Reciprocal Optimal Control Problems," J. Math. Appl., 14, 1966, pp. 141-152.

[K-37] G. Kron, "A Set of Principles to Interconnect the Solution of Physical Systems," J. Appl. Phys., 24, 1953, pp. 965-980.

[K-38] G. Kron, "Tearing and Interconnecting as a Form of Transformation," Quart. Appl. Math., 13, 155, pp. 147-159.

[K-39] V. F. Krotov, "Methods of Solving Variational Problems on the Basis of the Sufficient Conditions Governing the Absolute Minimum," Automat. Telemekh. 23, 1962, pp. 1473-1484.

[K-40] E. Kruger-Thiemer and R. R. Levine, "The Solution of Pharmacological Problems with Computers," Drug. Res., 18, 1968, pp. 1575-1579.

[L-1] R. B. Laning, Studies Basic to the Consideration of Artificial Heart Research and Development Program, Final Rep. PB 169-831, US Dept. of Commerce, National Bureau of Standards, Inst. for Applied Technology, Washington, D.C., 1961.

[L-2] P. A. Larkin, "The Possible Shape of Things to Come," SIAM Rev., 11, 1969, pp. 1-6.

[L-3] R. E. Larson, F. J. Rees and J. P. Nichols, "Dynamic Programming for Expansion Planning of Nuclear Power Generation Systems," American Nuclear Society Transactions, Vol. 13, No. 1, June-July 1970, pp. 37-38.

[L-4] R. E. Larson, Dynamic Programming with Continuous Independent Variable, Stanford Electronics Laboratory, TR 6302-6, Stanford, Calif., April 1964.

[L-5] R. E. Larson, "Dynamic Programming with Reduced Computational Requirements," IEEE Trans. on Automatic Control, Vol. AC-10 No. 2, April 1965, pp. 135-143.

[L-6] R. E. Larson, "An Approach to Reducing the High Speed Memory Requirement of Dynamic Programming," J. Math. Anal. and Appl., Vol. 11, Nos. 1-3, July 1965, pp. 519-537.

[L-7] R. E. Larson and J. Peschon, "A Dynamic Programming Approach to Trajectory Estimation," IEEE Trans. Auto. Contr., Vol. AC-10, No. 3, July 1966, pp. 537-540.

[L-8] R. E. Larson, "A Survey of Dynamic Programming Computational Procedures," IEEE Trans. Auto, Contr., Vol. AC-12, No. 6, December 1967, pp. 767-774.

[L-9] R. E. Larson and W. G. Keckler, "Applications of Dynamic Programming to the Control of Water Resource Systems," Automatica, Vol. 5, No. 1, January 1969, pp. 15-26.

[L-10] R. E. Larson and P. J. Wong, "Optimization of Natural-Gas Pipeline Systems Via Dynamic Programming," IEEE Trans.on Auto. Contr., Vol. AC-13, No. 5, October 1968, pp. 475-481.

[L-11] R. E. Larson, R. M. Dressler, and R. S. Ratner, "Precomputation of the Weighting Matrix in an Extended Kalman Filter," 1967 Joint Automatic Control Conference, University of Pennsylvania, Philadelphia, Pennsylvania, June 1967.

[L-12] R. E. Larson, L. Meier, and A. J. Tether, "Dynamic Programming for Stochastic Control of Discrete Systems," IEEE Trans. on Auto. Contr., Vol. AC-16, No. 6, December 1971.

[L-13] R. E. Larson and E. Tse, "Parallel Processing Algorithms for the Optimal Control of Nonlinear Dynamic Systems," IEEE Trans. on Computers, Vol. C-22, No. 8, August, 1973.

[L-14] R. E. Larson, State Increment Dynamic Programming, Elsevier, New York, 1968.

[L-15] R. E. Larson, Y. Bar-Shalom, and M. A. Grossberg, "Application of Stochastic Control Theory to Resource Allocation Under Uncertainty," IEEE Trans. on Auto. Contr., Vol. 19, No. 1, February 1974.

[L-16] R. E. Larson and W. G. Keckler, "Optimum Adaptive Control in an Unknown Environment," IEEE Trans. on Auto. Contr., Vol. AC-13, No. 4, August 1968, pp. 438-439.

[L-17] J. P. LaSalle, "Stability and Control," J. SIAM Control, 1, 1962, pp. 3-15.

[L-18] R. Lattes and J. L. Lions, The Method of Quasi-reversibility: Applications to Partial Differential Equations, Elsevier, New York, 1969.

[L-19] E. L. Lawler, "Partitioning Methods for Discrete Optimization Problems," in Recent Mathematical Advances in Operations Research, Univ. of Michigan, Ann Arbor, 1964.

[L-20] P. D. Lax, "Hyperbolic Systems of Conservation Laws II," Comm. Pure Appl. Math., 10, 1957, pp. 537-566.

[L-21] E. S. Lee, "Dynamic Programming, Quasilinearization and Dimensionality Difficulty," J. Math. Anal. Appl., 27, 1968, pp. 303-322.

[L-22] R. C. K. Lee, Optimal Estimation, Identification and Control, MIT Press, 1964.

[L-23] A. Lew, "Approximation Techniques in Discrete Dynamic Programming," TR70-10, Univ. Southern Calif., Dept. of Elec. Eng., January 1970.

[L-24] A. Lew, "Reduction of Dimensionality by Approximation Techniques: Diffusion Processes," J. Math. Anal. Appl., 1972.

[L-25] D. V. Lindley, Probability and Statistics, Vol. 2, Cambridge Univ. Press, Cambridge, England, 1964, pp. 59-72.

[L-26] L. D. Liozner, "Regeneration of Lost Organs," Izdatel's stvu Akad. Nauk SSR, Moscow, 1962, pp. 1-141; Engl. Trans., FTD-TT 63-576, Div. Foreign Technology, Air Force Systems Command, Wright-Patterson AFB, September 1963.

[L-27] G. G. Lorentz, "Metric Entropy and Approximation," Bull. Amer. Math. Soc., 72, 1966, pp. 903-937.

[L-28] A. G. Lubowe, "Optimal Functional Approximation Using Dynamic Programming," AIAA J., 2, 1964, pp. 376-377.

[L-29] R. D. Luce and H. Raiffa, Games and Decisions: Introduction and Critical Survey, John Wiley and Sons, New York, 1957.

[L-30] D. L. Lukes, "Optimal Regulation of Nonlinear Dynamical Systems," J. SIAM Control, 7, 1969, pp. 75-100.

[L-31] H. Luss and Z. Kander, "Inspection Policies when Duration of Checking is Non-Negligible," Operat. Res. Quart., 25, June 1974, pp. 299-309.

[M-1] S. MacLane, "Hamiltonian Mechanics and Geometry," Amer. Math. Monthly, 77, 1970, pp. 570-585.

[M-2] J. MacQueen, "A Test for Suboptimal Actions in Markovian Decision Problems," Opns. Res., 15, May 1967, pp. 559-561.

[M-3] R. B. Maffei, "Planning Advertising Expenditures by Dynamic Programming Methods," Management Technology, 1, 1960, pp. 94-100.

[M-4] M. Magazine, Optimal Policies for Queueing Systems with Periodic Review, Ph.D. Thesis, Univ. of Fla., 1969.

[M-5] O. Mangasarian, Nonlinear Programming, McGraw-Hill, New York, 1969.

[M-6] D. Mangeron, "The Bellman Equations of Dynamic Programming Concerning a New Class of Boundary Value Problems with Total Derivatives," J. Math Anal. Appl., 8, 1964, pp. 141-146.

[M-7] D. Mangeron, "Sur une classe d'equations fonctionnelles attaches aux problems extremaux correspondant aux systems polyvibrants," C. R. Acad. Sci. Paris, 266, 1968, pp. 1121-1124.

[M-8] A. S. Manne, "Linear Programming and Sequential Decisions," Management Science, 6, 1960, pp. 259-267.

[M-9] P. Masse, Les reserves et la regulation de l'avenir dans la vie economique; Vol. I: Avenir determine; Vol. II; Avenir aleataire, Hermann et Cie, Paris, 1946.

[M-10] S. R. McReynolds and A. E. Bryson, "A Successive Sweep Method for Solving Optimal Programming Problems," Joint Auto. Control Conf., Troy, N.Y., 1965.

[M-11] S. R. McReynolds, "The Successive Sweep Method and Dynamic Programming," J. Math. Anal. Appl., 18, 1967, pp. 355-364.

[M-12] L. Meier, "Combined Optimum Control Estimation," paper presented at 1965 Allerton Conference, University of Illinois, pp. 109-120.

[M-13] V. W. Merriam, "A Class of Optimum Control Systems," J. Franklin Inst., 267, 1959, pp. 367-381.

[M-14] C. F. Meyer and R. J. Newett, "Dynamic Programming for Feedlot Optimization," Man. Sci.: Application, 16, Feb.6, 1970, pp. B-410.

[M-15] A. Miele (ed.), Theory of Optimum Aerodynamic Shapes, Academic Press, New York, 1965.

[M-16] H. Mine and S. Osaki, Markovian Decision Problems, Elsevier, New York, 1970.

[M-17] G. J. Minty, "Monotone Networks," Proc. Roy Soc. London, Ser. A, 257, 1960, pp. 194-212.

[M-18] L. G. Mitten, "An Analytic Solution to the Least Cost Testing Sequence Problem," J. of Ind. Eng., 11, 17, 1960.

[M-19] L. G. Mitten and G. L. Nemhouser, "Multistage Optimization," Chemical Engineering Progress, 59, 1963, pp. 52-60.

[M-20] L. G. Mitten and G. L. Nemhauser, "Optimization of Multistage Separation Processes by Dynamic Programming," Can. J. of Chemical Eng., 41, 1963, pp. 187-194.

[M-21] L. G. Mitten, "Composition Principles for Synthesis of Optimal Multi-stage Processes," Opns. Res., 12, 1964, pp. 610-619.

[M-22] L. G. Mitten, "Preference Order Dynamic Programming," Man. Sci., 21, Sept. 1974, pp. 43-46.

[M-23] N. N. Moiseev, "Methods of Dynamic Programming in the Theory of Optimal Controls (IV)," USSR Computa. Math. Phys., 5, 1965, pp. 58-75.

[M-24] J. Monavek, "A Note on Minimal Path Problems," JMAA, 1970.

[M-25] B. Mond and M. A. Hanson, "Duality for Variational Problems," J. Math. Anal. Appl., 18, 1967, pp. 355-364.

[M-26] M. Morse, Calculus of Variations in the Large, American Mathematical Society, 1935.

[M-27] J. J. Moreau, "Convexity and Duality," Functional Analysis and Optimization, Academic Press, New York, 1966, pp. 145-169.

[M-28] J. J. Moreau, "Fonctions convexes duales et points proximaux dans un espace Hilbertien," C. R. Acad. Sci., Paris 255, 1962, pp. 2897-2899.

[M-29] J. J. Moreau, "Analyse Fonctionelle, Inf.-convolution des fonctions numeriques sur un espace vectoriel," C.R. Acad. Sci., 256, pp. 5047-5049.

[M-30] J. J. Moreau, "Proprietes des applications 'prox'," C. R. Acad. Sci., Paris, 256, 1963, pp. 1069-1071.

[M-31] T. L. Morin, A. O. Esogbue, "The Imbedded State Space Approach to Reducing Dimensionality in Dynamic Programs of Higher Dimensions," J. Math Analysis and Appl., 148, 1974, pp. 801-811.

[M-32] R. Morton, "On the Dynamic Programming Approach to Pontriagin's Maximum Principles," J. Appl. Prob., 5, 1968, pp. 679-692.

[M-33] T. E. Morton, "On the Asymptotic Convergence Rage of Cost Differences for Markovian Decision Processes," Opns. Res., 19, Jan. 1971, pp. 244-248.

[M-34] J. Moser, "A New Technique for the Construction of Solutions of Nonlinear Differential Equations," Proc. Nat. Acad. Sci. USA, 47, 1961, pp. 1824-1831.

[N-1] G. L. Nemhauser, "Applications of Dynamic Programming in the Process Industries," Am. Inst. Ind. Eng. Proceedings., 1963, pp. 279-292.

[N-2] G. L. Nemhauser, "Decomposition of Linear Programs by Dynamic Programming," Naval Research Logistics Quart., 11, 1964, pp. 191-196.

[N-3] G. L. Nemhauser and Z. Ullman, "Discrete Dynamic Programming and Capital Allocation," Manag. Sci., 15, 1969, pp. 494-505.

[N-4] G. L. Nemhauser, Introduction to Dynamic Programming, John Wiley and Sons, 1966.

[N-5] G. L. Nemhauser, "Scheduling Local and Express Service," Trans. Sci., 3, May 1969, pp. 164-175.

[N-6] G. L. Nemhauser and P. LuYu, "A Problem in Bulk Service Scheduling," Opns. Res., 20, July 1972, pp. 813-819.

[N-7] E. Noether, "Invariant Variationsprobleme," Nachr. Ges., Gottingen, 1918, pp. 235-237.

[O-1] A. Odoni, "On Finding the Maximal Gain for Markovian Decision Processes," Opns. Res., 17, Sept. 1969, pp. 857-860.

[O-2] C. Olech, "Existence Theorems for Optimal Problems with Vector-Valued Cost Functions," Trans. Amer. Math. Soc., 136, 1969, pp. 159-180.

[O-3] H. Osborn, "Euler Equations and Characteristics," Chpt. 7 in Dynamic Programming of Continuous Processes, (R. Bellman, ed.) the RAND Corp., R-271, 1954.

[O-4] A. Ostrowsky, "On Two Problems in Abstract Algebra Connected with Horner's Rule," Studies in Mathematics and Mechanics Presented to Richard von Mises, Academic Press, New York, 1954, pp. 40-48.

[P-1] V. G. Pavlov and V. P. Cheprasov, "Constructing Certain Invariant Solutions of Bellman's Equation," Automat. Remot. Control., January 1968, pp. 31-36.

[P-2] J. D. Peason, "Reciprocity and Duality in Control Programming Problems," J. Math. Anal. Appl., 1965, pp. 385-408.

[P-3] V. Pereyra, "On Improving an Approximate Solution of A Functional Equation by Deferred Corrections," Numer. Mat. 8, 1966, pp. 376-391.

[P-4] J. Peschon, Disciplines and Techniques of Systems Control, Blaisdell Publishing Co., New York, 1965.

[P-5] J. Peschon and R. E. Larson, "Analysis of an Intercept System," SRI, Menlo Park, Calif., Final Report to NIKE-X Project Office, Redstone Arsenal, Alabama, Contract DA-01-021-AMC-900005(y), SRI Project 5188-7, Dec. 1965.

[P-6] E. R. Peters, "A Dynamic Programming Model for the Expansion of Electric Power Systems," Man. Sci., 10, Dec. 1973 pp. 656-664.

[P-7] R. Petrovic, "Optimization of Resource Allocation in Project Planning," Opns. Res., 16, May 1968, pp. 559-568.

[P-8] G. Polya and G. Szego, Augfaben and Lehrsatze aus der Analysis, Dover, New York, 1945.

[P-9] M. Pollack and W. Weibenson, "Solution of the Shortest Route Problem - A Review," Operations Research, 8, 1960, pp. 224-230.

[P-10] G. C. Pomraming, "Reciprocal and Canonical Forms of Variational Problems Involving Linear Operators," J. Math. Phys., XLVII, 1968.

[P-11] L. S. Pontryagin et al., The Mathematical Theory of Optimal Processes, John Wiley & Sons, Inc., New York, 1962.

[P-12] E. L. Porteus, "Some Bounds for Discounted Sequential Decision Progress," Man. Sci., 1971, pp. 7-11.

[P-13] F. Prochan and T. A. Bray, "Optimal Redundancy Under Multiple Constraints," Opns. Res., 13, 1965, pp. 800-814.

[P-14] B. A. Powell, "Optimal Elevator Banking Under Heavy Up-Traffic," Trans. Sci., 5, May 1971, pp. 109-121.

[P-15] M. J. Protter, "Difference Methods and Soft Solutions," Nonlinear Partial Differential Equations, Academic Press, New York, 1963, pp. 161-170.

[P-16] E. B. Pyle, III, B. Douglas, G. W. Ebright, W. J. Westlake, A. D. Bender, "Scientific Manpower Allocation to New Drug Screening Programs," Man. Sci., 19, Aug. 1973, pp. 1433-1443.

[R-1] R. Radner, "Paths of Economic Growth that are Optimal with Regard Only to Final States: A 'Turnpike Theorem'," Rev. Econ. Stud., 28, 1961, pp. 98-104.

[R-2] L. B. Rall, "On Complementary Variational Principles," J. Math. Anal. Appl., 14, 1966, pp. 174-184.

[R-3] V. R. Rao, J. L. Thomas, "Dynamic Models for Sales Promotion Policies," Operat. Res. Quart., 24, Sept. 1973, pp. 403-417.

[R-4] R. S. Ratner, "Performance-Adaptive Renewal Policies for the Operation of Linear Systems Subject to Failure," Ph.D. Thesis, Dept. of Elec. Engr., Stanford Univ., Calif., June 1968.

[R-5] H. E. Rauch, F. Tung, and C. G. Streibel, "Maximum Likelihood Estimates of Linear Dynamic Systems," AIAA J., Vol. 3, No. 8, Aug. 1965, pp. 1445-1450.

[R-6] F. J. Rees and R. E. Larson, "Computer-Aided Dispatching and Operations Planning for an Electric Utility with Multiple Types of Generation," IEEE Transactions on Power Apparatus and Systems, Vol. PAS-90, No. 2, March-April 1971.

[R-7] D. W. Richardson, E. P. Hechtman, R. E. Larson, "Control Data Investigation for Optimization of Fuel on Supersonic Transport Vehicles," Phase II Report, Contract AF 33(657)-8822, Project 9056, Hughes Aircraft Co., Culver City, Calif., June 1963.

[R-8] R. D. Richtmyer, Difference Methods for Initial Value Problems, Interscience Publishers, Inc., New York, 1957.

[R-9] V. Riley and S. I. Gass, Linear Programming and Associated Techniques, The Johns Hopkins University Press, Baltimore 1958.

[R-10] S. M. Roberts, Dynamic Programming in Chemical Engineering and Process Control, Academic Press, New York, 1964.

[R-11] R. T. Rockafeller, "Conjugates and Legendre Transforms of Convex Functions," Can. J. Mat., 19, 1964, pp. 200-205.

[R-12] R. T. Rockafeller, "Duality and Stability in Extremum Problems Involving Convex Functions," Pacific J. Math., 21, 1967, pp. 200-205.

[R-13] R. T. Rockafeller, Convexity and Duality, Princeton Univ. Press, Princeton, N. J., 1968.

[R-14] R. J. Roman, "Mine-Mill Production Scheduling by Dynamic Programming," Operat. Res. Quart., 22, Dec. 1971, pp. 319-328.

[R-15] S. H. Ross, Applied Probability Models with Optimization Applications, Holden-Day, San Francisco, 1970.

[R-16] M. Rosseel, "Comments on a Paper by Romesh Saigal: 'A Constrained Shortest Route Problem'," Opns. Res., 16, 1968, pp. 1232-1234.

[R-17] C. J. Rose, "Dynamic Programming Processes within Dynamic Programming Processes, JMAA.

[R-18] P. Rosenblook, "The Method of Steepest Descent," Proc. Symps. Appl. Math., VI, 1956, pp. 127-176.

[R-19] R. Roth, "The Unscrambling of Data Studies in Segmental Differential Approximation," J. Math. Anal. Appl., 14, 1966, pp. 5-22.

[R-20] R. Roth, "An Application of Algebraic Topology: Kron's Method of Tearing," Quart. Appl. Math., 17, 1959, pp. 1-14.

[R-21] M. Rothstein, "An Airline Overbooking Model," Trans. Sci., 5, May 1971, pp. 180-192.

[R-22] L. I. Rozonoer, "The Maximum Principle of L. S. Pontryagin in Optimal System Theory," Automat. Remote Control, 20, 1960, pp. 1288-1302; 1517-1532.

[R-23] L. I. Rozonoer, "Variational Approach to the Problem of Invariance of Automatic Control Systems-I," Automat. Telemekh. 24, 1963, pp. 774-756.

[R-24] D. P. Rutenberg, "Design Commonality to Reduce Multi-Item Inventory: Optimal Depth of a Product Line," Opns. Res., 19, March 1971, pp. 491-509.

[S-1] T. L. Saaty, Mathematical Methods of Operations Research, McGraw-Hill, New York, 1959.

[S-2] P. Sannuto and P. V. Kokotovis, "Near Optimum Design of Linear Systems by a Singular Perturbation Method," IEEE Trans. Autom. Control, AC-14, 1969, pp. 15-21.

[S-3] L.J. Savage, The Foundations of Statistics, Wiley, New York, 1954.

[S-4] V. K. Savley, Integration of Equations of Parabolic Type by the Method of Nets, Trans. by G. S. Tee, Macmillan Co., New York, 1964.

[S-5] H. E. Scarf, D. M. Gilford and M. W. Shelly (eds.), Multi-Stage Inventory Models and Techniques, Stanford University Press, Stanford, Calif., 1963.

[S-6] A. Schronhage, "Multiplication grosser Zahlen," Computing (Arch. Electron, Rechnen), 1966, pp. 182-196.

[S-7] P. J. Schweitzer, "Multiple Policy Improvements in Undiscounted Markov Renewal Programming," Opns. Res., 19, May 1971, pp. 784-793.

[S-8] J. F. Shapiro, Shortest Route Methods for Finite State Space, Deterministic Dynamic Programming Problems, Tech. Rep. Operations Research Center, MIT, Cambridge, Mass., 1967.

[S-9] J. F. Shapiro, "Dynamic Programming Algorithms for the Integer Programming Problem-1: The Integer Programming Problem Viewed as a Knapsack Type Problem," Opns. Res., 16, Jan. 1968, pp. 103-121.

[S-10] J. F. Shapiro, "Turnpike Theorems for Integer Programming Problems," Opns. Res., 18, May 1970, pp. 432-440.

[S-11] M. W. Shelley, II, and G. L. Bryan (eds.), Human Judgements and Optimality, Wiley, New York, 1964.

[S-12] L. B. Slobodkin, "The Strategy of Evolution," Amer. Sci., 52, 1964, pp. 342-357.

[S-13] F. J. Smith, "An Algorithm for Summing Orthogonal Polynomial Series and Their Derivatives with Applications to Curve-Fitting and Interpolation," Math. Comp., 19, 1965, pp. 33-36.

[S-14] M. Sobral, "Sensitivity in Optimal Control Systems," Proc. IEEE, 45, 1968, pp. 1644-1752.

[S-15] W. R. Spillers and N. Hickerson, "Optimal Elimination for Sparse Symmetric Systems as a Graph Problem," Quart. Appl. Math., 26, 1968, pp. 425-432.

[S-16] V. P. Sreedharan, H. H. Wein, "A Stochastic, Multi-stage Investment Model," SIAM J. Appl. Math., 15, March 1967, pp. 347-358.

[S-17] D. V. Stewart, "Partitioning and Tearing Systems of Equations,' J. Soc. Indust. Appl. Math., Number Anal. 2, 1965, pp. 345-365.

[S-18] A. W. J. Stoddart, "Estimation of Optimality for Multidimensional Control Systems," J. Optimization Theory and Appl., 3, 1969, pp. 385-391.

[S-19] H. Stone, "Approximation of Curves by Line Segments," Math of Comp., 15, 1961, pp. 40-47.

[S-20] A. R. Stubberud and J. M. Swiger, "Minimum Energy Control of a Linear Plant with Magnitude Constraint on the Control Input Signals," 1965, JACC, pp. 308-406.

[S-21] J. Sugie, "An Extension of Fibonaccian Searching to the Multidimentional Cases," IEEE Trans. Automat. Control, AC-9, Jan. 1964.

[S-22] R. Sussman, "Optimal Control of Systems with Stochastic Disturbances," Technical Report Series 63, No. 20, Electronics Laboratory, Univ. of Calif., Berkeley, 1963.

[T-1] A. Tchamran, "A New Derivation of the Maximum Principle," J. Math. Anal. Appl., 25, 1969, pp. 350-361.

[T-2] L. Tisza, "The Conceptual Structure of Physics," Rev. Mod. Phys., 35, 1963, pp. 151-185.

[T-3] J. Todd, "Motivation for Working in Numerical Analysis," Comm. Pure Appl. Math, 8, 1955, pp. 97-116.

[T-4] J. Todd, "The Problem of Error in Digital Computation," Error Digital Computation, 1, 1965, pp. 3-41.

[T-5] J. T. Tou, Optimum Design of Digital Control Systems, Academic Press, New York, 1963.

[T-6] R. Tomovic and G. Boni, "An Adaptive Artificial Hand," IEEE Trans. on Autom. Control, AC-7, No. 3, April 1962, pp. 3-10.

[T-7] R. Tomovic and L. Radanopvic, "Homeostatic Control of Dynamic Systems," Proc. First Int. Symp. Optimization Adap. Control, April 1962, pp. 57-67.

[V-1] P. J. M. Van den Bogaard, A. M. Luque and C. Van de Panne, "A Study of the Implications of Alternative Layers in Quadratic Dynamic Programming," Rev. Fr. de Recherche Operationelle, 6, 1962, pp. 163-183.

[V-2] A. F. Veinott, Jr., "On Finding Optimal Policies in Discrete Dynamic Programming with No Discounting," Ann. Math'l. Stat., 37, 1966, pp. 1284-1295.

[V-3] A. F. Veinott, Jr., "Minimum Concave-Cost Solution of Leontief Substitution Models of Multi-Facility Inventory Systems," Opns. Res., 17, March 1969, pp. 262-291.

[V-4] T. L. Vincent and J. D. Mason, "Disconnected Optimal Trajectories," JOTA, 3, 1969, pp. 263-281.

[V-5] T. Von Karman, "The Engineer Grapples with Nonlinear Problems," Bull. Amer. Math. Soc., 46, 1940, pp. 615-683.

[V-6] J. Von Neumann and O. Morgensterm, Theory of Games and Economic Behavior, Princeton Univ. Press, Princeton, N. J., 1944.

[W-1] A. Wald, Sequential Analysis, Wiley, New York, 1947.

[W-2] A. Wald, Statistical Decision Functions, John Wiley & Sons, New York, 1950.

[W-3] P. K. C. Wang, "Optimal Control of Distributed Control Systems with Time Delays," IEEE Trans. Automat. Control, AC-9, 1964, pp. 13-22.

[W-4] P. K. C. Wang, "Control of Distributed Parameter Systems," in Advances in Control Systems, Vol. 1, Academic Press, New York, 1964, pp. 75-172.

[W-5] P. K. C. Wang and F. Tung, "Optimal Control of Distributed Parameter Systems," J. Basic Eng. Trans. ASME, SER D, 86, 1964, pp. 67-79.

[W-6] K. E. F. Watt, "Dynamic Programming, Look Ahead Programming and the Strategy of Insect Pest Control," Can. Entomol. 95, 1963, pp. 525-536.

[W-7] K. E. F. Watt, "Computers and the Evaluation of Resource Management Strategies," Amer. Sci., 62, 1964, pp. 408-419.

[W-8] H. M. Weingartner, "Capital Budgeting of Interrelated Problems: Survey and Synthesis," Man. Sci., 12, 1966, pp. 485-516.

[W-9] H. M. Weingartner, "Municipal Bond Coupon Schedules with Limitations on the Number of Coupons," Man. Sci., 19, Dec. 1972, pp. 369-378.

[W-10] J. H. Westcott, J. J. Florentin and J. D. Pearson, "Approximation Methods in Optimal Adaptive Control," Proc. IFAC Sec. Congr., Basel, 1963.

[W-11] D. M. White, "Dynamic Programming and Probabilistic Constraints," Opns. Res., 22, May 1974, pp. 654-664.

[W-12] P. Whittle, "The Deterministic Stochastic Transition in Control Processes and the Use of Maximum and Integral Transforms," in Proc. Fifth Berkeley Symp. Math. Stat. Prob., Vol. II, Univ. of Calif. Press, Berkeley, 1967.

[W-13] P. Whittle, "A View of Stochastic Control Theory," J. Royal Stat. Soc., 131, 1969, pp. 320-335.

[W-14] C. Wilkinson, S. K. Gupta, "Allocating Promotional Effort to Competing Activities: A Dynamic Programming Approach," Proc. of Fifth International Conf. on Opns. Res., June 1969, pp. 419-432.

[W-15] D. M. Wiberg, "Feedback Control of Linear Distributed Systems," Trans. ASME, 89, 1967, pp. 379-384.

[W-16] N. Wiener, "Problems of Sensory Prosthesis," Bull. Amer. Math. Soc., 57, No. 1, Part 1, Jan. 1951.

[W-17] D. J. Wilde, Optimum Seeking Methods, Prentice-Hall, Englewood Cliffs, New Jersey, 1964.

[W-18] D. J. Wilde and C. S. Beightler, Foundations of Optimization, Prentice-Hall, New York, 1967.

[W-19] D. M. G. Wishart, "A Survey on Control Theory," J. Royal Stat. Soc., 132, 1969, pp. 293.

[W-20] L. A. Wolsey, "Generalized Dynamic Programming Methods in Integer Programming," Math. Prog., 4, April 1973, pp. 222-232.

[W-21] P. J. Wong, "Dynamic Programming Using Shift Vectors," Ph.D. Thesis, Dept. of Elec. Engr., Stanford University, Stanford, Calif., Sept. 1967.

[W-22] P. J. Wong and D. G. Luenberger, "Reducing the Memory Requirements of Dynamic Programming," Oper. Res., 16, 1968, pp. 1115-1125.

[W-23] P. J. Wong, "A New Decomposition Procedure for Dynamic Programming," Oper. Res., 18, 1970, pp. 119-131.

[W-24] P. J. Wong, "An Approach to Reducing the Computing Time for Dynamic Programming," Opns. Res., 19, Jan. 1970, pp. 181-185.

[W-25] W. M. Wonham, "Random Differential Equations in Control Theory," in Probabilistic Methods in Applied Mathematics, Vol. 2, Academic Press, 1970, pp. 131-212.

[W-26] R. D. Woodman, "Replacement Policies for Components that Deteriorate," Operat. Res. Quart., 18, Sept. 1967, pp. 267-281.

[W-27] A. Wouk, "Approximation and Allocation," J. Math. Anal. Appl., 8, 1964, pp. 135-143.

[W-28] A. Wouk, "An Extension of the Caratheodory-Kalman Variational Method," J. Optimization Theory Appl. 3, 1968, pp. 2-34.

[Y-1] S. Yakowitz, Mathematics of Adaptive Control Processes, American Elsevier, Co., New York, 1969.

[Y-2] Y. B. Yasinsky and S. Kaplan, "On the Use of Dual Variational Principles for the Estimation of Error in Approximate Solutions of Diffusion Problems," Nuc. Sci. Eng., 31, 1968, pp. 80-90.

[Y-3] D. R. Young, "Scheduling a Fixed-Schedule, Common Carrier Passenger Transportation System," Trans. Sci., 4, Aug. 1970, pp. 243-269.

[Y-4] L. C. Young, Lectures on the Calculus of Variations and Optimal Control Theory, Saunders, Philadelphia, 1968.

[Z-1] J. Zaborsky and D. Gorman, "Control by Functional Lagrange Expansions," IFAC, 1966.

[Z-2] S. Zacks, "A Two-Echelon Multi-Station Inventory Model for Navy Applications," Naval Research Logistics Quarterly, 17, March 1970, pp. 79-85.

[Z-3] L. Zadeh, "What is Optimal?," IEEE Trans. Inform. Theory, IT-4, 1958, p. 3.

[Z-4] L. Zadeh, C. A. Doesoer, Linear System Theory: The State Space Approach, McGraw-Hill, New York, 1963.

SUBJECT INDEX